*Springer Monographs in Mathematics*

Springer-Verlag London Ltd.

J.A. Leach    D.J. Needham

# Matched Asymptotic Expansions in Reaction-Diffusion Theory

**With 52 Figures**

 Springer

*J.A. Leach, BSc, PhD*
*D.J. Needham, BSc, PhD*
Department of Mathematics
The University of Reading
Reading
Berkshire RG6 6AX
UK

British Library Cataloguing in Publication Data
Leach, J. A.
    Matched asymptotic expansions in reaction-diffusion theory.
    (Springer monographs in mathematics)
    1.Asymptotic expansions 2.Reaction-diffusion equations
    I.Title II.Needham, D. J.
    511.4
    ISBN 978-1-4471-1054-5        ISBN 978-0-85729-396-1 (eBook)
    DOI 10.1007/978-0-85729-396-1

Library of Congress Cataloging-in-Publication Data
A catalog record for this book is available from the Library of Congress.

---

Mathematics Subject Classification (1991): 35K57, 34E05

---

Springer Monographs in Mathematics ISSN 1439-7382
ISBN 978-1-4471-1054-5
http://www.springer.co.uk

Typesetting: Electronic text files prepared by authors

12/3830-543210  Printed on acid-free paper  SPIN 10943380

# Preface

The idea for this monograph was born out of the desire to collate the results from two distinct strands of the authors' research with the common theme of the application of the method of matched asymptotic expansions to problems arising in reaction-diffusion theory.

In Part I, the method of matched asymptotic expansions (MAE) is used to obtain the complete structure of the solution to reaction-diffusion equations of the Fisher-Kolmogorov type for large-$t$ (dimensionless time), which exhibit the formation of a permanent form travelling wave (PTW) structure. In particular, the wave speed for the large-$t$ PTW, the correction to the wave speed and the rate of convergence of the solution onto the PTW are obtained. The primary focus of Chapters 2-4 is the scalar Fisher-Kolmogorov equation with either the generalized Fisher nonlinearity or the $m$th-order ($m > 1$) Fisher nonlinearity while in Chapter 5 the analysis is extended by consideration of a system of Fisher-Kolmogorov equations. The methodology developed is flexible and has wide applicability to scalar and systems of Fisher-Kolmogorov equations in one or higher spatial dimensions. The method of matched asymptotic expansions has also been used successfully to give information about the structure and propagation speed of accelerating phase wave (PHW) structures which can evolve in reaction-diffusion equations (see Needham and Barnes [56]) and nonlinear diffusion equations of Fisher-Kolmogorov type. The approach presented in this part of the monograph is based on the results obtained in the series of papers by Leach and Needham [32],[33],[34], Leach, Needham and Kay [35],[37],[36] and Smith, Needham and Leach [65].

In Part II we analyze a class of singular (in the sense that the nonlinearities are not Lipschitz continuous) reaction-diffusion equations. These reaction-diffusion equations can display a wide range of behaviour including:

(i) Solutions which decay to zero with contracting support in finite $t$ (say $t_c$).
(ii) Spatially uniform solutions which grow algebraically in $t$.
(iii) Permanent form travelling waves which are excitable (rather than of Fisher-Kolmogorov type).

A detailed analysis is presented of the permanent form travelling wave theory in Chapter 7 and of the initial boundary value problem in Chapters 8 and 9. This analysis requires substantial modifications to the standard theory developed for regular reaction-diffusion equations. In particular, we develop, via the method of MAE, the asymptotic structure as $t \to 0$ and as $t \to \infty$ (or $t \to t_c^-$) over all parameter values. A system of singular reaction-diffusion equations is considered in Chapter 10, with particular emphasis on the asymptotic development of the solution as $t \to 0$. The approach presented in this part of the monograph is based on the results obtained in the series of papers by McCabe, Leach and Needham [38], [39],[40],[41] and [42].

This monograph contains a wealth of results and methodologies which are applicable to a wide range of related problems arising in reaction-diffusion theory. In particular, the regions (with the analysis of their associated boundary-value problems) that constitute the asymptotic structures presented can be considered as the building blocks of the asymptotic structures of other related problems. Hence this monograph can be viewed both as a handbook and as a detailed description of methodology.

Throughout we use the nomenclature of the theory of matched asymptotic expansions, as given in Van Dyke [70]. The monograph assumes a general knowledge of perturbation methods (see for example Lagerstrom and Casten [30], Nayfeh [53], Georgescu [18] and Hinch [26]), dynamical systems theory (see for example Perko [59] and Wiggins [72]) and reaction-diffusion theory (see for example Britton [10], Fife [14] and Volpert *et al* [71]).

Both authors would like to acknowledge the contribution made to the results of Part II of this monograph by their friend and former research student Dr. Philip McCabe, C&B, and express their appreciation to Dr. Alison Kay for the numerical simulations presented in Chapters 3 and 4 and to Stephanie Smith who as a M.Sc. student worked on the PTW theory given in Chapter 5. The authors would also like to acknowledge the publishers: Birkhauser Publishing Ltd, Elsevier, Oxford University Press and the Society for Industrial and Applied Mathematics who made this monograph possible by granting permission to reproduce material from the papers mentioned above.

Reading, England,                                        *John Leach*
December, 2002                                          *David Needham*

# Contents

# The Evolution of Travelling Waves in Scalar Fisher-Kolmogorov Equations

# 1

## Introduction

In Part I of this monograph, we develop, via the method of matched asymptotic expansions (MAE), a rational approach to obtaining the complete large-$t$ (dimensionless time) structure of the solution to initial-boundary value problems (IBVPs) and initial value problems (IVPs) for reaction-diffusion equations of the Fisher-Kolmogorov type, which exhibit the formation of a permanent form travelling wave (PTW) structure. In particular, this approach allows the wave speed for the large-$t$ PTW, the correction to the wave speed and the rate of convergence of the solution of the IBVP or IVP onto the PTW to be determined. This large-$t$ structure is obtained by careful consideration of the asymptotic structures as $t \to 0$ ($0 \leq x < \infty$) (where $x$ is the dimensionless distance) and as $x \to \infty$ ($t \geq O(1)$).

We exemplify this approach by considering in detail two classes of scalar reaction-diffusion equations, namely,

$$u_t = u_{xx} + F(u), \quad -\infty < x < \infty, \quad t > 0, \tag{1.1}$$

where the reaction function, $F(u)$, is given either by:

(A) The generalized Fisher nonlinearity. Where the reaction function, $F(u)$, satisfies the normalized conditions $(F1)-(F5)$, as described in Section 1.1.
(B) The $m$th-order ($m > 1$) Fisher nonlinearity. In this case the reaction function, $F(u)$, is given by $F(u) = u^m(1 - u)$ (known as the Zeldovich nonlinearity when $m = 2$).

Equation (1.1) (with either nonlinearity (A) or (B)) is to be solved subject to the initial condition

$$u(x, 0) = u_0(x), \quad -\infty < x < \infty,$$

and the boundary condition

$$u(x, t) \to 0 \quad \text{as} \quad |x| \to \infty.$$

Sketches of the reaction function, $F(u)$, when $F(u)$ is given by (A) and (B) are given in Figures 1.1 and 1.2 respectively. We note that these nonlinearities have similar qualitative behaviour near $u = 1$ but differ significantly near $u = 0$. In particular, the $m$th-order $(m > 1)$ Fisher nonlinearity has $F(u) \sim u^m$ as $u \to 0^+$ with $F'(0) = 0$ (and zero derivatives up to order $m - 1$ for $m \geq 2$), whereas the generalized Fisher nonlinearity has $F(u) \sim u$ as $u \to 0^+$ with $F'(0) = 1$.

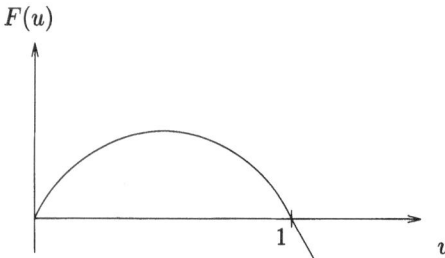

**Fig. 1.1.** The generalized Fisher nonlinearity.

Reaction-diffusion equations of the form (1.1) with associated nonlinearities (A) or (B) arise in many diverse scientific areas, for example chemistry (e.g. chemical kinetics) and biology (e.g. population dynamics and genetics). The dependant variable $u(x,t)$ may accordingly represent, for example the concentration of a chemical reactant or the population density of a biological species. A comprehensive review of the literature regarding mathematical models, based on (1.1), which arise from chemical and biological systems is given in Section 1 of McCabe, Leach and Needham [38], with a review of the basic properties of equation (1.1) being found in Section 2 of Xin [74]. For a general introduction to the mathematical modelling of chemical and biological systems see Gray and Scott [21], Murray [52] and Winfree [73].

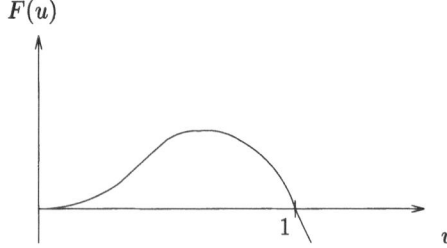

**Fig. 1.2.** The $m$th-order $(m > 1)$ Fisher nonlinearity.

We note throughout that the initial data, $u_0(x)$, is assumed to be continuous everywhere and analytic within the closure of its support. However, the more general case when $u_0(x)$ is simply piecewise differentiable can be treated in entirely the same manner but may require the inclusion of additional passive asymptotic regions in the structure of the solution as $t \to 0$. These additional regions have no influence on the final large-$t$ structure and for simplicity in what follows we restrict our attention to the former case.

Further, we assume in general that the initial data is symmetric about $x = 0$ and impose a symmetry condition, via the Neumann boundary condition

$$u_x(0, t) = 0, \quad t > 0,$$

and restrict attention to solving (1.1) in $x, t > 0$. In a chemical context this is appropriate to model the situation where the reaction proceeds on the domain $x \geq 0$ with an impermeable wall positioned at $x = 0$. This is not a technical restriction and is adopted merely for the convenience of presentation.

The methodology developed is applicable to a wide range of problems of the Fisher-Kolmogorov type provided:

(i)    (1.1) is parabolic.
(ii)   (1.1) is semilinear.
(iii)  The equilibrium state $u(x, t) \equiv 0$ of (1.1) is temporally unstable.
(iv)   A detailed knowledge of the PTW theory for (1.1) is available.

Thus we expect this method to be applicable to a wide variety of problems of the Fisher-Kolmogorov type, but not to excitable (bistable) problems where the equilibrium state $u(x, t) \equiv 0$ is temporally stable. We further note that the method is readily adaptable to parabolic systems of Fisher-Kolmogorov type and to problems in higher spatial dimensions. We conclude Part I by considering the extension of the presented method to a system of Fisher-Kolmogorov equations which arise as a simple model for an ionic autocatalytic system.

## 1.1 Generalized Fisher Nonlinearity

In this section we introduce the following initial-boundary-value problem for a scalar reaction-diffusion equation,

$$u_t = Du_{xx} + R(u), \quad x, t > 0, \tag{1.2}$$

$$u(x, 0) = \begin{cases} u_0 g(x), & 0 \leq x \leq \sigma, \\ 0, & x > \sigma, \end{cases} \tag{1.3}$$

$$u_x(0, t) = 0, \quad t > 0, \tag{1.4}$$

$$u(x, t) \to 0 \quad \text{as} \quad x \to \infty, \quad t \geq 0. \tag{1.5}$$

Here $g : [0, \sigma] \to \mathbb{R}$ is positive, analytic, has $\max_{x \in [0,\sigma]} g(x) = 1$ and $g(x) \sim g_\sigma (\sigma - x)^r$ as $x \to \sigma^-$ (with $g_\sigma > 0$ and $r \in \mathbb{N}$), with the parameters $\sigma, D, u_0 > 0$. The function $R : (-\infty, \infty) \to \mathbb{R}$ has the following properties:

(R1)  $R(u)$ is continuous and differentiable for all $u \in (-\infty, \infty)$,
(R2)  $R(0) = R(u_s) = 0$   $(u_s > 0)$,
(R3)  $R'(0) > 0, R'(u_s) < 0$,
(R4)  $R(u) > 0$ for all $u \in (0, u_s)$,
(R5)  $R(u) < 0$ for all $u \in (u_s, \infty)$.

The problem (1.2)-(1.5) can be simplified by introducing the scaled variables

$$t' = R'(0)t, \quad x' = \left[ \frac{R'(0)}{D} \right]^{1/2} x, \quad u' = \frac{u}{u_s}. \tag{1.6}$$

In terms of the above scaled variables (1.2)-(1.5) may be rewritten as (dropping primes for convenience)

$$u_t = u_{xx} + F(u), \quad x, t > 0, \tag{1.7}$$

$$u(x, 0) = \begin{cases} u_0 g(x), & 0 \le x \le \sigma, \\ 0, & x > \sigma, \end{cases} \tag{1.8}$$

$$u_x(0, t) = 0, \quad t > 0, \tag{1.9}$$

$$u(x, t) \to 0, \quad \text{as} \quad x \to \infty, \quad t \ge 0, \tag{1.10}$$

where now $F : (-\infty, \infty) \to \mathbb{R}$ satisfies the normalized conditions,

(F1)  $F(u)$ is continuous and differentiable for $u \in (-\infty, \infty)$,
(F2)  $F(0) = 0, \quad F(1) = 0$,
(F3)  $F'(0) = 1, \quad F'(1) < 0$,
(F4)  $F(u) > 0$ for all $u \in (0, 1)$,
(F5)  $F(u) < 0$ for all $u \in (1, \infty)$.

Henceforth, we will refer to (1.7)-(1.10) as IBVP (which will be discussed in detail in Chapter 2). It is readily established that IBVP has a unique, global, solution (see, for example, Smoller [66], Chapter 14) with

$$0 < u(x, t) < \max[1, u_0]$$

for all $x, t > 0$.

The particular case of IBVP which arises when

$$F(u) = u(1 - u) \tag{1.11}$$

has been studied extensively (see, for example, Fisher [16], Kolmogorov *et al* [29], McKean [43], Bramson [9], Larson [31], and Merkin and Needham [44]), when equation (1.7) is referred to as the Fisher-Kolmogorov equation. The starting point in analyzing IBVP with (1.11), is to examine the existence of propagating, permanent form travelling waves (PTW) which may be supported by equation (1.7). Any such PTW should have a constant propagation speed $v > 0$, have $u$ non-negative throughout the wave profile, whilst achieving the unreacted state $u = 0$ ahead of the wave front and the fully reacted state $u = 1$ to the rear of the wave front. Introducing a travelling wave coordinate $z = x - vt$, the existence of a PTW requires the existence of a solution to the following nonlinear boundary value problem

$$\left. \begin{array}{l} u_{zz} + v\,u_z + u(1 - u) = 0, \quad -\infty < z < \infty, \\ u(z) \to \begin{cases} 1 & \text{as} \quad z \to -\infty, \\ 0 & \text{as} \quad z \to +\infty, \end{cases} \\ u(z) \geq 0 \quad \text{for all} \quad -\infty < z < \infty. \end{array} \right\} \quad \text{BVP1}$$

This (BVP1) may be thought of as a nonlinear eigenvalue problem for the propagation speed $v > 0$. BVP1 has been studied extensively, with a review of the main results being given by Fife [14] (Chapter 4). In the present context, we recall the main result, that BVP1 has a unique (up to translation) solution if and only if

$$v \geq 2 \tag{1.12}$$

(which, in the original variables, requires $v_d \geq 2\sqrt{R'(0)D}$, with $v_d$ being the dimensional propagation speed).

In relation to IBVP, we may now enquire as to whether or not the structure of the solution to IBVP for $t \gg 1$ involves the formation of a PTW, and if so, what is the propagation speed $v \geq 2$ of this evolving PTW. For the case when $g(x) \equiv 1$ and $u_0 = 1$ this has been analyzed rigorously by Kolmogorov et al [29] and McKean [43]. It was established that a PTW does evolve in the solution of IBVP as $t \to \infty$, and this PTW is the one with minimum propagation speed, that is, the PTW with $v = 2$. For this specific case, the analysis was extended by Bramson [9], who obtained the following asymptotic estimate of the propagation speed as $t \to \infty$, namely,

$$\dot{s}(t) = 2 - \frac{3}{2}t^{-1} + o\left(t^{-1}\right), \tag{1.13}$$

where $s(t)$ is a measure of the location of the PTW wave front at time $t$. For more general initial data, when $u_0 \ll 1$, a formal theory for IBVP has been developed by Needham [54] which is primarily based on linearization of IBVP for $t \ll 1, x = O(1)$ followed by $t \geq O(1), x \gg O(t)$. This theory reproduces (1.13) for the case when $g(x)$ has finite support (In addition it enables the cases when the initial data has exponential and algebraic tails to be analyzed, which can lead to the propagation of a PTW which has $v > 2$ when $t \gg 1$). For the purpose of this chapter and that of Chapter 2 it is useful to highlight this theory for the case when $g(x)$ has finite support. The linearized version of IBVP, with $F(u)$ given by (1.11) and $u_0 \ll 1$, is

$$u_t = u_{xx} + u, \quad x, t > 0, \tag{1.14}$$

together with conditions (1.8)-(1.10). The solution to the linearized problem may be written as

$$u(x,t) = e^t D(x,t), \quad x,t \geq 0, \tag{1.15}$$

where $D(x,t)$ is the solution of the corresponding pure diffusion problem. Due to the temporal, exponential growth in (1.15), the linearized theory fails when $t \gg 1$ and $x = O(1)$, when $u = O(1)$. However, we expect (1.15) to remain a valid approximation for $t \gg 1$ when $x \gg 1$ and $u \ll 1$ (for details see Needham [54]). Approximating $D(x,t)$ for $t \gg 1$ and $x \gg O(t)$ (via steepest descents) we obtain, via (1.15),

$$u(x,t) \sim u_0 t^{-1/2} \exp\left[-t\left(\frac{y^2}{4} - 1\right)\right] \tag{1.16}$$

for $t \gg 1$ where $y = xt^{-1} = O(1)$. The approximation (1.15) followed by (1.16) will only remain accurate when $u(x,t)$ remains small, $u \leq O(u_0)$. An examination of (1.16) then shows that, when $t \gg 1$, (1.16) may be expected to remain valid for $y > 2$ but will fail for fixed $y < 2$. The overall conclusion is that for small initial data, with finite support, the solution to IBVP will be well approximated by (1.16) for fixed $y > 2$, with $u$ decaying exponentially to zero as $t \to \infty$. However, the exponential growth in (1.16) when $y < 2$ indicates that the linearized approximation fails when $y < 2$ and that $u = O(1)$ as $t \to \infty$ when $y < 2$. Therefore we may expect that a transition occurs in the solution to IBVP when $t \gg 1$ and $y \sim 2 + o(1)$; that is, when $t \gg 1$ and $x \sim 2t + o(t)$. This transition, from $u = O(1)$ to $u \ll 1$, when $x \sim 2t + o(t)$ and $t \gg 1$, is interpreted as the large-$t$ development of the PTW with minimum speed $v = 2$ in IBVP. This argument is in agreement with the rigorous results discussed earlier, and indicates that the mechanism which leads to the development of a PTW in the solution to IBVP when $t \gg 1$, and in particular the mechanism which selects the propagation speed of the emerging PTW, is based on the linearized approximation (1.16); that is, the selection of propagation speed from those available ($v \geq 2$) is determined via the evolution when $x \gg 1$, $t \geq 0$ (in the "far field").

We now move onto the more general case of IBVP, when the only restrictions on $F(u)$ are those given by (F1)-(F5). A PTW in this case requires the existence of a solution to the nonlinear boundary value problem

$$\left.\begin{array}{ll} u_{zz} + v\,u_z + F(u) = 0, & -\infty < z < \infty, \\ u(z) \to \begin{cases} 1 & \text{as} \quad z \to -\infty, \\ 0 & \text{as} \quad z \to +\infty, \end{cases} \\ u(z) \geq 0 \quad \text{for all} \quad -\infty < z < \infty, \end{array}\right\} \text{BVP2}$$

with the same notation as in BVP1. A general theory for BVP2 has been developed (and will be discussed further in Chapter 2). For the present it is sufficient to note that again there exists a value $v^* > 0$ such that BVP2 has a unique solution if and only if

$$v \geq v^*. \tag{1.17}$$

In particular, in the present context, we note that when $F(u)$ satisfies the additional condition

$$F(u) \leq F'(0)u = u \quad \text{for all} \quad u \in [0,1] \tag{1.18}$$

then

$$v^* = 2, \tag{1.19}$$

as in the case of the Fisher-Kolmogorov equation, when $F(u)$ is given by (1.11). However, when condition (1.18) is not satisfied by $F(u)$ then it is possible that

$$v^* = 2 \quad \text{or} \quad v^* > 2. \tag{1.20}$$

As an example, we exhibit in Chapter 2 that when

$$F(u) = u(1-u)(1+\hat{\sigma}u) \tag{1.21}$$

with the parameter $\hat{\sigma} \geq 0$, then (1.20(a)) holds for $0 \leq \hat{\sigma} \leq 2$, whilst (1.20(b)) holds for $2 < \hat{\sigma} < \infty$. With reference to IBVP, we may again enquire as to whether or not a PTW emerges for $t \gg 1$, and if so, what is the propagation speed $v \geq v^*$ of this emerging PTW. If we again follow the linearized theory, we obtain from IBVP, together with condition (F3), the same linearized equation as for the Fisher-Kolmogorov equation, and so we may conclude again that the solution to IBVP has the structure (1.16) when $y = O(1)$ and $t \gg 1$. This again *suggests* that a wave front emerges in the solution to IBVP when $t \gg 1$, and the wave front location, $x \sim s(t)$, has speed $\dot{s}(t) \sim 2 + o(1)$ as $t \to \infty$. This is consistent with the PTW theory when $v^* = 2$ and indicates that the PTW of minimum speed emerges in the solution to IBVP when $t \gg 1$, and this selection is dictated by the linear mechanisms when $t \gg 1$ and $x \gg 1$. However, a "paradox" emerges when we consider the situation when $v^* > 2$: the linearized theory indicates that the solution to IBVP when $x \gg 1$ and $t \gg 1$ will develop, via linear mechanisms, an emerging PTW which has location $x \sim s(t)$ and speed $\dot{s}(t) \sim 2 + o(1)$; but this is not possible, for in this case no PTW exists with speed $v = 2$ (as the minimum propagation speed $v^* > 2$). If a PTW emerges at all then its propagation speed must have $v \geq v^* > 2$. Thus, in this case the linearized mechanism fails to determine the large-$t$ propagation speed of the emerging PTW. This "paradox" has important consequences. For many reaction-diffusion problems of the Fisher-Kolmogorov type, when rigorous results are not available, linearized arguments have been extensively used to predict asymptotic wave speeds (see, for example, Murray [52], Sherratt [63], Snita *et al* [67], Gray *et al* [22]). The above argument demonstrates that this approach must be treated with caution: for IBVP the linearized approach can only be guaranteed to be accurate when $F(u)$ satisfies the additional condition (1.18).

In Chapter 2 we present a rational approach to establishing the large-$t$ structure of IBVP via a full analysis using the theory of matched asymptotic expansions. The results of the linearized theory are reproduced and extended when IBVP is such that $v^* = 2$. The situation when $v^* > 2$ is fully developed, and it is established that in this case a travelling wave of speed $v = v^*$ emerges in IBVP for $t \gg 1$, and the apparent paradox between this situation and that as presented by the linearized theory is resolved. In each case correction terms to the asymptotic PTW wave speed are obtained, which agree with the correction of Bramson when $F(u)$ has the form (1.11), and show that this correction remains the same whenever $F(u)$ satisfies (1.18). However, the correction term changes when $F(u) > u$ for some $u \in [0, 1]$, and the correction terms in this case when $v^* = 2$ and $v^* > 2$ are presented. It is particularly important to note that the theory developed in Chapter 2 is flexible, and may be applied to coupled systems of Fisher-Kolmogorov type equations, and to problems in higher spatial dimensions. At the end of Chapter 2, we illustrate the theory when applied to IBVP with $F(u)$ given by (1.21).

## 1.2 $m$th-Order ($m > 1$) Fisher Nonlinearity

In this section we introduce the following initial-boundary value problem for a scalar reaction-diffusion equation (which will be discussed in detail in Chapters 3 and 4),

$$
\left.
\begin{array}{lll}
u_t = u_{xx} + u^m(1-u), & x, t > 0, & \text{(P1)} \\
u(x,0) = u_0(x), & x \geq 0, & \text{(P2)} \\
u_x(0,t) = 0, & t > 0, & \text{(P3)} \\
u(x,t) \to 0 \text{ as } x \to \infty, & t \geq 0, & \text{(P4)}
\end{array}
\right\} \quad [\mathbf{P, m}]
$$

where the reaction order $m > 1$, and $u_0(x)$ is a continuous, piecewise analytic, non-negative and monotone decreasing function in $x \geq 0$, with $u_0(x) \to 0$ as $x \to \infty$. In particular, we consider the following classes of initial data:

(i) $u_0(x)$ is positive, analytic and has exponential decay rate as $x \to \infty$, with

$$
u_0(x) \sim
\begin{cases}
u_\infty e^{-\sigma x} + O[e^{-f(x)}] & \text{as } x \to \infty, \text{ (g1)} \\
\tilde{u}_0 + \sum_{l=1}^{\infty} \tilde{u}_l x^l & \text{as } x \to 0^+, \quad \text{ (g2)}
\end{cases}
$$

for some $f(x) > O(x)$ as $x \to \infty$, where $u_\infty, \sigma, \tilde{u}_o > 0$ and $\tilde{u}_l$ are constants.

(ii) $u_0(x)$ has compact support. In this case

$$
u_0(x) =
\begin{cases}
u_0 g(x), & 0 \leq x \leq \sigma, \\
0, & x > \sigma,
\end{cases} \quad \text{(d1)}
$$

where g:$[0,\sigma] \to \mathbb{R}$ is positive in $[0,\sigma)$, non-negative and analytic in $[0,\sigma]$, has $g(0) = 1$ and

$$
g(x) \sim
\begin{cases}
g_\sigma(\sigma - x)^r, & \text{as } x \to \sigma^-, \\
1 + g_{\tilde{m}} x^{\tilde{m}}, & \text{as } x \to 0^+.
\end{cases}
$$

Here r, $\tilde{m} \in \mathbb{N}$ with constants $g_{\tilde{m}} \neq 0$ and $g_\sigma, g_0 > 0$. The parameters $\sigma, u_0$ are positive.

(iii) $u_0(x)$ is positive, analytic and has algebraic decay rate as $x \to \infty$, with

$$u_0(x) \sim \begin{cases} u_\infty x^{-\alpha} + \text{EST}(x) & \text{as} \quad x \to \infty, \\ \tilde{u}_0 + \sum_{l=1}^{\infty} \tilde{u}_l x^l & \text{as} \quad x \to 0^+, \end{cases} \quad \text{(d2)}$$

for some $\alpha \geq \frac{1}{(m-1)}$, where $u_\infty > 0$ and $\text{EST}(x)$ denotes exponentially small terms in $x$ as $x \to \infty$.

In all cases the global existence and uniqueness of a solution to [P,m] follows directly via the comparison theorem for parabolic operators (see, for example, Smoller Chapter 14 [66]) with

$$0 < u(x,t) < \max[1, u_0] \quad (1.22)$$

for all $x, t > 0$.

The case when m=1, when (P1) is the Fisher-Kolmogorov equation, has been studied extensively (see Fisher [16], Kolmogorov *et al* [29], McKean [43], Bramson [9], Larson [31], Merkin and Needham [44]). On considering [P,1] it has been shown that travelling waves of permanent form (PTWs), which connect the equilibrium state $u = 0$ (ahead) to the equilibrium state $u = 1$ (at the rear), develop for initial data $u_0(x)$ such that

$$u_0(x) \leq O\left(e^{-\lambda x}\right) \quad \text{as} \quad x \to \infty, \quad (\lambda > 0), \quad (1.23)$$

with these PTWs having speed

$$v = \begin{cases} 2, & \lambda \geq 1, \\ \lambda + \frac{1}{\lambda}, & 0 < \lambda < 1, \end{cases} \quad (1.24)$$

(see Larson [31], McKean [43], Needham [54], Billingham and Needham [8]). However, when $u_0(x)$ is such that

$$u_0(x)e^{\lambda x} \to \infty \quad \text{as} \quad x \to \infty, \quad (1.25)$$

for all $\lambda > 0$, there are no PTW solutions which may develop in [P,1] as $t \to \infty$. In particular, Bramson [9] considered [P,1] when the initial data has compact support with a step function initial profile as $u_0(x)$ and determined that the PTW propagation speed is given by $v(t) \sim 2 - \frac{3}{2}\frac{1}{t}$ as $t \to \infty$. This result was also obtained formally by Billingham and Needham [8] via the method of matched asymptotic expansions. Further, Billingham and Needham [8] obtained (via the method of matched asymptotic expansions) the large time solution to [P,1] for initial data with compact support and unbounded support with algebraic and exponential decay rates as $x \to \infty$. They established the wave speed of the PTW in the cases when the initial data has compact support and unbounded support with exponential decay rate as $x \to \infty$ and that the asymptotic correction to the wave speed is of $O\left(\frac{1}{t}\right)$ as $t \to \infty$ in these cases. Moreover, when the initial data has unbounded support with algebraic decay rate as $x \to \infty$ (when via (1.25) no PTW exists) the large time solution exhibits an accelerating phase wave **PHW** structure (Needham and Barnes [56]). An alternative approach for $m = 1$ has been presented by Ebert and Van Sarloos [13]. However this approach does not generalize to the degenerate case $m > 1$.

On considering [**P,m**] for $m > 1$ fixed, it has been shown that travelling waves of permanent form, travelling with constant speed $v \geq v^*(m)(> 0)$ exist (see Billingham and Needham [5], Merkin and Needham [45] and Barnes [4]) which connect the equilibrium state $u = 0$ (ahead) to the equilibrium state $u = 1$ (at the rear). The PTW with minimum propagation speed $v = v^*(m)$ has exponential decay ahead of the wave front. However, each PTW with speed $v > v^*(m)$ decays algebraically ahead of the wavefront, with degree $(m - 1)^{-1}$.

The initial-boundary value problem [**P,m**] (with $m > 1$) has recently been considered by Needham and Barnes [56] in the complementary case to (iii) when $u_0(x)$ is positive, analytic and has algebraic decay rate as $x \to \infty$, given by (d2) when now $\alpha < \frac{1}{(m-1)}$ and $u_\infty > 0$. It was established that, when

$$u_0(x)x^{\frac{1}{(m-1)} - \delta} \to \infty \quad \text{as} \quad x \to \infty, \tag{1.26}$$

for some $\delta > 0$, then no PTW structure develops in the solution to [**P,m**] ($m > 1$) as $t \to \infty$ but an accelerating phase wave (**PHW**) structure develops as $t \to \infty$.

The asymptotic theory we develop for [**P,m**] with ($m > 1$), is similar in spirit to that developed by Billingham and Needham [8] for a system of reaction-diffusion equations which correspond to [**P,m**] with $m = 1$. For $m > 1$, this approach needs considerable adaptation due to the degenerate linearization of (P1) about $u = 0$, which leads to nonlinear effects being dominant for $x \gg 1, t \geq O(1)$ when $m > 1$, whereas linear effects are dominant when $m = 1$.

In Chapter 3 we obtain, using the method of matched asymptotic expansions, the full structure of the large-$t$ solution to [**P,m**] (with $m > 1$) for the cases when the initial data $u_0(x)$ has unbounded support with exponential decay rate (given by (g1), (g2), case (i)) as $x \to \infty$ and when the initial data has compact support (given by (d1), case (ii)). We establish that in both cases a PTW develops as $t \to \infty$ in [**P,m**] ($m > 1$). Further, we establish in both cases the wave speed of this PTW (this being the minimum available speed, $v = v^*(m)$), its asymptotic correction as $t \to \infty$, together with the rate of convergence of the solution to [**P,m**] onto the PTW as $t \to \infty$.

In Chapter 4 we consider [**P,m**] with initial data of the form (iii) with $\alpha \geq \frac{1}{(m-1)}$. It is demonstrated, via the method of matched asymptotic expansions, that $\alpha = \frac{1}{(m-1)}$ is a bifurcation point between the development of a PTW or a **PHW** in the solution to **[P,m]** $(m > 1)$ as $t \to \infty$, with the critical decay rate $\alpha = \frac{1}{(m-1)}$ falling into the PTW case. Moreover, we are able to determine the dependence of the propagation speed of the PTW upon the parameters $u_\infty, \alpha$ and $m$, together with its asymptotic correction as $t \to \infty$. As may be expected in the critical case $\alpha = \frac{1}{(m-1)}$, the propagation speed of the PTW, and its correction depends sensitively on the parameters $m$ and $u_\infty$.

# 2

---

# Generalized Fisher Nonlinearity

In this chapter we consider the following parabolic initial-boundary value problem, which is of reaction-diffusion type, namely,

$$u_t = u_{xx} + F(u), \quad x, t > 0, \tag{2.1}$$

$$u(x, 0) = \begin{cases} u_0 g(x), & 0 \le x \le \sigma, \\ 0, & x > \sigma, \end{cases} \tag{2.2}$$

$$u_x(0, t) = 0, \quad t > 0, \tag{2.3}$$

$$u(x, t) \to 0 \quad \text{as} \quad x \to \infty, \quad t \ge 0. \tag{2.4}$$

Here, $g[0, \sigma] \to \mathbb{R}$ is positive and analytic, has $\max_{x \in [0, \sigma]} g(x) = 1$ and has $g(x) \sim g_\sigma (\sigma - x)^r$ as $x \to \sigma^-$ (with $g_\sigma > 0$ and $r \in \mathbb{N}$), with the parameters $u_0, \sigma > 0$. The reaction function $F : (-\infty, \infty) \to \mathbb{R}$ is considered to be of generalized Fisher type, with, in particular,

(F1) $F(u)$ is continuous and differentiable for $u \in (-\infty, \infty)$,
(F2) $F(0) = 0, \quad F(1) = 0,$
(F3) $F'(0) = 1, \quad F'(1) < 0,$
(F4) $F(u) > 0 \quad$ for all $\quad u \in (0, 1),$
(F5) $F(u) < 0 \quad$ for all $\quad u \in (1, \infty).$

The initial-boundary value problem (2.1)-(2.4) is discussed in detail in Section 1.1 and henceforth we will refer to (2.1)-(2.4) as IBVP. As noted in Section 1.1, it is readily established that IBVP has a unique, global solution, with $0 < u(x, t) < \max[1, u_0]$ for all $x, t > 0$.

## 2.1 Permanent Form Travelling Waves

In this section we review the main results concerning the existence and structure of permanent form travelling waves (PTWs) which may occur in the solution to IBVP as $t \to \infty$. On introducing the travelling coordinate $z = x - vt$

(with $v > 0$ being the constant wave speed) a PTW is a solution to the following nonlinear boundary value problem

$$u'' + vu' + F(u) = 0, \quad -\infty < z < \infty, \tag{2.5}$$

$$u(z) \geq 0, \quad -\infty < z < \infty, \tag{2.6}$$

$$u(z) \to 0, \quad \text{as} \quad z \to +\infty, \tag{2.7}$$

$$u(z) \to 1, \quad \text{as} \quad z \to -\infty, \tag{2.8}$$

where $F(u)$ satisfies the properties (F1)-(F5), as laid down in the introduction. The nonlinear boundary value problem (2.5)-(2.8) can be regarded as an eigenvalue problem for the travelling wave propagation speed $v$ ($> 0$) and we have denoted this problem by BVP2. Any solution to BVP2 with $v > 0$ provides a permanent form travelling wave solution which could develop as the primary large-$t$ structure in the solution to the initial-boundary value problem IBVP. The nonlinear eigenvalue problem BVP2 has received considerable attention, and it is convenient to summarize the main results in the following theorem.

**Theorem 2.1.** *BVP2 has a unique PTW solution (say $u = u_T(z; v)$ with translational invariance fixed so that $u_T(0; v) = \frac{1}{2}$) for each $v \in [v^*, \infty)$, with $v^* \geq 2$. Moreover,*

*(a) when $F(u) \leq u$ $\forall u \in [0, 1]$, then $v^* = 2$ and*

$$u_T(z; v) \sim \begin{cases} (A^* z + B^*)e^{-z} & \text{as} \quad z \to \infty, \ v = v^*, \\ \bar{A}e^{\lambda_+(v)z} & \text{as} \quad z \to \infty, \ v > v^*; \end{cases}$$

*(b) when $F(u) \not\leq u$ $\forall u \in [0, 1]$ and $v^* = 2$, then*

$$u_T(z; v) \sim \begin{cases} (A^* z + B^*)e^{-z} & \text{as} \quad z \to \infty, \ v = v^*, \\ \bar{A}e^{\lambda_+(v)z} & \text{as} \quad z \to \infty, \ v > v^*; \end{cases}$$

*(c) when $F(u) \not\leq u$ $\forall u \in [0, 1]$ and $v^* > 2$, then*

$$u_T(z; v) \sim \begin{cases} A^* e^{\lambda_-(v^*)z} & \text{as} \quad z \to \infty, \ v = v^*, \\ \bar{A}e^{\lambda_+(v)z} & \text{as} \quad z \to \infty, \ v > v^*, \end{cases}$$

*where*

$$\lambda_\pm = -\frac{v}{2} \pm \frac{1}{2}\sqrt{v^2 - 4}.$$

*Further, in each of the above cases,*

$$u_T(z; v) \sim 1 - c^* e^{\lambda_m(v)z} \quad \text{as} \ z \to -\infty,$$

*where*

$$\lambda_m(v) = -\frac{v}{2} + \frac{1}{2}\sqrt{v^2 - 4F'(1)} \quad (> 0).$$

*Proof.* See, for example, Fife [14] or Hadeler and Rothe [24].     □

In the above, $A^*$, $B^*$ and $\bar{A}$ are constants ($A^*$ is non-negative, with $B^* > 0$ when $A^* = 0$, whilst $\bar{A}$ is positive) which can, in principle, be determined. In what follows we denote $u_T(z, v^*)$ by $u^*(z)$. Thus, for any $F(u)$ satisfying (F1)-(F4), travelling wave solutions always exist. More specifically, there is always a travelling wave of minimum speed $v = v^*$, together with faster travelling waves for each $v > v^*$. In particular, if the "curvature"of $F(u)$ on $[0, 1]$ is not too large, then $v^* = 2$. However, for sufficiently large curvature of $F(u)$ in $[0, 1]$, it is possible that $v^* > 2$.

Hence there are three possibilities which may arise from BVP2, namely,

(I) $v^* = 2$ and

$$u_T(z; v) \sim \begin{cases} A^* z e^{-z} & \text{as} \quad z \to \infty, \ v = v^*, \\ \bar{A} e^{\lambda_+(v)z} & \text{as} \quad z \to \infty, \ v > v^*; \end{cases}$$

(II) $v^* = 2$, and

$$u_T(z; v) \sim \begin{cases} B^* e^{-z} & \text{as} \quad z \to \infty, \ v = v^*, \\ \bar{A} e^{\lambda_+(v)z} & \text{as} \quad z \to \infty, \ v > v^*; \end{cases}$$

(III) $v^* > 2$, and

$$u_T(z; v) \sim \begin{cases} A^* e^{\lambda_-(v^*)z} & \text{as} \quad z \to \infty, \ v = v^*, \\ \bar{A} e^{\lambda_+(v)z} & \text{as} \quad z \to \infty, \ v > v^*, \end{cases}$$

where

$$\lambda_\pm(v) = -\frac{v}{2} \pm \frac{1}{2}\sqrt{v^2 - 4}.$$

We recall that a sufficient condition for case (I) to arise, is $F(u) \le u$ for all $u \in [0, 1]$ and hence that a necessary condition for case (II) or (III) to arise is $F(u) > u$ for some $u \in [0, 1]$. In cases (I) and (II), the travelling wave of minimum speed $v = v^* = 2$ has been referred to as a "pulled wave" whilst in case (III) the travelling wave of minimum speed has been referred to as a "pushed wave" (for a full discussion of terms "pulled wave" and "pushed wave" see Stokes [69]). This nomenclature is in reference to the observation that the linearized version of IBVP (replacing $F(u)$ by $u$ in (2.1)) will correctly predict the wave speed which develops in IBVP in cases (I) and (II), but fails to do so in case (III). This apparent paradox is discussed in detail in Section 1.1.

## 2.2 Asymptotic Solution to IBVP as $t \to \infty$

In this section we develop the asymptotic structure of the solution to IBVP as $t \to \infty$, for the three cases (I), (II) and (III) outlined in Section 2.1.

## 2.2.1 Case (III): $v^* > 2$

We must begin by examining the asymptotic structure of the solution to IBVP as $t \to 0$.

## (a) Asymptotic Solution as $t \to 0$

We first consider region **I**, in which $0 \le x \le \sigma - O(1)$ and $u = O(1)$ as $t \to 0$. Since $u(x,0) > 0$ and analytic in region **I**, with $u = O(1)$ as $t \to 0$, we expand $u(x,t)$ as a regular power series in $t$. After substitution into equation (2.1), equating powers of $t$ to zero, and applying initial condition (2.2), we obtain

$$u(x,t) = u_0 g(x) + t\left[u_0 g''(x) + F\left(u_0 g(x)\right)\right] + O(t^2) \tag{2.9}$$

as $t \to 0$ with $0 \le x \le \sigma - O(1)$. Now when $0 < (\sigma - x) \ll 1$, expansion (2.9) becomes

$$u(x,t) \sim u_0 g_\sigma \left[(\sigma - x)^r + \ldots\right] + t\left[u_0(r-1)r\, g_\sigma(\sigma-x)^{r-2} + \ldots\right.$$
$$\left. + u_0 g_\sigma(\sigma - x)^r + \ldots\right] + \ldots \tag{2.10}$$

as $t \to 0$, and it is clear from (2.10) that a non-uniformity develops in expansion (2.9) when $x = \sigma \pm O\left(t^{1/2}\right)$, when we observe that $u = O(t^{r/2})$. We must therefore introduce a further region, which we refer to as region **II**, in which $x = \sigma \pm O\left(t^{1/2}\right)$ as $t \to 0$.

To examine region **II**, we first introduce the scaled co-ordinate $\eta = (x - \sigma)t^{1/2}$ and look for an asymptotic expansion of the form

$$u(\eta,\, t) = t^{r/2}\check{u}(\eta) + o\left(t^{r/2}\right) \quad \text{as} \quad t \to 0, \tag{2.11}$$

with $\eta = O(1)$. On substitution of (2.11) into equation (2.1) (when written in terms of $\eta$ and $t$) we obtain at leading order

$$\check{u}_{\eta\eta} + \frac{\eta}{2}\check{u}_\eta - \frac{r}{2}\check{u} = 0 \quad -\infty < \check{u} < \infty, \tag{2.12}$$

which is to be solved subject to matching with region **I** as $\check{u} \to -\infty$, and the initial condition (2.2) as $t \to 0$, that is,

$$\check{u}(\eta) \sim u_0 g_\sigma(-\eta)^r \quad \text{as} \quad \eta \to -\infty, \tag{2.13}$$

$$\check{u}(\eta) = o\left(\eta^r\right) \quad \text{as} \quad \eta \to +\infty. \tag{2.14}$$

The solution to the boundary value problem (2.12)–(2.14) is unique and is readily obtained as

$$\check{u}(\eta) = \begin{cases} \dfrac{u_0 g_\sigma r!}{\left(\frac{1}{2}r\right)!\,\kappa_1} A(\eta) \displaystyle\int_\eta^\infty \dfrac{e^{-s^2/4}}{A^2(s)}\,ds, & r \text{ even}, \\[4mm] \dfrac{u_0 g_\sigma r!}{\left(\frac{1}{2}(r-1)\right)!\,\kappa_2}\left[\dfrac{A(\eta)}{\eta} - A(\eta)\displaystyle\int_\eta^\infty \left\{\dfrac{1}{s^2} - \dfrac{e^{-s^2/4}}{A^2(s)}\right\}\,ds\right], & r \text{ odd}, \end{cases} \tag{2.15}$$

with

$$
A(\eta) = \begin{cases} \displaystyle\sum_{p=0}^{\frac{r}{2}} \frac{\left[\frac{1}{2}r\right]!\,\eta^{2p}}{(2p)!\,\left[\frac{1}{2}r - p\right]!}, & r \text{ even}, \\[3ex] \displaystyle\sum_{p=0}^{\frac{1}{2}(r-1)} \frac{\left[\frac{1}{2}(r-1)\right]!\,\eta^{2p+1}}{(2p+1)!\,\left[\frac{1}{2}(r-1) - p\right]!}, & r \text{ odd} \end{cases} \tag{2.16}
$$

and

$$
\kappa_1 = \int_{-\infty}^{\infty} \frac{e^{-s^2/4}}{A^2(s)}\, ds, \tag{2.17}
$$

$$
\kappa_2 = \int_{-\infty}^{\infty} \left[\frac{1}{s^2} - \frac{e^{-s^2/4}}{A^2(s)}\right] ds. \tag{2.18}
$$

We note from (2.15) that $\breve{u}(\eta)$ is positive and monotone decreasing for all $-\infty < \eta < \infty$. From (2.15) we observe that

$$
\breve{u}(\eta) \sim C_\infty \frac{1}{\eta^{r+1}}\, e^{-\eta^2/4} \quad \text{as} \quad \eta \to \infty \tag{2.19}
$$

with

$$
C_\infty = \begin{cases} \dfrac{2u_0 g_\sigma (r!)^2}{\kappa_1 \left[\left(\frac{1}{2}r\right)!\right]^2}, & r \text{ even}, \\[3ex] \dfrac{2u_0 g_\sigma (r!)^2}{\kappa_2 \left[\left(\frac{1}{2}(r-1)\right)!\right]^2}, & r \text{ odd}. \end{cases} \tag{2.20}
$$

As $\eta \gg 1$ we move out of region **II** and we are left to introduce a final region, region **III**, where $x = \sigma + O(1)$ and $u(x,t)$ is exponentially small as $t \to 0$, via (2.11) and (2.19). The structure of the solution in region **II** as $\eta \to \infty$ (given by (2.11) and (2.19)) suggests that in region **III** we expand as

$$
u(x,t) = e^{-\frac{F(x,t)}{t}} \quad \text{as} \quad t \to 0, \tag{2.21}
$$

with

$$
F(x,t) = F_0(x) + F_1(x)t \ln t + F_2(x)t + O(t^2), \tag{2.22}
$$

where $x = \sigma + O(1)$ and $F(x,t) > 0$ for all $x > \sigma$. Substitution of (2.21) and (2.22) into equation (2.1) gives (on solving at each order in turn)

$$
u(x,t) = \exp\left\{ -\frac{(x+C)^2}{4t} - A \ln t - \left(\frac{1}{2} - A\right) \ln(x+C) - B + O(t) \right\} \tag{2.23}
$$

as $t \to 0$, with $x = \sigma + O(1)$, where $A$, $B$ and $C$ are arbitrary constants. Clearly (2.23) satisfies the initial condition in region **III** ($u(x,t) \to 0$ as $t \to 0$), together with the boundary condition as $x \to \infty$ (given by (2.4)). It remains to match expansion (2.23) in region **III** as $x \to \sigma^+$ with expansion (2.11) in region **II** as $\eta \to \infty$. Matching follows directly, giving

$$A = -\left(r + \frac{1}{2}\right) \quad B = -\ln C_\infty \quad C = -\sigma. \tag{2.24}$$

Furthermore, we conclude from (2.23) that this expansion remains uniform for $x \gg 1$ as $t \to 0$.

Finally, we note that, with $g(x) \sim g_0 + g_{\tilde{m}} x^{\tilde{m}} + \dots$ as $x \to 0^+$, where $g_0 > 0, g_{\tilde{m}} \neq 0$ and $\tilde{m} \in \mathbb{N}$ (as $g(x)$ is analytic in $0 \le x \le \sigma$), the expansion (2.9) in region **I** does not, in general, satisfy boundary condition (2.3) at $x = 0$, and a further passive region is required in the neighbourhood of $x = 0$ as $t \to 0$. We denote this as region $\mathbf{I_0}$ and it is readily deduced that in this region $x = O\left(t^{1/2}\right)$ as $t \to 0$. The appropriate expansion in this region (which satisfies the appropriate initial condition, boundary condition (2.3) and matches to region **I** for $x \gg t^{1/2}$) is readily obtained as, when $\tilde{m} = 1$,

$$u(\eta^*, t) = u_0 g_0 + t^{1/2} u_0 g_1 \eta^* \left[1 + \frac{1}{C} \int_{\eta^*}^\infty \frac{e^{-s^2/4}}{s^2} ds\right] + O(t) \tag{2.25}$$

as $t \to 0$, with $\eta^* = xt^{-1/2} = O(1)$ and

$$C = \int_0^\infty \frac{1 - e^{-s^2/4}}{s^2} ds. \tag{2.26}$$

For $\tilde{m} \ge 2$, we obtain

$$u(x, t) = u_0 g_0 + t \left[F(u_0 g_0) + 2\delta_{2,\tilde{m}} u_0 g_2 \left(1 + \frac{\eta^{*2}}{2}\right)\right] + O(t^2) \tag{2.27}$$

as $t \to 0$ where $\eta^* = O(1)$ and $\delta_{2,\tilde{m}}$ is the Kronecker delta.

The asymptotic structure as $t \to 0$ is now complete, with the expansions in regions $\mathbf{I_0}$, **I**, **II** and **III** providing a uniform approximation to the solution of IBVP as $t \to 0$. We next use this information to develop the asymptotic structure of the solution to IBVP, as $x \to \infty$ with $t = O(1)$.

### (b) Asymptotic Solution as $x \to \infty$

We now investigate the structure of the solution to IBVP as $x \to \infty$ with $t = O(1)$. The form of expansion (2.23) for $x \gg 1$ as $t \to 0$ suggests that in this region, which we will label as region **IV**, we expand as

$$u(x, t) = e^{-\check{H}(x,t)} \quad \text{as} \quad x \to \infty, \tag{2.28}$$

with

$$\check{H}(x, t) = H_0(t)x^2 + H_1(t)x + H_2(t)\ln x + H_3(t) + H_4(t)x^{-1} + O\left(x^{-2}\right), \tag{2.29}$$

where $t = O(1)$ as $x \to \infty$. On substituting from (2.28) and (2.29) into equation (2.1) and solving at each order in turn, we find (after matching with (2.23) as $t \to 0$) that

$$u(x,t) = \exp\left\{-\frac{x^2}{4t} + \frac{\sigma x}{2t} - (r+1)\ln x + \left[\left(r+\tfrac{1}{2}\right)\ln t + t + \ln C_\infty - \frac{\sigma^2}{4t}\right]\right.$$
$$\left. +O(x^{-1})\right\} \tag{2.30}$$

as $x \to \infty$ with $t = O(1)$. Expansion (2.30) will remain uniform for $t \gg 1$ provided that $x \gg t$, but fails to provide an asymptotic approximation when $x = O(t)$ as $t \to \infty$.

### (c) Asymptotic Solution as $t \to \infty$

As $t \to \infty$, the asymptotic expansion (2.30) of region **IV** continues to remain uniform for $x \gg t$. However, as already noted, a non-uniformity develops when $x = O(t)$. To proceed, we introduce a new region, region **V**. To examine region **V** we introduce the scaled coordinate $y = \frac{x}{t}$, where $y = O(1)$ as $t \to \infty$, and look for an expansion of the form (as suggested by (2.30))

$$u(y,t) = e^{-t\check{F}(y,t)} \quad \text{as} \quad t \to \infty, \tag{2.31}$$

with

$$\check{F}(y,t) = f_0(y) + f_1(y)\frac{\ln t}{t} + f_2(y)\frac{1}{t} + O\left(t^{-2}\right), \tag{2.32}$$

where $y = O(1)(> 0)$ as $t \to \infty$, and $f_0(y) > 0$. It is instructive to consider first the leading order problem in region **V**. On substituting from (2.31) and (2.32) into equation (2.1) (when written in terms of $y$ and $t$) we obtain the leading order problem as

$$f_{0y}^2 - y\,f_{0y} + f_0 + 1 = 0 \quad y > 0, \tag{2.33}$$

$$f_0(y) > 0 \quad y > 0, \tag{2.34}$$

$$f_0(y) \sim \frac{y^2}{4} - 1 \quad \text{as} \quad y \to \infty. \tag{2.35}$$

The final condition, (2.35), arises from matching expansion (2.31) ($y \gg 1$) with expansion (2.30) ($x = O(t)$). Equation (2.33) has a one-parameter family of linear solutions,

$$f_0(y) = c_0(y - c_0) - 1 \quad y > 0, \tag{2.36}$$

for any $c_0 \in \mathbb{R}$, together with the associated envelope solution

$$f_0(y) = \frac{y^2}{4} - 1, \quad y > 0. \tag{2.37}$$

Combinations of (2.36) and (2.37) which remain continuous and differentiable also provide solutions to (2.33) (envelope touching solutions). Applying the condition (2.35) requires us to select the solution

$$f_0(y) = \frac{y^2}{4} - 1, \quad y > 0, \tag{2.38}$$

or

$$f_0(y) = \begin{cases} \frac{y^2}{4} - 1, & y > y_0, \\ \frac{1}{2}y_0(y - \frac{1}{2}y_0) - 1, & 0 < y \le y_0, \end{cases} \tag{2.39}$$

for any $y_0 > 2$. We next check condition (2.34). We immediately observe that neither (2.38) nor (2.39) can fully satisfy this condition: (2.38) vanishes as $y \to 2^+$ whilst (2.39) vanishes as $y \to \left(\frac{2}{y_0} + \frac{y_0}{2}\right)^+$ ($> 2$ for $y_0 > 2$). We conclude that a non-uniformity occurs in expansion (2.31), (2.32) as $y \to y_c$ ($\ge 2$) where

$$y_c = \frac{2}{y_0} + \frac{y_0}{2} \begin{cases} = 2, & y_0 = 2, \\ > 2, & y_0 > 2, \end{cases} \tag{2.40}$$

for some $y_0 \ge 2$ (when $y_0 = 2$, $f_0(y)$ is given by (2.38), whilst when $y_0 > 2$, $f_0(y)$ is given by (2.39)). A consideration of further terms in (2.32) demonstrates that this nonuniformity occurs when

$$y = y_c + O\left(t^{-1}\right) \quad \text{with} \quad u = O(1) \tag{2.41}$$

as $t \to \infty$. We must introduce a further region, which we denote as region **TW**. In this region we write, via (2.41),

$$y = y_c + \frac{z}{t} \tag{2.42}$$

with $z = O(1)$ as $t \to \infty$, and expand

$$u(z,t) = u_c(z) + o(1), \quad z = O(1) \tag{2.43}$$

as $t \to \infty$. On substituting into equation (2.1) we obtain the leading order problem as

$$u_c'' + y_c\, u_c' + F(u_c) = 0, \quad -\infty < z < \infty, \tag{2.44}$$

$$u_c(z) > 0, \quad -\infty < z < \infty, \tag{2.45}$$

$$u_c(z) \to 0 \quad \text{as} \quad z \to +\infty, \tag{2.46}$$

$$u_c(z) \quad \text{bounded as} \quad z \to -\infty. \tag{2.47}$$

The condition (2.46) arises from matching expansion (2.43) ($z \to \infty$) with expansion (2.31), (2.32) (as $y \to y_c^+$). Moreover, a phase plane analysis (Poincaré-Bendixson theorem) of equation (2.44) with conditions (2.45) and (2.46) allows boundary condition (2.47) to be replaced by

$$u_c(z) \to 1 \quad \text{as} \quad z \to -\infty \tag{2.48}$$

(all other possibilities leading to unbounded $u_c(z)$ as $z \to -\infty$). We now recognize the nonlinear boundary value problem (2.34)-(2.48) as being precisely BVP2, its solutions representing permanent form travelling wave structures. We can now appeal to Theorem 2.1: (2.44)-(2.48) has a unique solution

$u_c(z) = u_T(z, y_c)$ for each $y_c \in [v^*, \infty)$. In particular, in this case $v^* > 2$. Therefore, we must have

$$y_c > 2, \tag{2.49}$$

and so, in region **V**, via (2.40), $y_0 > 2$ and $f_0(y)$ given by the associated form of (2.39). We next match expansion (2.31), (2.32) to expansion (2.43) at next order. It is convenient to match $U = \log u$ rather than $u$ itself. We follow the matching principle of Van Dyke [70] in matching the expansion (2.31), (2.32), for $U$, to $O(t)$, with the expansion (2.43) for $U$, to $O(1)$. On taking expansion (2.31), (2.32) for $U$, expanding in region **TW** up to $O(1)$ gives the expression

$$U_{t0} = -\frac{1}{2} y_0 \, z = \lambda_-(y_c) z \tag{2.50}$$

after using the relation (2.40) and the expression (2.39) for $f_0(y)$. Conversely, on taking expansion (2.43) for $U$, to $O(1)$, and expanding in region **V** up to $O(t)$ gives the expression

$$U_{0t} = \begin{cases} \lambda_-(y_c)z, & y_c = v^*, \\ \lambda_+(y_c)z, & y_c > v^*. \end{cases} \tag{2.51}$$

Van Dyke's matching principle requires $U_{0t} \equiv U_{t0}$. Comparing (2.50) and (2.51), matching then requires that

$$y_c = v^*, \tag{2.52}$$

and the travelling wave solution of minimum speed $v^* (> 2)$ is selected in region **TW**. With $y_c$ now fixed in (2.52), we can invert expression (2.40) to obtain $y_0$ as

$$y_0 = v^* + \left[ (v^*)^2 - 4 \right]^{1/2} = -2\lambda_-(v^*)(> 2) \tag{2.53}$$

and the expansions in both regions **V** and **TW** are now complete at leading order.

A sketch of $f_0(y)$ is given in Figure 2.1, where we label solutions (2.38) and (2.39) (with (2.52) and (2.53)) as $\mathcal{A}$ and $\mathcal{B}$ respectively. For completeness we note in this case (when $v^* > 2$ with $f_0(y)$ given by $\mathcal{B}$) that, although $f_0(y)$ and $f_0'(y)$ are continuous for $y \geq v^* (= y_c)$, the second derivative $f_0''(y)$ is discontinuous at the point $y = y_0 (> v^*)$. This indicates that a thin transition region, region **VI**, exists in the neighbourhood of the point $y = y_0$, in which second derivatives are retained at leading order to smooth out this discontinuity in curvature. Hence, region **V** is replaced by three regions, region **V(a)** $(y_0 + o(1) < y < \infty)$, region **VI** (transition region) and region **V(b)** $(v^* \leq y < y_0 - o(1))$. We consider each of these regions in turn. We begin with region **V(a)**, where $-2\lambda_-(v^*) + o(1) < y < \infty$. Substitution of (2.31), (2.32) into equation (2.1) (when written in terms of $y$ and $t$), gives on solving at each order in turn and matching to expansion (2.30) as $y \to \infty$ that

$$u(y,t) = \exp \left\{ -t \left[ \left( \frac{y^2}{4} - 1 \right) + \frac{1}{2} \frac{\ln t}{t} + \frac{H(y)}{t} + O\left( t^{-2} \right) \right] \right\} \tag{2.54}$$

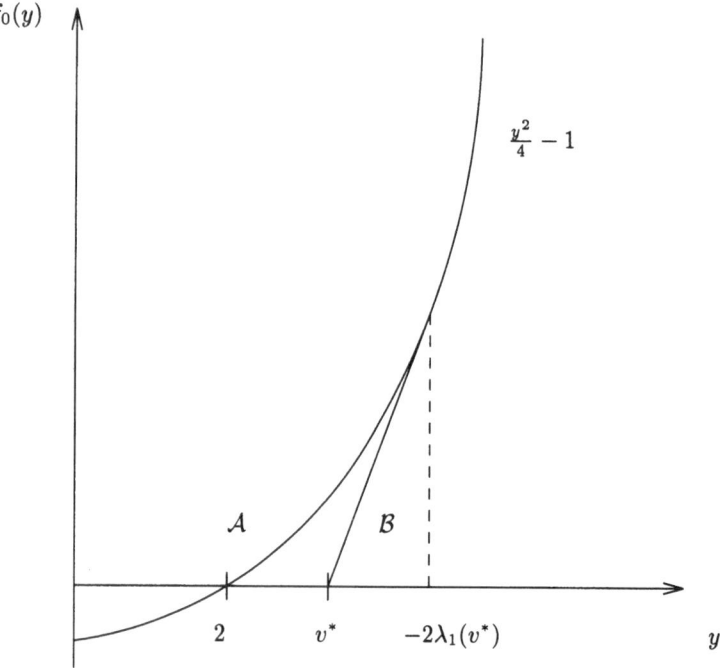

**Fig. 2.1.** A sketch of $f_0(y)$.

as $t \to \infty$ with $-2\lambda_-(v^*) + o(1) < y < \infty$. The function $H(y)$ remains undetermined. This is a consequence of using the far field asymptotics $x, t \gg 1$ rather than information from the bulk region $x, t = O(1)$ as a basis for the large time asymptotic structure. However, we can determine the asymptotic properties of $H(y)$ on matching with the far field (as $y \to \infty$) and matching to region **VI** (as $y \to -2\lambda_-(v^*)^+$). Matching with expansion (2.30) ($x = O(t)$) as $y \to \infty$ requires that

$$H(y) \sim (r+1)\ln y - \ln C_\infty - \frac{\sigma y}{2} \quad \text{as } y \to \infty. \tag{2.55}$$

In region **V(b)** we have

$$u(y,t) = \exp\left\{-t\left[-\lambda_-(v^*)\left(y + \lambda_-(v^*)\right) - 1\right] + \alpha \ln t \right. \\ \left. + \alpha \ln\left(-2\lambda_-(v^*) - y\right) + \beta + o(1)\right\} \tag{2.56}$$

as $t \to \infty$ with $v^* + o(1) < y < -2\lambda_-(v^*) - o(1)$. Here the constants $\alpha$ and $\beta$ are, as yet, undetermined. As $y \to -2\lambda_-(v^*)^+$ we move into the transition region, region **VI**. An examination of expansion (2.54) as $y \to -2\lambda_-(v^*)$ reveals that in this region, $y = -2\lambda_-(v^*) + O\left(t^{-1/2}\right)$ as $t \to \infty$. Thus in region **VI** (via (2.54) and (2.56)) we introduce the scaled co-ordinate $\eta = \left(y + 2\lambda_-(v^*)\right) t^{1/2}$, and expand as

$$u(\eta, t) = F(\eta) \exp\left\{-t\left[\lambda_-^2(v^*) - 1\right] + \lambda_-(v^*)\eta\, t^{1/2} + o(1)\right\} \qquad (2.57)$$

as $t \to \infty$ with $F(\eta) > 0$, $\eta = O(1)$. Substitution of (2.57) into equation (2.1) (when written in terms of $\eta$ and $t$ ) gives at leading order,

$$F_{\eta\eta} + \frac{1}{2}\eta F_\eta = 0 \quad -\infty < \eta < \infty. \qquad (2.58)$$

On matching (2.57) to (2.56), we require

$$\alpha = 0 \qquad (2.59)$$

and

$$F(\eta) \to e^\beta \quad \text{as} \quad \eta \to -\infty. \qquad (2.60)$$

The solution of (2.58) which satisfies (2.60) is

$$F(\eta) = A \operatorname{erfc}\left(\frac{1}{2}\eta\right) + B, \qquad (2.61)$$

where $A$, $B \geq 0$ and $2A + B = e^\beta$. Moreover, matching expansion (2.56) (as $\eta \to \infty$) with expansion (2.54) (as $y \to -2\lambda_-(v^*)^+$) then requires that

$$H(y) \sim \ln\left(y + 2\lambda_-(v^*)\right) + \ln\frac{2A}{\sqrt{\pi}} \quad \text{as} \quad y \to -2\lambda_-(v^*)^+, \qquad (2.62)$$

and

$$B = 0. \qquad (2.63)$$

The expansion in region **VI** is now given via (2.57) and (2.59) as

$$u(\eta, t) = A \operatorname{erfc}\left(\frac{1}{2}\eta\right) \exp\left\{-t\left[\lambda_-^2(v^*) - 1\right] + \lambda_-(v^*)\eta t^{1/2} + o(1)\right\} \quad (2.64)$$

as $t \to \infty$ with $\eta = O(1)$. The expansion in region **V(b)** is now given as

$$u(y, t) = \exp\left\{-t\left[-\lambda_-(v^*)\left(y + \lambda_-(v^*)\right) - 1\right] + \ln 2A + o(1)\right\} \qquad (2.65)$$

as $t \to \infty$ with $v^* + o(1) < y < -2\lambda_-(v^*) - o(1)$. Now as $y \to (v^*)^+$ we move into the wave front region, region **TW**, where $y = v^* + O\left(\frac{1}{t}\right)$ as $t \to \infty$. In region **TW**, $x \sim s(t)$ and $u$ (when written in terms of the travelling wave co-ordinate, $z$ ) has the form (via (2.43) and (2.52)),

$$u(z, t) = u_T(z, v^*) + o(1) \qquad (2.66)$$

as $t \to \infty$ with $z = O(1)$, where $z = x - s(t)$ and we now have, via (2.52), that

$$s(t) = v^* t + v_1 \phi(t) + \phi_0 + o(1). \qquad (2.67)$$

Here $\phi(t) = o(t)$ is as yet an undetermined gauge function (to be fixed on matching with region $\mathbf{V(b)}$ as $z \to \infty$), $\phi_0$ is a constant, whilst $u_T(z, v^*)$ represents the minimum speed $(v = v^*)$ permanent form travelling wave solution. We recall from Section 2.1 the following asymptotic properties of $u_T(z, v^*)$,

$$u_T(z, v^*) \sim \begin{cases} A^* e^{\lambda_-(v^*)z} & \text{as} \quad z \to +\infty, \\ 1 - C^* e^{\lambda_m(v^*)z} & \text{as} \quad z \to -\infty, \end{cases} \tag{2.68}$$

where $A^*$, $C^*$ are positive constants which are, in principle, determined. Matching expansion (2.66) (as $z \to \infty$) with expansion (2.65) of region $\mathbf{V(b)}$ (as $y \to (v^*)^+$) fixes $\phi(t) \equiv 0$ and requires that $\phi_0 = \frac{1}{\lambda_-(v^*)} \ln \left( \frac{A^*}{2A} \right)$. Hence we have established that the velocity of propagation of the wave front is given by

$$\dot{s}(t) = v^* + o\left(t^{-1}\right) \tag{2.69}$$

as $t \to \infty$. Further, to obtain a balance at the next order in equation (2.1) (when written in terms of $z$ and $t$) we require the correction to (2.66) to be $o\left(t^{-1}\right)$, giving

$$u(z, t) = u_T(z, v^*) + o\left(t^{-1}\right) \quad \text{as} \quad t \to \infty. \tag{2.70}$$

The correction terms to expansions (2.69) and (2.70) can readily be determined and we consider these now before completing the asymptotic structure as $t \to \infty$ in $0 \leq y < v^* - o(1)$.

We begin in region $\mathbf{VI}$. We develop the leading order expansion (2.64) as

$$u(\eta, t) = E(t, \eta) \left[ h_0(\eta) + \hat{\psi}(\eta, t) \right] \tag{2.71}$$

with $\eta = O(1)$ as $t \to \infty$. Here

$$E(t, \eta) = \exp \left\{ -t \left[ \lambda_-^2(v^*) - 1 \right] + \lambda_-(v^*) \eta\, t^{1/2} \right\}, \tag{2.72}$$

$$h_0(\eta) = A \text{erfc} \left( \frac{1}{2} \eta \right), \tag{2.73}$$

$$\hat{\psi}(\eta, t) = o(1) \quad \text{as} \quad t \to \infty, \quad \text{with} \quad \eta = O(1). \tag{2.74}$$

On substitution from (2.71)-(2.73) into the full equation (2.1) (when written in terms of $\eta$ and $t$), we arrive at

$$\hat{\psi}_{\eta\eta} + \frac{1}{2} \eta\, \hat{\psi}_\eta = t\hat{\psi}_t + O\left[t\, h_0^2(\eta)\, E(t, \eta)\right]. \tag{2.75}$$

A non-trivial balance in this equation (as $t \to \infty$ with $\eta = O(1)$) requires that $\hat{\psi}(\eta, t)$ has the form

$$\hat{\psi}(\eta, t) = \tilde{\psi}(\eta)\, E(t, \eta). \tag{2.76}$$

The problem for $\tilde{\psi}(\eta)$ need not be solved here, but we have established that in region **VI** that

$$u(\eta, t) = A\text{erfc}\left(\frac{1}{2}\eta\right) E(t, \eta) + O\left[E(t, \eta)^2\right] \quad \text{as} \quad t \to \infty, \quad (2.77)$$

with $\eta = O(1)$. We now examine the form of expansion (2.77) for $(-\eta) \gg 1$ (as we move into region **V(b)**). From (2.77) we have,

$$u(\eta, t) \sim 2A\, E(t, \eta) + \frac{2AE(t, \eta)}{\sqrt{\pi}\eta} e^{-\frac{\eta^2}{4}} + O\left[E(t, \eta)^2\right] \quad \text{as} \quad t \to \infty, \quad (2.78)$$

with $(-\eta) \gg 1$. When written in terms of $y$, (2.78) becomes

$$u(y, t) \sim 2Ae^{-t[-\lambda_-(v^*)(y+\lambda_-(v^*))-1]} + \frac{2A}{\sqrt{\pi}[y+2\lambda_-(v^*)]t^{\frac{1}{2}}} e^{-t\left[\frac{y^2}{4}-1\right]}$$
$$+ O\left[e^{-2t[-\lambda_-(v^*)(y+\lambda_-(v^*))-1]}\right]. \quad (2.79)$$

Thus, we must look for an expansion in region **V(b)** in the form

$$u(y, t) = 2A\, e^{-t[-\lambda_-(v^*)(y+\lambda_-(v^*))-1]} + \frac{2A}{\sqrt{\pi}} \hat{H}(y)\, t^{-1/2} e^{-t\left[\frac{y^2}{4}-1\right]}$$
$$+ O\left[e^{-2t[-\lambda_-(v^*)(y+\lambda_-(v^*))-1]}\right] \quad (2.80)$$

as $t \to \infty$, with $v^* + O\left(t^{-1}\right) < y < -2\lambda_-(v^*) - O\left(t^{-1/2}\right)$ as $t \to \infty$. On substitution from (2.80) into the full equation (2.1) (when written in terms of $y$ and $t$), we find that $\hat{H}(y)$ is indeterminate at this order. However, matching to region **VI** (as $y \to -2\lambda_-(v^*)$) requires that

$$\hat{H}(y) \sim \frac{1}{[y + 2\lambda_-(v^*)]} \quad \text{as} \quad y \to (-2\lambda_-(v^*))^-. \quad (2.81)$$

We now move into region **TW**, where

$$y = \frac{s(t)}{t} + \frac{z}{t} \quad (2.82)$$

with $z = O(1)$ as $t \to \infty$, and

$$s(t) = v^* t + \phi_0 + \chi(t) \quad \text{as} \quad t \to \infty. \quad (2.83)$$

Here, via (2.67), $\chi(t) = o(1)$ as $t \to \infty$. To determine the correction to expansion (2.70) in region **TW**, we first examine the form of expansion (2.80) as we move into region **TW** ($y \to (v^*)^+$). We obtain from (2.80), when written in terms of $z$, that

$$u(z, t) \sim 2Ae^{[\lambda_-(v^*)(z+\phi_0)+\lambda_-(v^*)\chi]} + \frac{2A}{\sqrt{\pi}} \hat{H}t^{-\frac{1}{2}} e^{-t\left[\frac{(v^*)^2}{4}-1\right]}$$
$$\times e^{-\frac{1}{2}v^*[z+\phi_0+\chi]} + O\left[e^{2\lambda_-(v^*)(z+\phi_0)}\right] \quad (2.84)$$

as $t \to \infty$ with $z \gg 1$. We now make the assumption (which we will verify as consistent) that

$$\hat{H}(y) \sim c(y - v^*)^{\hat{\gamma}} \quad \text{as} \quad y \to (v^*)^+ \tag{2.85}$$

for some constants $c$ and $\hat{\gamma}$. With (2.85), we can reduce (2.84) to

$$u(z,t) \sim \left\{ 2Ae^{\lambda_-(v^*)(z+\phi_0)} + O\left[ e^{2\lambda_-(v^*)(z+\phi_0)} \right] \right\} + \left\{ 2A\lambda_-(v^*)\chi e^{\lambda_-(v^*)(z+\phi_0)} \right.$$

$$\left. + \dots \right\} + \left\{ \frac{2A}{\sqrt{\pi}} t^{-\frac{1}{2}-\hat{\gamma}} e^{-t\left[ \frac{(v^*)^2}{4} - 1 \right]} e^{-\frac{1}{2}v^*(z+\phi_0)} cz^{\hat{\gamma}} + \dots \right\} + \dots \tag{2.86}$$

as $t \to \infty$ with $z \gg 1$. We conclude from (2.86) that in region **TW** we must have $u(z,t) = u_T(z,v^*) + O\left( \chi(t) \right)$ as $t \to \infty$ (we note that the translational invariance of $u_T(z,v^*)$ with respect to $z$ is fixed by requiring $u_T(0,v^*) = \frac{1}{2}$). Now equation (2.1) when written in terms of $z$ and $t$ becomes

$$u_t - v^* u_z - u_{zz} - F(u) = \chi'(t)u_z. \tag{2.87}$$

Whilst $u(z,t) = u_T(z,v^*) + O\left( \chi(t) \right)$. Thus, to obtain a non-trivial balance in (2.87) at $O\left( \chi'(t) \right)$ we require

$$\chi'(t) = O\left( \chi(t) \right) \quad \text{as} \quad t \to \infty. \tag{2.88}$$

We conclude from (2.88) that $\chi(t)$ must be of exponentially small order as $t \to \infty$. Thus we set

$$\chi(t) = \hat{v} t^{\gamma} e^{-\hat{\sigma} t} \left[ 1 + o(1) \right] \tag{2.89}$$

as $t \to \infty$, with the constants $\hat{v}, \gamma$ and $\hat{\sigma}$ ($> 0$) to be determined. We now expand in region **TW** as

$$u(z,t) = u_T(z,v^*) + \chi'(t)u_1(z) + o\left[ \chi'(t) \right] \tag{2.90}$$

as $t \to \infty$ with $z = O(1)$. At $O\left( \chi'(t) \right)$ we obtain the equation

$$u_1'' + v^* u_1' + \left[ \hat{\sigma} + F'\left( u_T(z,v^*) \right) \right] u_1 = -u_T'(z,v^*) \quad -\infty < z < \infty. \tag{2.91}$$

Now for $z \gg 1$, $F'\left( u_T(z,v^*) \right) \to 1$, and $u_T'(z,v^*) \sim \lambda_-(v^*)A^* e^{\lambda_-(v^*)z}$. Therefore, (2.91) becomes

$$u_1'' + v^* u_1' + \left[ \hat{\sigma} + 1 \right] u_1 \sim -\lambda_-(v^*)A^* e^{\lambda_-(v^*)z} \tag{2.92}$$

for $z \gg 1$. Thus we have, from (2.92),

$$u_1(z) \sim Ee^{\alpha_+ z} + De^{\alpha_- z} - \frac{\lambda_-(v^*)}{\hat{\sigma}} A^* e^{\lambda_-(v^*)z} \tag{2.93}$$

for $z \gg 1$, where $E, D$ are constants and

$$\alpha_{\pm} = \frac{-v^* \pm \left[ (v^*)^2 - 4(\hat{\sigma} + 1) \right]^{1/2}}{2}. \tag{2.94}$$

Matching with (2.86) will require no oscillations in $u_1(z)$, and we may then conclude immediately that

$$\hat{\sigma} \le \frac{1}{4} \left( (v^*)^2 - 4 \right). \tag{2.95}$$

Finally, we have in region **TW**, via (2.90) and (2.93), that

$$\begin{aligned}
u(z,t) \sim \ & \left[ A^* e^{\lambda_-(v^*)z} + O\left( e^{2\lambda_-(v^*)z} \right) \right] - \hat{v}\hat{\sigma} t^\gamma e^{-\hat{\sigma}t} \\
& \times \left[ E e^{\alpha_+ z} + D e^{\alpha_- z} - \frac{\lambda_-(v^*)A^*}{\hat{\sigma}} e^{\lambda_-(v^*)z} \right] + \cdots
\end{aligned} \tag{2.96}$$

as $t \to \infty$ with $z \gg 1$. It now remains to match expansion (2.86) from region **V(b)** with expansion (2.96) from region **TW**. Matching terms of $O(1)$ has already been considered and requires that

$$\phi_0 = \frac{1}{\lambda_-(v^*)} \ln \left[ \frac{A^*}{2A} \right]. \tag{2.97}$$

Next we match the exponentially small terms. These will match only if

$$\alpha_+ = \alpha_- = -\frac{1}{2} v^*, \tag{2.98}$$

which gives, via (2.94), that

$$\hat{\sigma} = \frac{1}{4} \left( (v^*)^2 - 4 \right). \tag{2.99}$$

However, the form of (2.96) now needs to be adjusted. With (2.99) and (2.98), the homogeneous part of (2.93) changes, so that (2.93) is now

$$u_1(z) \sim \hat{E} z e^{-\frac{1}{2} v^* z} - \frac{\lambda_-(v^*)}{\hat{\sigma}} A^* e^{\lambda_-(v^*)z}, \tag{2.100}$$

for $z \gg 1$, with $\hat{E}$ constant. The corresponding form of (2.96) is

$$\begin{aligned}
u(z,t) \sim \ & \left[ A^* e^{\lambda_-(v^*)z} + O\left( e^{2\lambda_-(v^*)z} \right) \right] - \hat{v}\hat{\sigma} t^\gamma e^{-\hat{\sigma}t} \\
& \times \left[ \hat{E} z e^{-\frac{1}{2} v^* z} - \frac{\lambda_-(v^*)A^*}{\hat{\sigma}} e^{\lambda_-(v^*)z} \right] + \cdots.
\end{aligned} \tag{2.101}$$

Comparison of the exponentially small terms in (2.101) and (2.86) now requires

$$\hat{\gamma} = 1, \quad \gamma = -\frac{1}{2} - \hat{\gamma} = -\frac{3}{2}, \hat{v} = \frac{-2Ac}{\sqrt{\pi}\,\hat{E}\left[ \frac{1}{4}(v^*)^2 - 1 \right]} e^{-\frac{1}{2} v^* \phi_0}, \tag{2.102}$$

and matching is complete. Note that $\hat{v}$ is given in terms of $A$ and $\hat{E}$, which, although both non-zero $(A > 0$, in fact), are indeterminate at this order in our asymptotic theory. Finally, we have in region **TW**,

$$u(z,t) = u_T(z,v^*) + O\left[t^{-3/2}e^{-\frac{1}{4}((v^*)^2-4)t}\right] \qquad (2.103)$$

as $t \to \infty$ with $z = O(1)$, and

$$s(t) = v^*t + \phi_0 + \hat{v}t^{-3/2}e^{-\frac{1}{4}((v^*)^2-4)t} + o\left[t^{-3/2}e^{-\frac{1}{4}((v^*)^2-4)t}\right] \qquad (2.104)$$

as $t \to \infty$.

We now complete the asymptotic structure as $t \to \infty$ by developing the asymptotic structure of the solution of IBVP in $0 \le y < v^* - o(1)$.

As $z \to -\infty$, we move into region **VII**, where $0 \le y < v^* - o(1)$. The structure of expansion (2.103) as $z \to -\infty$ (obtained via (2.68)), suggests that in region **VII** we expand as

$$u(y,t) = 1 - C^* e^{[f(y)t+o(1)]} t^{-\frac{\lambda_m(v^*)}{2\lambda_-(v^*)}} \qquad (2.105)$$

as $t \to \infty$ with $0 \le y < v^* - o(1)$ and $f(y) < 0$. On substituting (2.105) into equation (2.1) (when written in terms of $y$ and $t$) we obtain at leading order the problem

$$f_y^2 - yf_y + f + F'(1) = 0, \quad 0 < y < v^*, \qquad (2.106)$$

$$f(y) \sim \lambda_m(v^*)(y - v^*) \quad \text{as} \quad y \to (v^*)^-, \qquad (2.107)$$

$$f(y) < 0, \quad 0 \le y < v^*, \qquad (2.108)$$

with condition (2.107) arising from matching expansion (2.105) (as $y \to (v^*)^-$) with expansion (2.103) (as $z \to -\infty$). The solution of (2.106)-(2.108) is readily obtained as

$$f(y) = \lambda_m(v^*)[y - v^*], \quad 0 \le y < v^*. \qquad (2.109)$$

Therefore we have, via (2.105) and (2.109), that the expansion in region **VII** has the form

$$u(y,t) \sim 1 - C^* e^{-\lambda_m(v^*)t\left\{\left[\frac{s(t)}{t} - y\right] + o(1)\right\}} \qquad (2.110)$$

as $t \to \infty$ with $0 \le y < v^* - o(1)$ and $s(t)$ given by (2.104). We note that expansion (2.110) becomes non-uniform when $y = O\left(\frac{1}{t}\right)$ [that is, $x = O(1)$] and the boundary condition (2.3) at $y = 0$ fails to be satisfied.

To complete the asymptotic structure as $t \to \infty$, we introduce the final region, region **VIII**, where $x = O(1)$ as $t \to \infty$. The structure of expansion (2.110) in region **VII** when $y = O\left(\frac{1}{t}\right)$ as $t \to \infty$ suggests that we should look for a solution in region **VIII** of the form $u \sim 1 + O\left(e^{-\lambda_m(v^*)s(t)}\right)$ as $t \to \infty$. Thus in region **VIII** we expand as

$$u(x,t) = 1 + G(x)e^{-\lambda_m(v^*)s(t)} + o\left(e^{-\lambda_m(v^*)s(t)}\right) \qquad (2.111)$$

as $t \to \infty$ with $x = O(1)$. On substituting expansion (2.111) into equation (2.1) we obtain at leading order the problem

$$G_{xx} - \lambda_m^2(v^*)G = 0, \quad 0 < x < \infty, \tag{2.112}$$

$$G_x(0) = 0, \tag{2.113}$$

$$G(x) \sim -C^* e^{\lambda_m(v^*)x} \quad \text{as} \quad x \to \infty. \tag{2.114}$$

Condition (2.113) ensures that boundary condition (2.3) on $x = 0$ is satisfied, whilst the final condition (2.114) arises from matching expansion (2.111) (as $x \to \infty$) with expansion (2.110) $\left(y = O\left(\frac{1}{t}\right)\right)$. The solution of (2.112)-(2.114) is readily obtained as

$$G(x) = -2C^* \cosh(\lambda_m(v^*)x), \quad x \geq 0, \tag{2.115}$$

giving, via (2.111), that in region **VIII**,

$$u(x,t) = 1 - 2C^* \cosh(\lambda_m(v^*)x) e^{-\lambda_m(v^*)s(t)} + o\left(e^{-\lambda_m(v^*)s(t)}\right) \tag{2.116}$$

as $t \to \infty$ with $x = O(1)$.

This completes the asymptotic structure as $t \to \infty$ in this case. In particular, with $u(x,t)$ the solution to IBVP, we have established (via regions **V-VIII**) that

$$u(z + s(t), t) \to u_T(z, v^*) \quad \text{as} \quad t \to \infty, \tag{2.117}$$

uniformly in $z$ (and through terms exponentially small in $t$ as $t \to \infty$ ) with $s(t)$ as given in (2.104). In the terminology of Volpert et al [71] the solution to IBVP converges to $u_T(z, v^*)$ *uniformly* as $t \to \infty$. A schematic representation of the location and thickness of the asymptotic regions as $t \to \infty$ is given in Figure 2.2. We now move onto case (II).

## 2.2.2 Case (II): $v^* = 2$, $u_T(z, v^*) \sim B^* e^{-z}$ as $z \to \infty$

In this case, following Section 2.2.1, we only have region **V** before reaching region **TW**. We have, from Section 2.2.1:

**Region V** $y > 2 + O(t^{-1})$, $t \to \infty$

$$u(y,t) = \exp\left\{-t\left(\frac{y^2}{4} - 1\right) - \frac{1}{2}\ln t - \tilde{H}(y) + o(1)\right\}, \tag{2.118}$$

with $\tilde{H}(y)$ an undetermined function for $y \in (2, \infty)$, but having

$$\tilde{H}(y) \sim (r+1)\ln y - \ln C_\infty - \frac{1}{2}\sigma y \quad \text{as} \quad y \to \infty, \tag{2.119}$$

with the constant $C_\infty$ as given in Section 2.2.1.

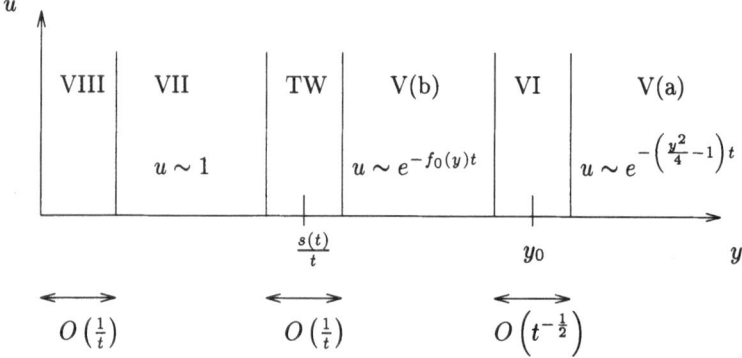

**Fig. 2.2.** Schematic representation of the location and thickness of the asymptotic regions as $t \to \infty$ in case (III). Note that in region **V(b)** $f_0(y) = -\lambda_-(v^*)(y + \lambda_-(v^*)) - 1$.

**Region TW**　$y = \frac{s(t)}{t} \pm O\left(t^{-1}\right), \quad t \to \infty$

$$u(z,t) = u_T(z,2) + O\left(s'(t) - 2\right) \tag{2.120}$$

with

$$s(t) = 2t + \tilde{\chi}(t) + \tilde{\phi}_0 + o(1) \quad \text{as} \quad t \to \infty. \tag{2.121}$$

Here $\tilde{\phi}_0$ is a constant and $1 \ll \tilde{\chi}(t) \ll t$ as $t \to \infty$.

It now remains to match expansion (2.120) ($z \to \infty$) with expansion (2.118) ($y \to 2$). The matching is readily performed (matching $U \equiv \ln u$ is most convenient) to give

$$\tilde{\chi}(t) = -\frac{1}{2} \ln t,$$
$$\tilde{\phi}_0 = -\ln B^*,$$
$$\tilde{H}(y) = o(1) \quad \text{as} \quad y \to 2^+. \tag{2.122}$$

Thus, we have in region TW,

$$u(z,t) = u_T(z,2) + O\left(t^{-1}\right) \quad \text{as} \quad t \to \infty, \tag{2.123}$$

with $z = O(1)$, and

$$s(t) = 2t - \frac{1}{2} \ln t + \tilde{\phi}_0 + o(1) \quad \text{as} \quad t \to \infty, \tag{2.124}$$

via (2.120), (2.121) and (2.122). Again, the structure for $t \gg 1$ is completed by regions **VII** and **VIII**, and these follow directly from Section 2.2.1. Now, with $u(x,t)$ the solution to IBVP, we have established (via regions **V-VIII**) that in this case,

$$u\left(z + s(t), t\right) \to u_T(z, 2) \quad \text{as} \quad t \to \infty, \tag{2.125}$$

uniformly in $z$ (and through terms of $O\left(t^{-1}\right)$ as $t \to \infty$), with $s(t)$ as given in (2.124). In the terminology of Volpert *et al* [71] the solution to IBVP converges to $u_T(z, 2)$ *in form* as $t \to \infty$.

### 2.2.3 Case (I): $v^* = 2$, $u_T(z, v^*) \sim A^* z e^{-z}$ as $z \to \infty$

The details in this case follow, after minor modifications, those given in Section 2.2.2. Again we only have region **V** before reaching region **TW**. We have from Section 2.2.1 that:

**Region V** $\quad y > 2 + O(t^{-1}), \quad t \to \infty$

$$u(y, t) = \exp\left\{-t\left(\frac{y^2}{4} - 1\right) - \frac{1}{2}\ln t - \hat{H}(y) + o(1)\right\}, \tag{2.126}$$

with $\hat{H}(y)$ an undetermined function for $y \in (2, \infty)$, but having

$$\hat{H}(y) \sim (r + 1)\ln y - \ln C_\infty - \frac{1}{2}\sigma y \quad \text{as} \quad y \to \infty, \tag{2.127}$$

with the constant $C_\infty$ as given in Section 2.2.1.

**Region TW** $\quad y = \frac{s(t)}{t} \pm O\left(t^{-1}\right), \quad t \to \infty$

$$u(z, t) = u_T(z, 2) + O\left(s'(t) - 2\right) \tag{2.128}$$

with

$$s(t) = 2t + \hat{\chi}(t) + \tilde{\phi}_1 + o(1) \quad \text{as} \quad t \to \infty. \tag{2.129}$$

Here $\tilde{\phi}_1$ is a constant and $1 \ll \hat{\chi}(t) \ll t$ as $t \to \infty$.

It now remains to match expansion (2.128) ($z \to \infty$) with expansion (2.126) ($y \to 2$). The matching is readily performed (matching $U \equiv \ln u$ is most convenient) and fixes

$$\hat{\chi}(t) = -\frac{3}{2}\ln t,$$

and requires that

$$\hat{H}(y) \sim -\ln(y - 2) \quad y \to 2^+.$$

Thus, we have in region **TW** that

$$u(z, t) = u_T(z, 2) + O\left(t^{-1}\right) \quad \text{as} \quad t \to \infty, \tag{2.130}$$

with $z = O(1)$, and

$$s(t) = 2t - \frac{3}{2}\ln t + \tilde{\phi}_1 + o(1) \quad \text{as} \quad t \to \infty, \tag{2.131}$$

where $\phi_1$ is a constant. Again, the structure for $t \gg 1$ is completed by regions **VII** and **VIII**, and these follow directly from Section 2.2.1. Now, with $u(x,t)$ the solution to IBVP, we have established (via regions **V-VIII**) that in this case,

$$u(z + s(t), t) \to u_T(z, 2) \quad \text{as} \quad t \to \infty, \tag{2.132}$$

uniformly in $z$ (and through terms of $O(t^{-1})$ as $t \to \infty$), with $s(t)$ as given in (2.131). In the terminology of Volpert *et al* [71] the solution to IBVP converges to $u_T(z, 2)$ *in form* as $t \to \infty$.

### 2.2.4 Conclusions

Using the method of matched asymptotic expansions, we have analyzed IBVP, a generalized Fisher problem. In particular, by careful consideration of the asymptotic structures when $t \to 0$ $(0 \le x < \infty)$ and $x \to \infty$ $(t \ge O(1))$ we have been able to develop the complete asymptotic structure to IBVP as $t \to \infty$, uniformly in $0 \le x < \infty$. This approach confirms the validity of using linearized approximations to determine the asymptotic structure and wave speed as $t \to \infty$ in the case when $v^* = 2$, but more importantly, for the case when $v^* > 2$, it resolves the paradox which occurs in using the linearized approach and allows the correct asymptotic structure, wave speed and correction to wave speed to be determined as $t \to \infty$. We have demonstrated that the solution to IBVP, $u(x,t)$, has

$$u(z + s(t), t) = u_T(z, v^*) + O[\dot{s}(t) - v^*] \tag{2.133}$$

as $t \to \infty$, uniformly in $z$, with:

**Case (I)**

$$s(t) = 2t - \frac{3}{2}\ln t + \phi_1 + o(1) \quad \text{as} \quad t \to \infty. \tag{2.134}$$

**Case (II)**

$$s(t) = 2t - \frac{1}{2}\ln t + \phi_2 + o(1) \quad \text{as} \quad t \to \infty. \tag{2.135}$$

**Case (III)**

$$s(t) = v^* t + \phi_3 + \hat{v}\, t^{-3/2} e^{-\frac{1}{4}((v^*)^2 - 4)t} + o\left[t^{-3/2} e^{-\frac{1}{4}((v^*)^2 - 4)t}\right] \quad \text{as} \quad t \to \infty. \tag{2.136}$$

Here $\phi_i (i = 1, 2, 3)$ and $\hat{v}$ are constants. The change in convergence rate in (2.133) from algebraic to exponential between cases (I), (II) and (III) is best understood from regions **V** and region **TW**. Whenever region **TW** requires the envelope-tangent solution in regions **V (a, b)**, then matching to region **TW** ultimately leads to exponential corrections. However, whenever region **TW** requires the full envelope solution in region **V**, then matching to region **TW** leads to algebraic corrections. A sketch of $f_0(y)$ vs $y$ illustrating the envelope-tangent (case (III)) and full envelope (cases (I) and (II)) solutions is given in Figure 2.3. Similarly, the change of coefficient in the $\ln t$ terms in $s(t)$ between cases (I) and (II) is induced by the change in structure of $u_T(z, 2)$ as $z \to \infty$ between these two cases; that is

$$u_T(z, 2) \sim \begin{cases} A^* z e^{-z}, & z \to \infty \quad \text{case (I)}, \\ B^* e^{-z}, & z \to \infty \quad \text{case (II)}. \end{cases} \tag{2.137}$$

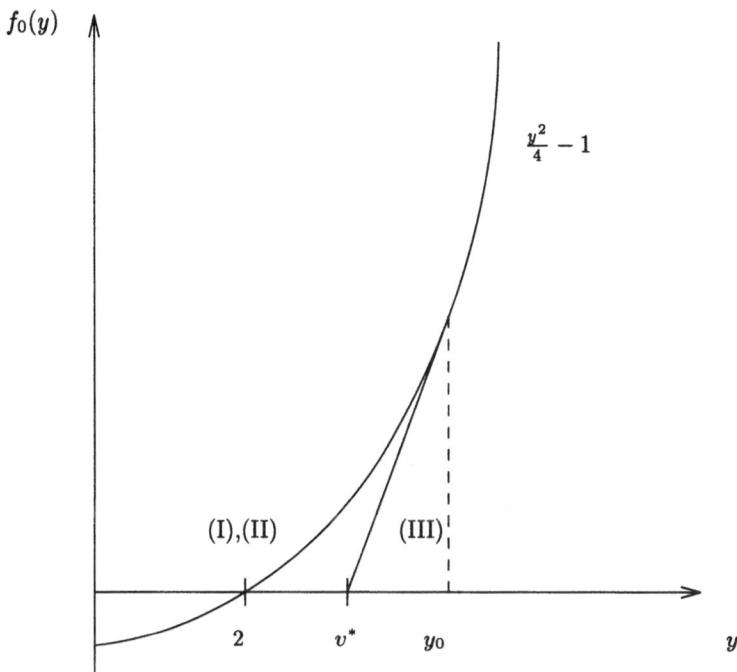

**Fig. 2.3.** A sketch of $f_0(y)$ illustrating the envelope-tangent and full envelope solutions.

## 2.3 Example: Fisher's General Genetic Population Model

Fisher's general model for the migration of advantageous genes (Fisher [16]) can be transformed (see Hadeler and Rothe [24]) into

$$u_t = u_{xx} + F(u, \sigma), \tag{2.138}$$

with

$$F(u, \sigma) = u(1 - u)(1 + \sigma u), \quad -\infty < u < \infty, \tag{2.139}$$

where $\sigma \in [0, \infty)$ is a dimensionless parameter and it is readily checked that $F(u, \sigma)$ satisfies properties (F1)-(F5) for each $\sigma \in [0, \infty)$. In particular, $F'(0, \sigma) = 1$ and $F'(1, \sigma) = -(1 + \sigma) < 0$ for each $\sigma \in [0, \infty)$ (note that (2.138) and (2.139) reduce to the classical Fisher model when $\sigma = 0$). In a chemical context the scalar equation (2.138) with (2.139) can also be considered as a model for mixed quadratic and cubic autocatalysis, where now $\sigma$ measures the ratio of the quadratic to the cubic reaction rates (see, for example, Gray *et al* [20]). It is straightforward to verify that (2.138) has the explicit travelling wave solution

$$u_e(z) = \left[1 + e^{\sqrt{\frac{\sigma}{2}}z}\right]^{-1}, \quad -\infty < z < \infty, \tag{2.140}$$

for any $\sigma > 0$, with propagation speed

$$v = v_e = \frac{\sqrt{2}}{\sigma^{\frac{1}{2}}} + \frac{\sigma^{\frac{1}{2}}}{\sqrt{2}} \quad \sigma > 0. \tag{2.141}$$

We observe immediately that

$$v_e \begin{cases} > 2, & \sigma \in (0, 2) \cup (2, \infty), \\ = 2, & \sigma = 2, \end{cases} \tag{2.142}$$

whilst

$$u_e(z) \sim e^{-\sqrt{\frac{\sigma}{2}}z} \quad \text{as} \quad z \to \infty. \tag{2.143}$$

The form (2.143) demonstrates that, as $z \to \infty$,

$$u_e(z) \sim \begin{cases} e^{\lambda_+(v_e)z}, & 0 < \sigma < 2, \\ e^{-z}, & \sigma = 2, \\ e^{\lambda_-(v_e)z}, & \sigma > 2. \end{cases} \tag{2.144}$$

We can now use Theorem 2.1. We observe immediately from Theorem 2.1 that, for this model (when $\sigma > 0$),

$$2 \leq v^* \leq v_e. \tag{2.145}$$

On using (2.142) together with (2.145) we conclude that

$$v^* = 2 \quad \text{when } \sigma = 2. \tag{2.146}$$

Moreover, it is straightforward to establish that, in this case, $F(u) \leq u$ if and only if $\sigma \in [0, 1]$. Thus we obtain via Theorem 2.1 that

$$v^* = 2 \quad \text{when } \sigma \in [0, 1]. \tag{2.147}$$

However, using (2.144) together with Theorem 2.1 establishes that

$$v^* = v_e = \frac{\sqrt{2}}{\sigma^{\frac{1}{2}}} + \frac{\sigma^{\frac{1}{2}}}{\sqrt{2}} \quad (> 2) \quad \text{when} \quad \sigma \in (2, \infty). \tag{2.148}$$

It remains to consider $\sigma \in (1, 2)$. Equation (2.144) together with Theorem 2.1 requires that

$$2 \leq v^* < v_e \quad \text{when } \sigma \in (1, 2). \tag{2.149}$$

However, it has been demonstrated by Hadeler and Rothe [24] that in fact

$$v^* = 2 \quad \text{when} \quad \sigma \in (1, 2). \tag{2.150}$$

Combining these results, we have, for the general Fisher model, that

$$v^* = \begin{cases} 2, & \sigma \in [0, 2], \\ \left( \dfrac{\sqrt{2}}{\sigma^{\frac{1}{2}}} + \dfrac{\sigma^{\frac{1}{2}}}{\sqrt{2}} \right) (> 2), \sigma \in (2, \infty) . \end{cases} \tag{2.151}$$

Thus, with regards to Theorem 2.1, this model represents case (a) when $\sigma \in [0, 1]$, case (b) when $\sigma \in (1, 2]$ and case (c) when $\sigma \in (2, \infty)$. We have established that, for this example,

$$\begin{array}{llll} v^* = 2, & \text{case (I),} & \text{for all } \sigma \in [0, 2), \\ v^* = 2, & \text{case (II),} & \text{for } \sigma = 2, \\ v^* = \dfrac{\sqrt{2}}{\sigma^{\frac{1}{2}}} + \dfrac{\sigma^{\frac{1}{2}}}{\sqrt{2}} > 2, \text{case (III),} & \text{for all } \sigma \in (2, \infty) \end{array} \tag{2.152}$$

and

$$\lambda_-(v^*) = -\frac{\sigma^{\frac{1}{2}}}{\sqrt{2}} \quad \text{for all} \quad \sigma \in (2, \infty). \tag{2.153}$$

Thus, via the theory developed in Section 2.2, we have that the asymptotic location of the emerging permanent form travelling wave is at $x = s(t)$, where

$$\dot{s}(t) = \begin{cases} 2 - \frac{3}{2}t^{-1} + o(t^{-1}), & \sigma \in [0, 2), \\[2mm] 2 - \frac{1}{2}t^{-1} + o(t^{-1}), & \sigma = 2, \\[2mm] \left( \dfrac{\sqrt{2}}{\sigma^{\frac{1}{2}}} + \dfrac{\sigma^{\frac{1}{2}}}{\sqrt{2}} \right) - \dfrac{\hat{v}(\sigma-2)^2}{8\sigma} t^{-\frac{3}{2}} e^{-\frac{(\sigma-2)^2 t}{8\sigma}} + o\left[ t^{-\frac{3}{2}} e^{-\frac{(\sigma-2)^2 t}{8\sigma}} \right], \\ \qquad\qquad \sigma \in (2, \infty) \end{cases} \tag{2.154}$$

as $t \to \infty$.

---

# $m$th-Order ($m > 1$) Fisher Nonlinearity: Initial Data with Exponential Decay Rates or Compact Support

In this chapter we consider the following initial-boundary value problem for a scalar reaction-diffusion equation, namely,

$$\left.\begin{array}{lll} u_t = u_{xx} + u^m(1-u), & x, t > 0, & \text{(P1)} \\ u(x,0) = u_0(x), & x \geq 0, & \text{(P2)} \\ u_x(0,t) = 0, & t > 0, & \text{(P3)} \\ u(x,t) \to 0 \ \text{ as } \ x \to \infty, & t \geq 0, & \text{(P4)} \end{array}\right\} \ \ [\mathbf{P,m}]$$

where the reaction order $m > 1$, and $u_0(x)$ is a continuous, non-negative and monotone decreasing function in $x \geq 0$, with $u_0(x) \to 0$ as $x \to \infty$. In particular, we consider the following classes of initial data:

(i) $u_0(x)$ is positive, analytic and has exponential decay rate as $x \to \infty$, with

$$u_0(x) \sim \begin{cases} u_\infty e^{-\sigma x} + O[e^{-f(x)}] & \text{as } x \to \infty, & \text{(g1)} \\ \tilde{u}_0 + \sum_{l=1}^{\infty} \tilde{u}_l x^l & \text{as } x \to 0^+, & \text{(g2)} \end{cases}$$

for some $f(x) > O(x)$ as $x \to \infty$, where $u_\infty, \sigma, \tilde{u}_0 > 0$ and $\tilde{u}_l$ are constants.

(ii) $u_0(x)$ has compact support. In this case

$$u_0(x) = \begin{cases} u_0 g(x), & 0 \leq x \leq \sigma, \\ 0, & x > \sigma, \end{cases} \quad \text{(d1)}$$

where g:$[0,\sigma] \to \mathbb{R}$ is positive in $[0,\sigma)$, non-negative and analytic in $[0,\sigma]$, has $g(0) = 1$ and

$$g(x) \sim \begin{cases} g_\sigma(\sigma - x)^r & \text{as } x \to \sigma^-, \\ 1 + g_{\tilde{m}} x^{\tilde{m}} & \text{as } x \to 0^+. \end{cases}$$

Here r, $\tilde{m} \in \mathbb{N}$ with constants $g_{\tilde{m}} \neq 0$ and $g_\sigma, g_0 > 0$. The parameters $\sigma, u_0$ are positive.

We note that the more general case when (g1) is replaced by

$$u_0(x) \sim u_\infty x^{-n} e^{-\sigma x} \quad \text{as} \quad x \to \infty, \tag{3.1}$$

where $-\infty < n < \infty$ has been considered by Leach et al [36]. The modifications required in this case to the theory presented in this chapter along with the main reults are summerised in Section 3.5.

As noted in Section 1.2, it is readily established that [**P,m**] has a unique, global solution, with $0 < u(x,t) < \max[1, u_0]$ for all $x, t > 0$.

## 3.1 Permanent Form Travelling Waves

In this section we review the main results concerning the existence and structure of PTWs which may occur in the solution to [**P,m**] ($m > 1$) as $t \to \infty$. On introducing the travelling coordinate $z = x - vt$ (with $v > 0$ being the constant wave speed) a PTW is a solution to the following nonlinear boundary value problem,

$$\begin{aligned}
&u'' + vu' + u^m(1-u) = 0, \quad -\infty < z < \infty, \\
&u(z) \geq 0, \quad -\infty < z < \infty, \\
&u(z) \to 0, \quad \text{as} \quad z \to +\infty, \\
&u(z) \to 1, \quad \text{as} \quad z \to -\infty.
\end{aligned} \tag{3.2}$$

Concerning (3.2) we have the following fundamental result:

**Theorem 3.1.** *For each $m > 1$, there exists a unique PTW solution of (3.2) (say $u = u_T(z; v)$ with translational invariance fixed so that $u_T(0; v) = \frac{1}{2}$) for each $v \geq v^*(m)(> 0)$, whilst no PTW solution of (3.2) exists for each $0 < v < v^*(m)$. Moreover, as $z \to \infty$,*

$$u_T(z; v) \sim \begin{cases} B^* e^{-v^*(m)z}, & v = v^*(m), \\ \left(\frac{v}{(m-1)z}\right)^{\frac{1}{(m-1)}}, & v > v^*(m), \end{cases}$$

*and, as $z \to -\infty$,*

$$u_T(z; v) \sim 1 - A^* e^{\lambda_m(v)z},$$

*where $A^*, B^*$ are fixed positive constants and,*

$$\lambda_m(v) = -\frac{v}{2} + \frac{1}{2}\sqrt{v^2 + 4} \quad (> 0).$$

*Proof.* See for example Billingham and Needham [5] and Barnes [4]. □

In addition it was established in Needham and Barnes [56] that $v^*(m)$ is continuous and monotone decreasing for $m > 1$, with

$$v^*(m) \sim \begin{cases} 2 - 2.33107(m-1)^{2/3} + \cdots, & m \to 1^+, \\ \frac{\sqrt{2}}{m} + \cdots, & m \to \infty. \end{cases}$$

We further note that $v^*(2) = \frac{1}{\sqrt{2}}$ (see Britton [10]).

## 3.2 Evolution of Travelling Waves in $[P, m]$

In this section we develop the asymptotic structure to [**P,m**] as $t \to \infty$ with particular attention to PTW formation. We must first begin by examining the asymptotic structure of the solution to [**P,m**] as $t \to 0$.

### 3.2.1 Asymptotic Solution as $t \to 0$ for $0 \le x < \infty$

### (a) Initial Data with Exponential Decay Rate as $x \to \infty$

We first consider region I, where $x = O(1)$ as $t \to 0$ and expand the solution to [**P,m**] as

$$u(x,t) = u_0(x) + tu_1(x) + O(t^2) \tag{3.3}$$

as $t \to 0$. On substituting into equation (P1) and applying initial condition (P2) we readily obtain

$$u(x,t) = u_0(x) + t[u_0''(x) + u_0^m(x)(1 - u_0(x))] + O(t^2) \tag{3.4}$$

as $t \to 0$, with $x = O(1)$. Now, for $x \gg 1$, expansion (3.4), with (g1), takes the form

$$u(x,t) \sim [u_\infty e^{-\sigma x} + \ldots] + t[\sigma^2 u_\infty e^{-\sigma x} + \ldots] + \ldots \tag{3.5}$$

as $t \to 0$, and we conclude that (3.4) remains uniform for $x \gg 1$ as $t \to 0$. However, for $x \ll 1$, (3.4) with (g2) takes the form

$$u(x,t) \sim (\tilde{u}_0 + \tilde{u}_1 x + \ldots) + t[(2\tilde{u}_2 + 6\tilde{u}_3 x + \ldots)$$
$$+ (\tilde{u}_0 + \tilde{u}_1 x + \ldots)^m (1 - (\tilde{u}_0 + \tilde{u}_1 x + \ldots)) + \ldots] + \ldots \tag{3.6}$$

as $t \to 0$, and in general, will not satisfy the boundary condition (P3) at $x = 0$. Therefore, we will require an inner region, region $\mathbf{I_0}$, when $x = o(1)$ as $t \to 0$, over which this condition is satisfied. To retain the highest spatial derivative in (P1) at leading order, we find that the inner region must have the scaling $x = O(t^{1/2})$ as $t \to 0$ and we introduce the inner coordinate $\eta = O(1)$ as $t \to 0$, where $\eta = xt^{-1/2}$. The form of (3.6) then suggests that we expand in the inner region as

$$u(\eta, t) = \tilde{u}_0 + t^{\frac{1}{2}}\bar{u}_1(\eta) + O(t) \tag{3.7}$$

as $t \to 0$, with $\eta = O(1)$. The leading order problem then becomes, on matching with (3.6),

$$\bar{u}'' + \frac{1}{2}\eta\bar{u}' - \frac{1}{2}\bar{u} = 0, \quad \eta > 0, \tag{3.8}$$

$$\bar{u}'(0) = 0, \tag{3.9}$$

$$\bar{u}(\eta) \sim \tilde{u}_1\eta \quad \text{as } \eta \to +\infty. \tag{3.10}$$

The solution to the boundary value problem (3.8)–(3.10) is readily obtained as

$$\bar{u}(\eta) = 2\sqrt{\pi}\, \tilde{u}_1\, {}_1F_1[1, 1/2; \eta^2/4]e^{-\frac{\eta^2}{4}}, \tag{3.11}$$

where ${}_1F_1[a,b;z]$ is the confluent hypergeometric function (Abramowitz and Stegun [1]). This completes the asymptotic structure as $t \to 0$, with expansions (3.4) (in region **I**) and (3.7) (in region **I₀**) providing a uniform approximation in $x \geq 0$ to the solution of **[P,m]** as $t \to 0$.

## (b) Initial Data with Compact Support

Following Chapter 2, the asymptotic structure of the solution to **[P,m]** (with $m > 1$) as $t \to 0$ has three regions. The details of these regions are for brevity summarized below (with the full details following, after minor modification, those given in Chapter 2 for a related problem).

**Region I**   $0 \leq x < \sigma - O\left(t^{1/2}\right), \quad t \to 0,$

$$u(x,t) = u_0 g(x) + t\left(u_0 g''(x) + u_0^m g^m(x)(1 - u_0 g(x))\right) + O\left(t^2\right). \tag{3.12}$$

**Region II**   $x = \sigma \pm O\left(t^{1/2}\right), \quad t \to 0,$

$$u(\hat{\eta}, t) = t^{r/2}\tilde{u}(\hat{\eta}) + o\left(t^{r/2}\right), \tag{3.13}$$

with $\hat{\eta} = (x - \sigma)t^{1/2} = O(1)$ as $t \to 0$.

Here the function $\tilde{u} : \mathbb{R} \to \mathbb{R}$ is defined in Chapter 2 (Section 2.2.1 ) and need not be repeated here.

**Region III**   $x > \sigma + O\left(t^{1/2}\right), \quad t \to 0,$

$$u(x,t) = \exp\left(-\frac{(x-\sigma)^2}{4t} + \left(r + \tfrac{1}{2}\right)\ln t - (r+1)\ln(x - \sigma)\right.$$
$$\left. + \ln C_\infty + O(t)\right), \tag{3.14}$$

where the constant $C_\infty$ is defined in Chapter 2.

This completes the asymptotic structure as $t \to 0$, with expansions (3.12) (in region **I**), (3.13) (in region **II**) and (3.14) (in region **III**) providing a uniform approximation in $x \geq 0$ to the solution of **[P,m]** as $t \to 0$.

We next consider the asymptotic structure of **[P,m]** as $x \to \infty$ with $t = O(1)$.

### 3.2.2 Asymptotic Solution as $x \to \infty$ for $t = O(1)$

#### (a) Initial Data with Exponential Decay Rate as $x \to \infty$

We now investigate the structure of the solution to $[\mathbf{P,m}]$ as $x \to \infty$ for $t = O(1)$. The form of (3.5) indicates that $u = O(e^{-\sigma x})$ as $x \to \infty$ with $t = O(1)$, which suggests that in this region, region $\mathbf{II}$, we write

$$u(x, t) = e^{-\phi(x,t)} \quad \text{as} \quad x \to \infty, \tag{3.15}$$

and expand $\phi(x, t)$ in the form

$$\phi(x, t) = \phi_0(t)x + \phi_1(t) + o(1) \quad \text{as} \quad x \to \infty, \tag{3.16}$$

with $t = O(1)$ . On substituting from (3.15) and (3.16) into equation (P1) and solving at each order in turn, we find (after matching with (3.5) as $t \to 0$) that

$$u(x, t) = \exp[-\sigma x + (\sigma^2 t + \ln u_\infty) + o(1)] \tag{3.17}$$

as $x \to \infty$ with $t = O(1)$. Expansion (3.17) will remain uniform for $t \gg 1$ *provided* that $x \gg t$. However, expansion (3.17) fails to provide an asymptotic approximation for $t \gg 1$ when $x \leq O(t)$.


#### (b) Initial Data with Compact Support

Following Chapter 2, we have

**Region IV**   $t = O(1)$   $x \to \infty$,

$$u(x, t) = \exp \left( -\tfrac{x^2}{4t} + \tfrac{\sigma x}{2t} - (r + 1) \ln x + \left( \left(r + \tfrac{1}{2}\right) \ln t \right. \right.$$
$$\left. \left. + t + \ln C_\infty - \tfrac{\sigma^2}{4t} \right) + O(x^{-1}) \right), \tag{3.18}$$

which matches with region $\mathbf{III}$ as $t \to 0$ ($x \gg 1$). We observe that the expansion in region $\mathbf{IV}$ continues to remain uniform when $t \gg 1$ *provided* that $x \gg t$, but fails to provide an asymptotic approximation when $x \leq O(t)$ as $t \to \infty$.


### 3.2.3 Asymptotic Solution as $t \to \infty$ in the Case of Initial Data with Exponential Decay Rate as $x \to \infty$

As $t \to \infty$, the asymptotic expansion (3.17) in region $\mathbf{II}$ continues to remain uniform for $x \gg t$. However, as already noted, a nonuniformity develops when $x = O(t)$. To proceed, we introduce a new region, region $\mathbf{III}$. To examine region **III** we introduce the scaled coordinate $Y = \frac{x}{t}$, where $Y = O(1)$ as $t \to \infty$, and look for an expansion of the form (as suggested by (3.17))

$$u(Y, t) = e^{-tF(Y,t)}, \tag{3.19}$$

where

$$F(Y, t) = f_0(Y) + f_1(Y)\frac{1}{t} + o(t^{-1}) \text{ as } t \to \infty, \tag{3.20}$$

with $Y = O(1)(> 0)$ as $t \to \infty$, and $f_0(Y) > 0$. It is instructive to consider first the leading order problem in region **III**. On substituting from (3.19) and (3.20) into (P1) (when written in terms of $Y$ and $t$) we obtain the leading order problem as

$$f_{0Y}^2 - Y f_{0Y} + f_0 = 0, \quad Y > 0, \tag{3.21}$$
$$f_0(Y) > 0, \quad Y > 0, \tag{3.22}$$
$$f_0(Y) \sim \sigma[Y - \sigma] \text{ as } Y \to +\infty. \tag{3.23}$$

The final condition, (3.23), arises from matching expansion (3.19) ($Y \gg 1$) with expansion (3.17) [$x = O(t)$]. Equation (3.21) has a one-parameter family of linear solutions

$$f_0(Y) = A[Y - A], \quad Y > 0, \tag{3.24}$$

for any $A \in \mathbb{R}$, together with the associated envelope solution

$$f_0(Y) = \frac{Y^2}{4}, \quad Y > 0. \tag{3.25}$$

Combinations of (3.24) and (3.25) which remain continuous and differentiable also provide solutions to (3.21) (envelope touching solutions). Applying the conditions (3.22) and (3.23) requires us to select one of the following solutions:

$$f_0(Y) = \sigma[Y - \sigma], \qquad Y > \sigma, \tag{3.26}$$

$$f_0(Y) = \begin{cases} \sigma[Y - \sigma], & Y > 2\sigma, \\ \frac{Y^2}{4}, & 0 < Y \leq 2\sigma, \end{cases} \tag{3.27}$$

$$f_0(Y) = \begin{cases} \sigma[Y - \sigma], & Y > 2\sigma, \\ \frac{Y^2}{4}, & 2A \leq Y \leq 2\sigma, \\ A[Y - A], & A < Y < 2A, \end{cases} \tag{3.28}$$

for $\sigma > 0$ and any $0 < A < \sigma$. We conclude that a nonuniformity occurs in expansion (3.19),(3.20) when $f_0(Y)$ is given by (3.26),(3.27) and (3.28) as $Y \to Y_c^+$, where $Y_c = \sigma, 0$ and $A$ respectively. A consideration of further terms in (3.20) demonstrates that this non-uniformity occurs when $Y = Y_c + O(t^{-1})$. Thus we write

$$Y = Y_c + \frac{z}{t} \tag{3.29}$$

with $z = O(1)$ as $t \to \infty$. We refer to this region as region **TW**. In region **TW** expansions (3.19) and (3.20) demonstrate that $u = O(1)$ as $t \to \infty$, and so we expand in the form

$$u(z, t) = u_c(z) + o(1) \quad \text{as } t \to \infty \tag{3.30}$$

with $z = O(1)$. On substituting into equation (P1) (when written in terms of $z$ and $t$) we obtain the leading order problem as

$$u_c'' + Y_c u_c' + u_c^m(1 - u_c) = 0, \quad -\infty < z < \infty, \tag{3.31}$$
$$u_c(z) \geq 0, \quad -\infty < z < \infty, \tag{3.32}$$
$$u_c(z) \to 0, \quad \text{as } z \to +\infty, \tag{3.33}$$
$$u_c(z) \text{ bounded as } z \to -\infty. \tag{3.34}$$

The condition (3.33) arises from matching expansion (3.30) (as $z \to \infty$), with expansion (3.19),(3.20) (as $Y \to Y_c^+$). Moreover, a phase plane analysis (Poincaré-Bendixson theorem) of equation (3.31) with conditions (3.32) and (3.33) allows boundary condition (3.34) to be replaced by

$$u_c(z) \to 1 \quad \text{as } z \to -\infty, \tag{3.35}$$

(all other possibilities leading to unbounded $u_c(z)$ as $z \to -\infty$). Thus, the complete boundary value problem for $u_c(z)$ may be stated as

$$\left. \begin{array}{l} u_c'' + Y_c u_c' + u_c^m(1 - u_c) = 0, \quad -\infty < z < \infty, \\ u_c(z) \geq 0, \quad -\infty < z < \infty, \\ u_c(z) \to 0, \quad \text{as } z \to +\infty, \\ u_c(z) \to 1, \quad \text{as } z \to -\infty. \end{array} \right\} \text{[TW]}$$

We observe that [**TW**] is precisely the boundary value problem (3.2) for PTWs with $v = Y_c$, and its solutions are classified in Theorem 3.1. It is then clear from (3.19),(3.20) and (3.26)–(3.28), together with Theorem 3.1, that the matching of expansion (3.30) (as $z \to \infty$) with expansion (3.19),(3.20) (as $Y \to Y_c^+$) requires $f_0(Y)$ to be chosen as follows:

(i) $\underline{\sigma > v^*}$
   $f_0(Y)$ must be chosen as (3.28) with $A = v^*$.

(ii) $\underline{\sigma = v^*}$
   $f_0(Y)$ must be chosen as (3.26).

(iii) $\underline{0 < \sigma < v^*}$
   In this case matching of expansions (3.30) as $z \to \infty$ with expansion (3.19),(3.20) as $Y \to Y_c$ cannot be achieved with any of the forms (3.26)-(3.28) for $f_0(Y)$. We conclude in this case that expansion (3.19), (3.20) becomes nonuniform *before* $Y = 2\sigma$, and further regions will be necessary in this case.

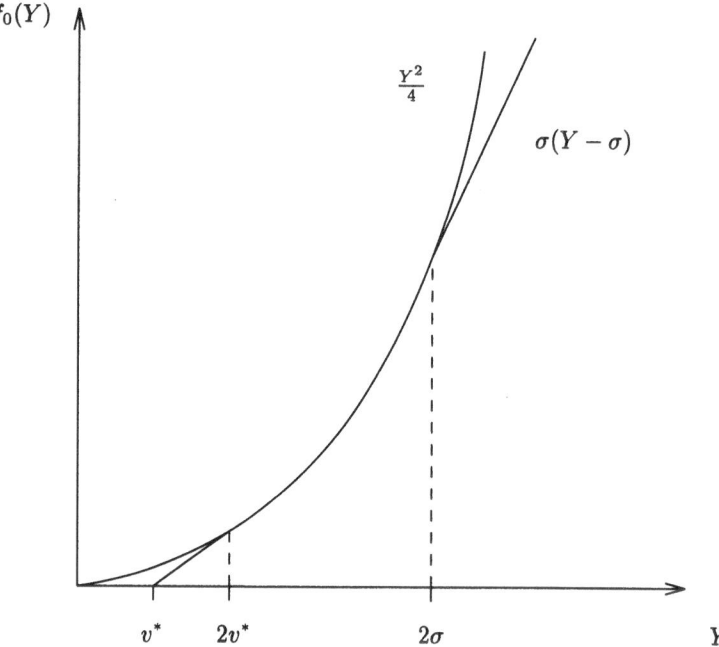

**Fig. 3.1.** A sketch of $f_0(Y)$ when $\sigma > v^*(m)$.

A sketch of $f_0(Y)$ when $\sigma > v^*(m)$ is given in Figure 3.1. We finally note that, although $f_0(Y)$ and $f_0'(Y)$ are continuous over the range of definition of the solutions (3.26) and (3.28), the second derivative $f_0''(Y)$ is discontinuous at the points at which the linear solutions meet the envelope solution $[Y = 2\sigma, 2A$ in (3.28)]. This indicates that thin transition regions exist in the neighbourhood of the points of contact of the linear solutions and the envelope in which second derivatives are retained at leading order to smooth out these discontinuities in curvature.

In summary, a consideration of the leading order problem in region **III** has established that there are three distinct cases to consider, namely $\sigma > v^*(m), \sigma = v^*(m)$ and $0 < \sigma < v^*(m)$. We now consider the details of each of these cases in turn.

**(a) $\sigma > v^*(m)$**

In this case, region **III** must be replaced by five regions, region **III(a)** $(2\sigma + o(1) < Y < \infty)$, region **TRA** (transition region), region **III(b)** $(2v^*(m) + o(1) < Y < 2\sigma - o(1))$, region **TRB** (transition region) and region **III(c)** $(v^*(m) + o(1) < Y < 2v^*(m) - o(1))$. We consider each of these regions in turn. We begin with region **III(a)**, where $2\sigma + o(1) < Y < \infty$. Substitution of (3.19), (3.20) into equation (P1) (when written in terms of $Y$ and

$t$) gives on solving at each order in turn and matching to expansion (3.17) $[x = O(t)]$ as $Y \to \infty$ that

$$u(Y, t) = \exp\left\{-\sigma t[Y - \sigma] + \ln u_\infty + o(1)\right\} \qquad (3.36)$$

as $t \to \infty$ with $2\sigma + o(1) < Y < \infty$. As $Y \to 2\sigma^+$ we move into the transition region, region **TRA**. An examination of expansion (3.36) (as $y \to 2\sigma^+$) reveals that in this region $Y = 2\sigma + O(t^{-1/2})$ as $t \to \infty$. To examine region **TRA** we introduce the scaled coordinate $\eta = (Y - 2\sigma)t^{1/2}$, where $\eta = O(1)$ as $t \to \infty$, and expand as

$$u(\eta, t) = [h(\eta) + o(1)]e^{-\sigma^2 t - \sigma\eta t^{1/2}} \qquad (3.37)$$

as $t \to \infty$ with $h(\eta) > 0$. On substitution of (3.37) into equation (P1) (when written in terms of $\eta$ and $t$), we obtain at leading order

$$h_{\eta\eta} + \frac{1}{2}\eta h_\eta = 0, \quad -\infty < \eta < \infty. \qquad (3.38)$$

The solution of (3.38) is readily obtained as

$$h(\eta) = A_0 \operatorname{erfc}\left\{\frac{\eta}{2}\right\} + B_0, \qquad (3.39)$$

where $A_0$ and $B_0$ are to be determined on matching. Matching expansion (3.37) (as $\eta \to \infty$) with expansion (3.36) (as $Y \to 2\sigma^+$) requires $B_0 = u_\infty$. As $\eta \to -\infty$ we move out of region **TRA** into region **III(b)**, where $2v^*(m) + o(1) < Y < 2\sigma - o(1)$. We now examine the form of expansion (3.37) for $(-\eta) \gg 1$ (as we move into region **III(b)**). From (3.37) and (3.39) we have

$$u(\eta, t) \sim \left\{2A_0 + u_\infty + \frac{2A_0}{\sqrt{\pi}} \frac{e^{-\eta^2/4}}{\eta}\right\} \exp[-(\sigma^2 t + \sigma\eta t^{1/2})], \qquad (3.40)$$

with $(-\eta) \gg 1$. When written in terms of $Y$, (3.40) becomes

$$u(Y, t) \sim (2A_0 + u_\infty) e^{-\sigma t(Y-\sigma)} + \frac{2A_0}{\sqrt{\pi}} \frac{t^{-1/2}}{(Y - 2\sigma)} e^{-\frac{Y^2}{4}t}. \qquad (3.41)$$

We recall from Section 3.2.3 (via (3.19),(3.20) and (3.28)) that in this case (when $\sigma > v^*(m)$) $u = O\left\{e^{-\frac{Y^2 t}{4}}\right\}$ as $t \to \infty$ in region **III(b)**. Hence, to enable matching between region **TRA** (as $\eta \to -\infty$) and region **III(b)** (as $Y \to 2\sigma^-$) we require that $A_0 = -\frac{u_\infty}{2}$. Thus, we look for an expansion in region **III(b)** in the form

$$u(Y, t) = (H(Y) + o(1))t^{-1/2}e^{-\frac{Y^2}{4}t} \qquad (3.42)$$

as $t \to \infty$ with $2v^*(m) + o(1) < Y < 2\sigma - o(1)$. On substitution of (3.42) into equation (P1) (when written in terms of $Y$ and $t$), we find that $H(Y)$ is indeterminate at this order (although (1.22) requires $H(Y) > 0$). However, matching to region **TRA** (as $Y \to 2\sigma^-$) requires that

$$H(Y) \sim \left\{ \frac{u_\infty}{\sqrt{\pi}(2\sigma - Y)} \right\} \tag{3.43}$$

as $Y \to 2\sigma^-$. As $Y \to 2v^*(m)^+$ we move from region **III(b)** into the transition region, region **TRB**. An examination of expansion (3.42) as $Y \to 2v^*(m)^+$ reveals that in this region, $Y = 2v^*(m) + O(t^{-1/2})$ as $t \to \infty$. To examine region **TRB** we introduce the scaled coordinate $\xi = (Y - 2v^*(m))t^{1/2}$, and expand as indicated by expansion (3.42), that is,

$$u(\xi, t) = (F(\xi) + o(1))e^{[-(v^*(m))^2 t - v^*(m)\xi t^{1/2}]} \tag{3.44}$$

as $t \to \infty$ with $F(\xi) > 0, \xi = O(1)$. Substitution of (3.44) into equation (P1) (when written in terms of $\xi$ and $t$) gives at leading order

$$F_{\xi\xi} + \frac{1}{2}\xi F_\xi = 0, \quad -\infty < \xi < \infty. \tag{3.45}$$

The solution to (3.45) is readily obtained as

$$F(\xi) = C_0 \text{erfc}\left\{ \frac{\xi}{2} \right\} + D_0, \tag{3.46}$$

where $C_0$ and $D_0$ are to be determined on matching. Matching expansion (3.42) (as $Y \to 2v^*(m)^+$) to expansion (3.44) (as $\xi \to \infty$) requires that

$$H(Y) \sim \frac{2C_0}{\sqrt{\pi}(Y - 2v^*(m))} \quad \text{as} \quad Y \to (2v^*(m))^+, \tag{3.47}$$

and

$$D_0 = 0. \tag{3.48}$$

The expansion in region **TRB** is thus given via (3.44) and (3.46) as

$$u(\xi, t) = C_0 \text{ erfc}(\xi/2) \exp[-((v^*(m))^2 t + v^*(m)\xi t^{1/2})]$$
$$+ O\left\{ \exp[-m((v^*(m))^2 t + v^*(m)\xi t^{1/2})] \right\} \tag{3.49}$$

as $t \to \infty$ with $\xi = O(1)$, where $C_0 > 0$ (via (1.22)) is undetermined. The correction to (3.49) was obtained by continuing expansion (3.44) to the next order, the details of which are omitted for brevity although similar details are given in Section 2.2. As $\xi \to -\infty$ we move out of region **TRB** into region **III(c)**, where $v^*(m) + o(1) < Y < 2v^*(m) - o(1)$. We now examine the form of (3.49) for $(-\xi) \gg 1$ (as we move into region **III(c)**). From (3.49) we have

$$u(\xi, t) \sim (2C_0 + \frac{2C_0}{\sqrt{\pi}} \frac{e^{-\xi^2/4}}{\xi}) \exp[-((v^*(m))^2 t + v^*(m)\xi t^{1/2})]$$
$$+ O\left\{ \exp[-m((v^*(m))^2 t + v^*(m)\xi t^{1/2})] \right\}, \tag{3.50}$$

with $(-\xi) \gg 1$. When written in terms of $Y$, (3.50) becomes

$$u(Y,t) \sim 2C_0 \, e^{-v^*(m)t(Y-v^*(m))} + \frac{2C_0}{\sqrt{\pi}} \frac{t^{-1/2}}{(Y-2v^*(m))} e^{-\frac{Y^2}{4}t}$$
$$+ O\left\{e^{-mv^*(m)t(Y-v^*(m))}\right\}. \tag{3.51}$$

Thus, we must look for an expansion in region **III(c)** in the form

$$u(Y,t) = 2C_0 \, e^{-v^*(m)t(Y-v^*(m))} + \frac{2C_0}{\sqrt{\pi}} \hat{H}(Y)t^{-1/2}e^{-\frac{Y^2}{4}t}$$
$$+ O\left\{e^{-mv^*(m)t(Y-v^*(m))}\right\} \tag{3.52}$$

as $t \to \infty$ with $v^*(m) + O(t^{-1}) < Y < 2v^*(m) - O(t^{-1/2})$ as $t \to \infty$. On substitution of (3.52) into equation (P1) (when written in terms of $Y$ and $t$), we find that $\hat{H}(Y)$ is indeterminate at this order. However, matching to region **TRB** (as $Y \to 2v^*(m)^-$) requires that

$$\hat{H}(Y) \sim \frac{1}{(Y - 2v^*(m))} \quad \text{as} \quad Y \to (2v^*(m))^-. \tag{3.53}$$

As $Y \to v^*(m)^+$ we move out of region **III(c)** into the wave front region, region **TW**, where $Y = v^*(m) \pm O(t^{-1})$ as $t \to \infty$. In region **TW**, $x \sim s(t)$ and $u$ (when written in terms of the travelling wave coordinate, $z$) has the form (via (3.30))

$$u(z,t) = u_T(z, v^*(m)) + o(1) \tag{3.54}$$

as $t \to \infty$ with $z = O(1)$, where $z = x - s(t)$ and $s(t) = v^*(m)t + \hat{\phi}(t) + \phi_0 + \hat{\psi}(t)$ as $t \to \infty$. Here $1 \ll \hat{\phi}(t) \ll t, \phi_0$ is a constant and $\hat{\psi}(t) = o(1)$ as $t \to \infty$ and are as yet undetermined gauge functions (to be fixed on matching with region **III(c)** as $z \to \infty$), whilst $u_T(z, v^*(m))$ represents the minimum speed permanent form travelling wave solution. We recall from Section 3.1 the following asymptotic properties of $u_T(z, v^*(m))$,

$$u_T(z, v^*(m)) \sim \begin{cases} B^* e^{-v^*(m)z} & \text{as } z \to \infty, \\ 1 - A^* e^{\lambda_m(v^*)z} & \text{as } z \to -\infty, \end{cases} \tag{3.55}$$

where $A^*, B^*$ are positive constants which are, in principle, determined. On examining expansion (3.52) in region **III(c)** as we move into region **TW** (as $Y \to v^*(m)^+$), we obtain from (3.52), when written in terms of $z$ (via $Y = \frac{s(t)}{t} + \frac{z}{t}$) that

$$u(z,t) \sim 2C_0 \exp\left\{-v^*(m)\hat{\phi}(t) - v^*(m)[z + \phi_0] - v^*(m)\hat{\psi}(t)\right\} + \frac{2C_0}{\sqrt{\pi}}\hat{H}t^{-1/2}$$
$$\times \exp\left\{-\frac{(v^*(m))^2}{4}t - \frac{v^*(m)}{2}\hat{\phi}(t) - \frac{v^*(m)}{2}[z + \phi_0] - \frac{v^*(m)}{2}\hat{\psi}(t)\right\}$$
$$+ \dots \tag{3.56}$$

as $t \to \infty$ with $z \gg 1$. We now make the assumption (which we will verify as consistent) that

$$\hat{H}(Y) \sim \hat{H}_c(Y - v^*(m))^{\hat{\gamma}} \quad \text{as} \quad Y \to v^*(m)^+, \tag{3.57}$$

for some constants $\hat{H}_c$ and $\hat{\gamma}$. With (3.57) we can rewrite (3.56) as

$$
\begin{aligned}
u(z,t) \sim & \left(2C_0 - 2C_0\, v^*(m)\hat{\psi}(t) + \dots\right) \exp\left\{-v^*(m)\hat{\phi}(t) - v^*(m)[z + \phi_0]\right\} \\
& + \left(\tfrac{2C_0}{\sqrt{\pi}}\hat{H}_c t^{-1/2-\hat{\gamma}}[\hat{\phi} + (z + \phi_0)]^{\hat{\gamma}} + \dots\right) \\
& \times \exp\left\{-\tfrac{(v^*(m))^2}{4}t - \tfrac{v^*(m)}{2}\hat{\phi}(t) - \tfrac{v^*(m)}{2}[z + \phi_0]\right\} \\
& + \dots
\end{aligned} \tag{3.58}
$$

as $t \to \infty$ with $z \gg 1$. Matching expansion (3.54) (as $z \to \infty$) up to $O(1)$ with expansion (3.58) fixes $\hat{\phi}(t) \equiv 0$ and requires that $\phi_0 = -\frac{1}{v^*(m)}\ln\left\{\frac{B^*}{2C_0}\right\}$. On rewriting (3.58) we obtain

$$
\begin{aligned}
u(z,t) \sim & \left\{2C_0 e^{-v^*(m)[z+\phi_0]} + O\left\{e^{-mv^*(m)[z+\phi_0]}\right\}\right\} \\
& + \left\{-2C_0\, v^*(m)\hat{\psi}(t)e^{\{-v^*(m)[z+\phi_0]\}} + \dots\right\} \\
& + \left\{\tfrac{2C_0}{\sqrt{\pi}}\hat{H}_c t^{-1/2-\hat{\gamma}}z^{\hat{\gamma}}e^{\left\{-\tfrac{(v^*(m))^2}{4}t - \tfrac{v^*(m)}{2}[z+\phi_0]\right\}} + \dots\right\} \\
& + \dots
\end{aligned} \tag{3.59}
$$

as $t \to \infty$ with $z \gg 1$. We conclude from (3.59) that in region **TW** we must have

$$u(z,t) = u_T(z, v^*(m)) + O(\hat{\psi}(t)) \tag{3.60}$$

as $t \to \infty$. The translational invariance of $u_T(z, v^*(m))$ with respect to $z$ is fixed by requiring $u_T(0, v^*(m)) = \frac{1}{2}$. On substitution of (3.60) into equation (P1) (when written in terms of $z$ and $t$) we require to obtain a nontrivial balance at $O\left\{\hat{\psi}'(t)\right\}$ that $\hat{\psi}'(t) = O\left\{\hat{\psi}(t)\right\}$ as $t \to \infty$ and we conclude that $\hat{\psi}(t)$ must be exponentially small in $t$, as $t \to \infty$, and hence set

$$\hat{\psi}(t) = v_1 t^{\tilde{\gamma}} e^{-\lambda t}[1 + o(1)] \tag{3.61}$$

as $t \to \infty$, where $v_1$, $\tilde{\gamma}$ and $\lambda(> 0)$ are to be determined. We now continue the expansion in region **TW** as

$$u(z,t) = u_T(z, v^*(m)) + \hat{u}_c(z)\hat{\psi}'(t) + o\left\{\hat{\psi}'(t)\right\} \tag{3.62}$$

as $t \to \infty$ with $z = O(1)$. On solving at $O\left\{\hat{\psi}'(t)\right\}$ for $\hat{u}_c(z)$, we obtain that,

$$
\hat{u}_c(z) \sim
\begin{cases}
E_0 e^{m-z} + F_0 e^{m+z} + \dfrac{v^*(m)B^*}{4}e^{-v^*(m)z} & \text{if } \lambda < \dfrac{[v^*(m)]^2}{4}, \\[2mm]
(E_0 + zF_0)e^{-\tfrac{v^*(m)z}{2}} + \dfrac{v^*(m)B^*}{\lambda}e^{-v^*(m)z} & \text{if } \lambda = \dfrac{[v^*(m)]^2}{4},
\end{cases} \tag{3.63}
$$

as $z \to \infty$, where $m_{\pm} = -\frac{v^*(m)}{2} \pm \frac{1}{2}\sqrt{[v^*(m)]^2 - 4\lambda}$ and $E_0, F_0$ are constants. We note that $\lambda > \frac{[v^*(m)]^2}{4}$ would lead to $\hat{u}_c(z)$ being oscillatory and is excluded to allow matching with (3.59). Matching expansion (3.62) (with (3.63)) (as $z \to \infty$) with expansion (3.59) up to exponentially small terms of $O\{t^{\tilde{\gamma}}e^{-\lambda t}\}$ as $t \to \infty$, requires that $F_0 \neq 0$, and

$$\lambda = \frac{[v^*(m)]^2}{4}, E_0 = 0, \tilde{\gamma} = -\frac{3}{2}, \hat{\gamma} = 1, v_1 = -\frac{8C_0\hat{H}_c}{\sqrt{\pi}[v^*(m)]^2 F_0}e^{-\frac{v^*(m)\phi_0}{2}},$$

$$(3.64)$$

after which matching is complete. Note that $v_1$ is given in terms of $\hat{H}_c$ and $C_0$, which, although both non-zero ($C_0 > 0$), are indeterminate at this order in our asymptotic theory. Finally, we have in region **TW**,

$$u(z, t) = u_T(z, v^*(m)) + O\left\{t^{-\frac{3}{2}}e^{-\frac{[v^*(m)]^2}{4}t}\right\} \qquad (3.65)$$

as $t \to \infty$ with $z = O(1)$, and

$$s(t) = v^*(m)t + \phi_0 + v_1 t^{-\frac{3}{2}}e^{-\frac{[v^*(m)]^2}{4}t} + o\left\{t^{-\frac{3}{2}}e^{-\frac{[v^*(m)]^2}{4}t}\right\} \qquad (3.66)$$

as $t \to \infty$, with the asymptotic speed of the PTW being given by

$$\dot{s}(t) \sim v^*(m) - v_1\frac{[v^*(m)]^2}{4}t^{-\frac{3}{2}}e^{-\frac{[v^*(m)]^2}{4}t} + \ldots \qquad (3.67)$$

as $t \to \infty$. We note that the correction to the propagation speed is exponential in $t$, as $t \to \infty$, being of $O\left\{t^{-\frac{3}{2}}e^{-\frac{[v^*(m)]^2}{4}t}\right\}$. We further note that the rate of convergence of the solution of **[P,m]** to the PTW as $t \to \infty$ is exponential in $t$, being of $O\left\{t^{-\frac{3}{2}}e^{-\frac{[v^*(m)]^2}{4}t}\right\}$. As $z \to -\infty$, we move out of the localized region **TW** into region **IV**, where $0 \leq Y < v^*(m) - o(1)$. The structure of expansion (3.65) as $z \to -\infty$ (obtained via (3.55)), suggests that in region **IV** we expand as

$$u(Y, t) = 1 - A^*e^{-\lambda_m(v^*(m))s(t)}e^{f(Y)t}[1 + o(1)] \qquad (3.68)$$

as $t \to \infty$, with $0 \leq Y < v^*(m) - o(1)$. On substituting (3.68) into equation (P1) (when written in terms of $Y$ and $t$) we obtain the leading order problem as

$$f_Y^2 + Yf_Y - f = 1 - \lambda_m(v^*(m))v^*(m), \quad 0 < Y < v^*(m), \qquad (3.69)$$

$$f(Y) \sim \lambda_m(v^*(m))Y \quad \text{as} \quad Y \to (v^*(m))^-, \qquad (3.70)$$

with condition (3.70) arising from matching expansion (3.68) (as $Y \to (v^*(m))^-$) with expansion (3.54) (as $z \to -\infty$). The solution of (3.69),(3.70) is obtained directly as

$$f(Y) = \lambda_m(v^*(m))Y, \quad 0 < Y < v^*(m), \tag{3.71}$$

with, via (3.68) and (3.71), that the expansion in region **IV** has the form

$$u(Y,t) = 1 - A^* e^{-\lambda_m(v^*(m))\{\frac{s(t)}{t} - Y\}t}(1 + o(1)) \tag{3.72}$$

as $t \to \infty$ with $0 < Y < v^*(m) - o(1)$. We note that expansion (3.72) becomes nonuniform when $Y = O(t^{-1})$ [that is, $x = O(1)$], and the boundary condition (P3) at $Y = 0$ is not satisfied. To complete the asymptotic structure as $t \to \infty$, we therefore introduce the final region, region **V**, where $x = O(1)$ as $t \to \infty$. The structure of expansion (3.68) (with (3.71)) in region **IV** when $Y = O(t^{-1})$ as $t \to \infty$ suggests that we should, in region **V**, look for a solution of the form

$$u(x,t) = 1 + G(x)e^{-\lambda(v^*(m))s(t)} + o\left\{e^{-\lambda(v^*(m))s(t)}\right\} \tag{3.73}$$

as $t \to \infty$, with $x = O(1)$. On substituting (3.73) into equation (P1) we obtain at leading order the problem

$$G_{xx} - \lambda_m^2(v^*(m))G = 0, \quad 0 < x < \infty, \tag{3.74}$$

$$G_x(0) = 0, \tag{3.75}$$

$$G(x) \sim -A^* e^{\lambda_m(v^*(m))x} \quad \text{as} \quad x \to \infty. \tag{3.76}$$

Condition (3.75) ensures that the boundary condition (P3) at $x = 0$ is satisfied, whilst the final condition (3.76) arises from matching expansion (3.73) (as $x \to \infty$) with expansion (3.72) [$Y = O(t^{-1})$]. The solution to (3.74)–(3.76) is readily obtained as

$$G(x) = -2A^* \cosh[\lambda_m(v^*(m))x], \quad x \ge 0, \tag{3.77}$$

giving, via (3.73), that in region **V**,

$$u(x,t) = 1 - 2A^* \cosh[\lambda_m(v^*(m))x]e^{-\lambda(v^*(m))s(t)} + o\left\{e^{-\lambda(v^*(m))s(t)}\right\} \tag{3.78}$$

as $t \to \infty$, with $x = O(1)$. In particular, we have from (3.78) that

$$u(0,t) \sim 1 - 2A^* e^{-\lambda(v^*(m))s(t)} + \dots \tag{3.79}$$

as $t \to \infty$.

This completes the asymptotic structure as $t \to \infty$ in this case. In particular, with $u(x,t)$ the solution to [P,m], we have established (via regions **III(a)**-**V**) that

$$u(z + s(t), t) \to u_T(z, v^*) \quad \text{as} \quad t \to \infty, \tag{3.80}$$

uniformly in $z$ (and through terms exponentially small in $t$ as $t \to \infty$) with $s(t)$ as given in (3.66). In the terminology of Volpert *et al* [71] the solution to [P,m], when $\sigma > v^*(m)$, converges to $u_T(z, v^*)$ *uniformly* as $t \to \infty$. A schematic representation of the location and thickness of the asymptotic regions as $t \to \infty$ is given in Figure 3.2.

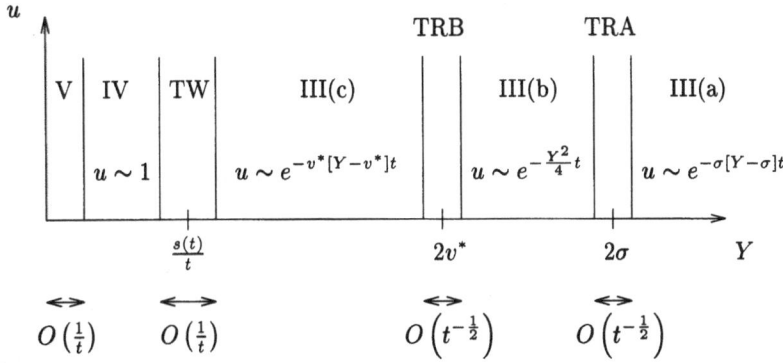

**Fig. 3.2.** Schematic representation of the location and thickness of the asymptotic regions as $t \to \infty$ when $\sigma > v^*(m)$.

**(b) $\sigma = v^*(m)$**

In this case substitution of (3.19),(3.20) into equation (P1) (when written in terms of $Y$ and $t$), gives on solving at each order in turn and matching to expansion (3.17) $[x = O(t)]$ as $Y \to \infty$ that the expansion in region **III** is given by

$$u(Y,t) = \exp\{-v^*(m)t[Y - v^*(m)] + \ln u_\infty + o(1)\} \qquad (3.81)$$

as $t \to \infty$ with $2v^*(m) + o(1) < Y < \infty$. Consideration of further terms in (3.81) indicates that a nonuniformity develops in (3.81) as $Y \to (2v^*(m))^+$. Hence, in this case, region **III** must be replaced by three regions: region **III(a)** $(2v^*(m) + o(1) < Y < \infty)$, region **TR** (transition region) and region **III(b)** $(v^*(m) + o(1) < Y < 2v^*(m) - o(1))$. As $Y \to [2v^*(m)]^+$ we move into the transition region, region **TR**. An examination of expansion (3.81) as $Y \to 2v^*(m)$ reveals that in this region, $Y = 2v^*(m) + O\left\{t^{-\frac{1}{2}}\right\}$ as $t \to \infty$. To examine region **TR** we introduce the scaled coordinate $\eta = (Y - 2v^*(m))t^{1/2}$, where $\eta = O(1)$ as $t \to \infty$, and expand as

$$u(\eta,t) = (h(\eta) + o(1))\exp[-(v^*(m))^2 t - v^*(m)\eta t^{\frac{1}{2}}] \qquad (3.82)$$

as $t \to \infty$ with $h(\eta) > 0$ and $\eta = O(1)$. On substitution of (3.82) into equation (P1) (when written in terms of $\eta$ and $t$), we obtain at leading order

$$h_{\eta\eta} + \frac{1}{2}\eta h_\eta = 0, \quad -\infty < \eta < \infty. \qquad (3.83)$$

The solution of (3.83) is readily obtained as

$$h(\eta) = A_0 \text{erfc}\left\{\frac{\eta}{2}\right\} + B_0, \qquad (3.84)$$

where $A_0$ and $B_0$ are to be determined on matching. Matching expansion (3.82) (as $\eta \to \infty$) with expansion (3.81) (as $Y \to 2v^*(m)^+$) requires $B_0 = u_\infty$. The expansion in region **TR** is thus given via (3.82) and (3.84) as

$$u(\eta, t) = (A_0 \operatorname{erfc}(\eta/2) + u_\infty) \exp[-((v^*(m))^2 t + v^*(m)\eta t^{1/2})]$$
$$+ O\left\{\exp[-m((v^*(m))^2 t + v^*(m)\eta t^{1/2})]\right\} \tag{3.85}$$

as $t \to \infty$ with $\eta = O(1)$, where $A_0 > 0$ is undetermined. The correction to (3.85) was obtained by continuing expansion (3.82) to next order, the details of which are omitted for brevity. As $\eta \to -\infty$ we move out of region **TR** into region **III(b)**, where $v^*(m) + o(1) < Y < 2v^*(m) - o(1)$. We now examine the form of expansion (3.82) for $(-\eta) \gg 1$ (as we move into region **III(b)**). From (3.82) we have,

$$u(\eta, t) \sim \left\{2A_0 + u_\infty + \frac{2A_0}{\sqrt{\pi}} \frac{e^{-\eta^2/4}}{\eta}\right\} \exp[-((v^*(m))^2 t + v^*(m)\eta t^{1/2})]$$
$$+ O\left\{e^{-m((v^*(m))^2 t + v^*(m)\eta t^{1/2})}\right\}, \tag{3.86}$$

with $(-\eta) \gg 1$. When written in terms of $Y$, (3.86) becomes,

$$u(Y, t) \sim (2A_0 + u_\infty) e^{-v^*(m)t(Y - v^*(m))} + \frac{2A_0}{\sqrt{\pi}} \frac{t^{-1/2}}{(Y - 2v^*(m))} e^{-\frac{Y^2}{4}t}$$
$$+ O\left\{e^{-mv^*(m)t[Y - v^*(m)]}\right\}. \tag{3.87}$$

We recall from Section 3.2.3 (via (3.19),(3.20) and (3.25)) that in this case (when $\sigma = v^*(m)$) $u = O\left\{e^{-v^*(m)t[y - v^*(m)]}\right\}$ as $t \to \infty$ in region **III(b)**. Thus, we must look for an expansion in region **III(b)** of the form

$$u(Y, t) = (2A_0 + u_\infty)e^{-v^*(m)t[Y - v^*(m)]} + \frac{2A_0 t^{-\frac{1}{2}} \breve{H}(Y)}{\sqrt{\pi}} e^{-\frac{Y^2}{4}t}$$
$$+ O\left\{e^{-mv^*(m)t[Y - v^*(m)]}\right\} \tag{3.88}$$

as $t \to \infty$ with $v^*(m) + o(1) < Y < 2v^*(m) - o(1)$. On substitution of (3.88) into equation (P1) (when written in terms of $Y$ and $t$), we find $\breve{H}(Y)$ is indeterminate at this order. However, matching to region **TR** (as $Y \to (2v^*(m))^-$) requires that

$$\breve{H}(Y) \sim \frac{1}{(Y - 2v^*(m))} \quad \text{as} \quad Y \to (2v^*(m))^-. \tag{3.89}$$

As $Y \to v^*(m)^+$ we move out of region **III(b)** into the wave front region, region **TW**, where $Y = v^*(m) + O(t^{-1})$ as $t \to \infty$. In region **TW**, $x \sim s(t)$ and $u$ (when written in terms of the travelling wave coordinate, $z$) has the form (via (3.30)),

$$u(z, t) = u_T(z, v^*(m)) + o(1) \tag{3.90}$$

as $t \to \infty$ with $z = O(1)$, where $z = x - s(t)$ and $s(t) = v^*(m)t + \hat{\phi}(t) + \phi_0 + \hat{\psi}(t)$ as $t \to \infty$. Here $1 \ll \hat{\phi}(t) \ll t$, $\phi_0$ is a constant and $\hat{\psi}(t) = o(1)$ as

$t \to \infty$ and are as yet undetermined gauge functions (to be fixed on matching with region **III(b)** as $z \to \infty$), whilst $u_T(z, v^*(m))$ represents the minimum speed permanent form travelling wave solution. We recall from Section 3.1 the following asymptotic properties of $u_T(z, v^*(m))$:

$$u_T(z, v^*(m)) \sim \begin{cases} B^* e^{-v^*(m)z} & \text{as } z \to \infty, \\ 1 - A^* e^{\lambda_m(v^*)z} & \text{as } z \to -\infty, \end{cases} \tag{3.91}$$

where $A^*, B^*$ are positive constants which are, in principle, determined. Matching expansion (3.88) (as $Y \to (v^*(m))^+$) of region **III(b)** to expansion (3.90) (as $z \to \infty$) of region **TW** follows after minor modifications ($C_0$ being replaced by $A_0$ in the expression for $v_1$ and $\phi_0 = -\frac{1}{v^*(m)} \ln \left\{ \frac{B^*}{2A_0 + u_\infty} \right\}$) that given in part (a) and is not repeated here. In summary, we have in region **TW**,

$$u(z, t) = u_T(z, v^*(m)) + O\left\{ t^{-\frac{3}{2}} e^{-\frac{[v^*(m)]^2}{4} t} \right\} \tag{3.92}$$

as $t \to \infty$ with $z = O(1)$, and

$$s(t) = v^*(m)t + \phi_0 + v_1 t^{-\frac{3}{2}} e^{-\frac{[v^*(m)]^2}{4} t} + o\left\{ t^{-\frac{3}{2}} e^{-\frac{[v^*(m)]^2}{4} t} \right\} \tag{3.93}$$

as $t \to \infty$, with the asymptotic speed of the PTW being given by

$$\dot{s}(t) \sim v^*(m) - v_1 \frac{[v^*(m)]^2}{4} t^{-\frac{3}{2}} e^{-\frac{[v^*(m)]^2}{4} t} + \ldots \tag{3.94}$$

as $t \to \infty$. We note that the correction to the propagation speed is exponential in $t$, as $t \to \infty$, being of $O\left\{ t^{-\frac{3}{2}} e^{-\frac{[v^*(m)]^2}{4} t} \right\}$. We further note that the rate of convergence of the solution of [**P,m**] to the PTW as $t \to \infty$ is exponential in $t$, being of $O\left\{ t^{-\frac{3}{2}} e^{-\frac{[v^*(m)]^2}{4} t} \right\}$. As $z \to -\infty$, we move out of region **TW** and the remaining structure in this case, regions **IV** and **V** follows that given in part (a).

This completes the asymptotic structure as $t \to \infty$ in this case. In particular, with $u(x, t)$ the solution to [**P,m**], we have established (via regions **III(a)-V**) that

$$u(z + s(t), t) \to u_T(z, v^*) \quad \text{as} \quad t \to \infty, \tag{3.95}$$

uniformly in $z$ (and through terms exponentially small in $t$ as $t \to \infty$) with $s(t)$ as given in (3.93). In the terminology of Volpert *et al* [71] the solution to [**P,m**], when $\sigma = v^*(m)$, converges to $u_T(z, v^*)$ *uniformly* as $t \to \infty$. A schematic representation of the location and thickness of the asymptotic regions as $t \to \infty$ is given in Figure 3.3.

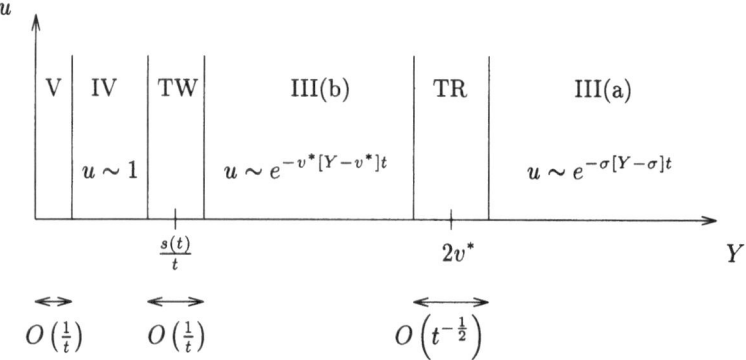

**Fig. 3.3.** Schematic representation of the location and thickness of the asymptotic regions as $t \to \infty$ when $\sigma = v^*(m)$.

## (c) $0 < \sigma < v^*(m)$

In this case the leading order problem [**TW**] of region **TW** has no solution, via Theorem 3.1. We must return to region **III** and conclude that the expansion in region **III** becomes nonuniform as $Y \to Y_T^+$, where $Y_T > 2\sigma$ is to be determined. Substitution of (3.19),(3.20) into equation (P1) (when written in terms of $Y$ and $t$), gives on solving at each order in turn and matching to expansion (3.17) [$x = O(t)$] as $Y \to \infty$ that

$$u(Y,t) = \exp\left\{-t\sigma(Y-\sigma) + \ln u_\infty + o(1)\right\} \qquad (3.96)$$

as $t \to \infty$, with $Y_T + o(1) < Y < \infty$. Thus we continue the asymptotic structure with a transition layer, region **TR**, in which $Y = Y_T + o(1)$. The distinguished limit in this case gives $Y = Y_T + O(1/t)$. To examine region **TR** we introduce the scaled coordinate $\xi = (Y - Y_T)t$, where $\xi = O(1)$ as $t \to \infty$, and expand as indicated from expansion (3.96), that is

$$u(\xi,t) = [F(\xi) + o(1)]e^{-\sigma t[Y_T - \sigma]} \qquad (3.97)$$

as $t \to \infty$ with $F(\xi) > 0$ and $\xi = O(1)$. Substitution of (3.97) into equation (P1) (when written in terms of $\xi$ and $t$) gives at leading order

$$F_{\xi\xi} + Y_T F_\xi + \sigma[Y_T - \sigma]F = 0, \quad -\infty < \xi < \infty. \qquad (3.98)$$

The solution to (3.98) is given directly as

$$F(\xi) = Ae^{-\sigma\xi} + Be^{-(Y_T-\sigma)\xi}, \quad -\infty < \xi < \infty, \qquad (3.99)$$

where $A, B > 0$ are constants to be determined. Matching expansion(3.97) (as $\xi \to \infty$) with expansion (3.96) (as $Y \to Y_T^+$) then requires

$$A = u_\infty. \qquad (3.100)$$

The constant $B$ will be fixed by matching expansion (3.97) (as $\xi \to \infty$) to expansion(3.96) (as $Y \to Y_T^+$), when expansion (3.96) is taken to the next order. The expansion in region **TR** is (via (3.97) and (3.99) (with(3.100)) then given by

$$u(\xi, t) = \left\{ u_\infty e^{-\sigma\xi} + B e^{-(Y_T - \sigma)\xi} \right\} e^{-\sigma t[Y_T - \sigma]}$$
$$+ o \left\{ e^{-\sigma t[Y_T - \sigma]} \right\} \tag{3.101}$$

as $t \to \infty$ with $\xi = O(1)$. As $\xi \to -\infty$, we move out of region **TR** into region **III(b)** (we relabel region **III** as **III(a)**), where $0 \le Y < Y_T - o(1)$. An examination of the form of expansion (3.101) for $(-\xi) \gg 1$ (as we move into region **III(b)**) suggests that in region **III(b)** we look for an expansion of the form

$$u(Y, t) = g_0(Y) e^{-(Y_T - \sigma)t(Y - (Y_T - \sigma))} + g_1(Y) e^{-\sigma t(Y - \sigma)} + o\left\{ e^{-\sigma t(Y - \sigma)} \right\} \tag{3.102}$$

as $t \to \infty$, with $g_0(Y) > 0$ and $Y_T - \sigma + o(1) < Y < Y_T - o(1)$. On substitution of (3.102) into equation (P1) (when written in terms of $Y$ and $t$) and solving at each order in turn, we find (after matching with (3.101) as $Y \to (Y_T)^-$) that

$$u(Y, t) = B e^{-(Y_T - \sigma)t(Y - (Y_T - \sigma))} + u_\infty e^{-\sigma t(Y - \sigma)} + o\left\{ e^{-\sigma t(Y - \sigma)} \right\} \tag{3.103}$$

as $t \to \infty$, where $Y_T - \sigma + o(1) < Y < Y_T - o(1)$. We note that expansion (3.103) becomes nonuniform as $Y \to (Y_T - \sigma)^+$. As $Y \to (Y_T - \sigma)^+$ we move into region **TW**, where $Y = (Y_T - \sigma) + O(t^{-1})$ as $t \to \infty$. In region **TW**, $u$ (when written in terms of the travelling wave coordinate, $z$) has the form (via (3.30)),

$$u(z, t) = u_c(z) + o(1) \quad \text{as} \quad t \to \infty, \tag{3.104}$$

with $z = O(1)$. The leading order problem in region **TW** is given by problem **[TW]** except now $Y_c$ is replaced by $(Y_T - \sigma)$. We know, via Theorem 3.1, that **[TW]** has a unique PTW solution (say $u_c(z) = u_T(z, Y_T - \sigma)$) for each $Y_T - \sigma \ge v^*(m)(> 0)$, whilst no PTW solution of **[TW]** exists for each $0 < Y_T - \sigma < v^*(m)$. Moreover, as $z \to \infty$,

$$u_T(z, Y_T - \sigma) \sim \begin{cases} B^* e^{-v^*(m)z}, & Y_T - \sigma = v^*(m), \\ \left\{ \frac{(Y_T - \sigma)}{(m-1)z} \right\}^{\frac{1}{m-1}}, & Y_T - \sigma > v^*(m), \end{cases} \tag{3.105}$$

where $B^* > 0$. Matching expansion (3.104) (as $z \to \infty$) to expansion (3.103) (as $Y \to (Y_T - \sigma)^+$), at leading order, requires that

$$Y_T = v^*(m) + \sigma \quad (> 2\sigma), \tag{3.106}$$

and so $u_c(z) = u_T(z, v^*(m))$. Region **TW** is the wave front region, where $x \sim s(t)$ and $u$ (when written in terms of the travelling wave coordinate, $z$) has the form (via (3.104)),

$$u(z,t) = u_T(z, v^*(m)) + o(1) \tag{3.107}$$

as $t \to \infty$ with $z = O(1)$, where $z = x - s(t)$ and $s(t) = v^*(m)t + \hat{\phi}(t) + \phi_0 + \hat{\psi}(t)$ as $t \to \infty$. Here $1 \ll \hat{\phi}(t) \ll t$, $\phi_0$ is a constant and $\hat{\psi}(t) = o(1)$ as $t \to \infty$ and are as yet undetermined gauge functions (to be fixed on matching with region III(b) as $z \to \infty$), whilst $u_T(z, v^*(m))$ represents the minimum speed permanent form travelling wave solution. We recall from Section 3.1 the following asymptotic properties of $u_T(z, v^*(m))$ :

$$u_T(z, v^*(m)) \sim \begin{cases} B^* e^{-v^*(m)z} & \text{as } z \to \infty, \\ 1 - A^* e^{\lambda_m(v^*)z} & \text{as } z \to -\infty, \end{cases} \tag{3.108}$$

where $A^*, B^*$ are positive constants which are, in principle, determined. On examining expansion (3.103) (with (3.106)) in region III(b) as we move into region TW (as $Y \to v^*(m)^+$), we obtain from (3.103) (with (3.106)), when written in terms of $z$ (via $Y = \frac{s(t)}{t} + \frac{z}{t}$), that

$$u(z,t) \sim B \exp\left\{-v^*(m)\hat{\phi}(t) - v^*(m)[\phi_0 + z] - v^*(m)\hat{\psi}(t)\right\}$$
$$+ u_\infty \exp\left\{-\sigma(v^*(m) - \sigma)t - \sigma\hat{\phi}(t) - \sigma[\phi_0 + z] - \sigma\hat{\psi}(t)\right\} \tag{3.109}$$

as $t \to \infty$ with $z \gg 1$. Matching expansion (3.107) (as $z \to \infty$) with expansion (3.109) up to leading order fixes $\hat{\phi}(t) \equiv 0$ and requires that $\phi_0 = -\frac{1}{v^*(m)} \ln\left\{\frac{B^*}{B}\right\}$. On rewriting (3.109) we obtain

$$u(z,t) \sim B e^{(-v^*(m)\phi_0)} e^{-v^*(m)z}[1 - v^*(m)\hat{\psi}(t) + \ldots]$$
$$+ u_\infty e^{-\sigma\phi_0} e^{-\sigma z} e^{(-\sigma(v^*(m)-\sigma)t)} + \ldots \tag{3.110}$$

as $t \to \infty$ with $z \gg 1$. We conclude from (3.110) that in region TW we must have

$$u(z,t) = u_T(z, v^*(m)) + O(\hat{\psi}(t)) \tag{3.111}$$

as $t \to \infty$. The translational invariance of $u_T(z, v^*(m))$ with respect to $z$ is fixed by requiring $u_T(0, v^*(m)) = \frac{1}{2}$. On substitution of (3.111) into equation (P1) (when written in terms of $z$ and $t$) we require to obtain a nontrivial balance at $O\left\{\hat{\psi}'(t)\right\}$ that $\hat{\psi}'(t) = O\left\{\hat{\psi}(t)\right\}$ as $t \to \infty$ and we conclude that $\hat{\psi}(t)$ must be exponentially small in $t$, as $t \to \infty$ and hence set

$$\hat{\psi}(t) = v_1 t^\gamma e^{-\lambda t}[1 + o(1)] \tag{3.112}$$

as $t \to \infty$, where $v_1$, and $\lambda(> 0)$ are to be determined. We now continue the expansion in region TW as

$$u(z,t) = u_T(z, v^*(m)) + \hat{u}_c(z)\hat{\psi}'(t) + o\left\{\hat{\psi}'(t)\right\} \tag{3.113}$$

as $t \to \infty$ with $z = O(1)$. The asymptotic behaviour of $\hat{u}_c(z)$ as $z \to \infty$ is given by (3.63). Matching expansion (3.113) (as $z \to \infty$) with expansion (3.110) up to exponentially small terms of $O\left\{e^{-\lambda t}\right\}$ as $t \to \infty$ requires that

$$\lambda = \sigma(v^*(m) - \sigma), E_0 = 0, \gamma = 0, v_1 = -\frac{B^*}{BF_0} u_\infty (v^*(m) - \sigma)^{-1}, \quad (3.114)$$

and matching is complete (we recall that $E_0$ and $F_0$ are from (3.63)). Finally, we have in region **TW**,

$$u(z, t) = u_T(z, v^*(m)) + O\left\{e^{-\sigma(v^*(m) - \sigma)t}\right\} \quad (3.115)$$

as $t \to \infty$ with $z = O(1)$, and

$$s(t) = v^*(m)t + \phi_0 + v_1 e^{-\sigma(v^*(m) - \sigma)t} + o\left\{e^{-\sigma(v^*(m) - \sigma)t}\right\} \quad (3.116)$$

as $t \to \infty$, with the asymptotic speed of the PTW being given by

$$\dot{s}(t) \sim v^*(m) - v_1 \sigma(v^*(m) - \sigma)e^{-\sigma(v^*(m) - \sigma)t} + \ldots \quad (3.117)$$

as $t \to \infty$. We note that the correction to the propagation speed is exponential in $t$, as $t \to \infty$, being of $O\left\{e^{-\sigma(v^*(m) - \sigma)t}\right\}$. We further note that the rate of convergence of the solution of $[P,m]$ to the PTW as $t \to \infty$ is exponential in $t$, being of $O\left\{e^{-\sigma(v^*(m) - \sigma)t}\right\}$. The remaining structure in this case, regions **IV** and **V**, follows that given in part (a).

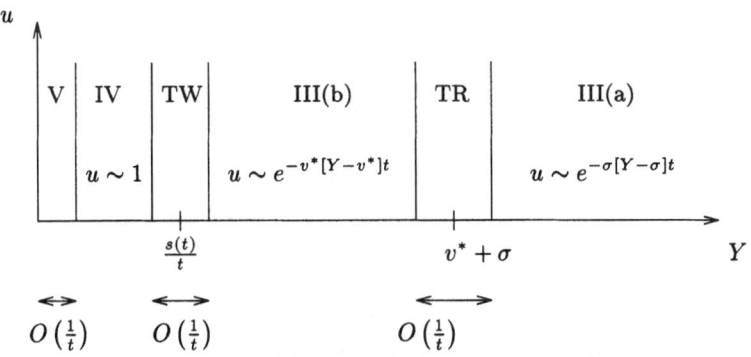

**Fig. 3.4.** Schematic representation of the location and thickness of the asymptotic regions as $t \to \infty$ when $0 < \sigma < v^*(m)$.

This completes the asymptotic structure as $t \to \infty$ in this case. In particular, with $u(x, t)$ the solution to $[P,m]$, we have established (via regions **III(a)-V**) that

$$u(z + s(t), t) \to u_T(z, v^*) \quad \text{as} \quad t \to \infty, \quad (3.118)$$

uniformly in $z$ (and through terms exponentially small in $t$ as $t \to \infty$) with $s(t)$ as given in (3.116). In the terminology of Volpert *et al* [71] the solution to [**P,m**], when $0 < \sigma < v^*(m)$, converges to $u_T(z, v^*)$ *uniformly* as $t \to \infty$. A schematic representation of the location and thickness of the asymptotic regions as $t \to \infty$ is given in Figure 3.4.

### 3.2.4 Asymptotic Solution as $t \to \infty$ in the Case of Initial Data with Compact Support

As $t \to \infty$, the asymptotic expansion (3.18) in region **IV** continues to remain uniform for $x \gg t$. However, as already noted, a nonuniformity develops when $x = O(t)$. To proceed, we introduce a new region, region **V**. To examine region **V** we introduce the scaled coordinate $Y = \frac{x}{t}$, where $Y = O(1)$ as $t \to \infty$, and look for an expansion of the form (3.19) where now (as suggested by (3.18))

$$F(Y, t) = f_0(Y) + f_1(Y)\frac{\ln t}{t} + f_2(Y)\frac{1}{t} + o\left(\frac{1}{t}\right). \tag{3.119}$$

As in Section 3.2.3 it is instructive to consider first the leading order problem in region **V**. On substituting (3.19) into equation (P1) (when written in terms of $Y$ and $t$) we obtain the leading order problem as

$$f_{0Y}^2 - Y f_{0Y} + f_0 = 0, \quad Y > 0, \tag{3.120}$$
$$f_0(Y) > 0, \quad Y > 0, \tag{3.121}$$
$$f_0(Y) \sim \frac{Y^2}{4} \quad \text{as } Y \to +\infty. \tag{3.122}$$

The final condition, (3.122), arises from matching expansion (3.19) ($Y \gg 1$) with expansion (3.18) [$x = O(t)$]. Equation (3.120) has a one-parameter family of linear solutions

$$f_0(Y) = A[Y - A], \quad Y > 0, \tag{3.123}$$

for any $A \in \mathbb{R}$, together with the associated envelope solution

$$f_0(Y) = \frac{Y^2}{4}, \quad Y > 0. \tag{3.124}$$

Combinations of (3.123) and (3.124) which remain continuous and differentiable also provide solutions to (3.120) (envelope touching solutions). Applying the conditions (3.121) and (3.122) requires us to select one of the following solutions

$$f_0(Y) = \frac{Y^2}{4}, \qquad Y > 0, \tag{3.125}$$

$$f_0(Y) = \begin{cases} \frac{Y^2}{4}, & Y > 2A, \\ A[Y - A], & A < Y \leq 2A, \end{cases} \tag{3.126}$$

for $A > 0$. We conclude, after consideration of higher order terms in (3.19), (3.119) that a nonuniformity occurs in expansion (3.19), when $Y \to Y_c^+ (> 0)$, where $Y_c = 0$ $[Y_c = A(> 0)]$ when $f_0(Y)$ is given by (3.125) [(3.126)] respectively. This nonuniformity occurs when

$$Y = Y_c + \frac{z}{t}, \tag{3.127}$$

with $z = O(1)$ as $t \to \infty$. We refer to this region as region **TW**. In region **TW** expansions (3.19)and (3.119) demonstrate that $u = O(1)$ as $t \to \infty$, and so we expand in the form

$$u(z,t) = u_c(z) + o(1), \quad \text{as } t \to \infty, \tag{3.128}$$

with $z = O(1)$. On substituting into equation (P1) (when written in terms of $z$ and $t$) we obtain the leading order problem as (3.31)-(3.34). As in Section 3.3.1 a phase plane analysis (Poincaré-Bendixson theorem) of equation (3.31) with conditions (3.32),(3.33) allows boundary condition (3.34) to be replaced by $u_c(z) \to 1$ as $z \to -\infty$, (all other possibilities leading to unbounded $u_c(z)$ as $z \to -\infty$). Thus, the complete boundary value problem for $u_c(z)$ may be stated as

$$\left.\begin{array}{l} u_c'' + Y_c u_c' + u_c^m(1 - u_c) = 0, \quad -\infty < z < \infty, \\ u_c(z) \geq 0, \quad -\infty < z < \infty, \\ u_c(z) \to 0, \quad \text{as } z \to +\infty, \\ u_c(z) \to 1, \quad \text{as } z \to -\infty. \end{array}\right\} \textbf{[TW]}$$

We observe that **[TW]** is precisely the boundary value problem (3.2) for PTWs with $v = Y_c$, and its solutions are classified in Theorem 3.1.

Matching expansion (3.128) (as $z \to \infty$) to expansion (3.19) (as $Y \to Y_c^+$) requires that $Y_c = A = v^*(m)$ and that $f_0(Y)$ is given by (3.126). Hence the large-$t$ structure in region **TW** is dominated by the evolution of the PTW with asymptotic wave speed $v = v^*(m)$ (the minimum available speed). Finally, we note that, although $f_0(Y)$ and $f_0'(Y)$ are continuous over the range of definitions of (3.126), the second derivative $f_0''(Y)$ is discontinuous at the point $Y = 2A$ $(2v^*(m))$, at which the linear solution meets the envelope solution. This indicates that a thin transition region exists in the neighbourhood of $Y = 2A$ $(2v^*(m))$, in which second order derivatives are retained at leading order to smooth out this discontinuity in curvature.

Hence, to accommodate this transition region, region **V** is replaced by three regions: region **Va** $(2v^*(m) + o(1) < Y < \infty)$, region **TR** (transition region) and region **V(b)** $(v^*(m)+o(1) < Y < 2v^*(m)-o(1))$. We consider each of these in turn. We begin with region **V(a)**, where $2v^*(m) + o(1) < Y < \infty$. Substitution of (3.19) and (3.119) into equation (P1) (when written in terms of $Y$ and $t$) gives on solving at each order in turn and matching to expansion (3.18) $[x = O(t)]$ as $Y \to \infty$ that

$$u(Y,t) = \exp\left\{-t\frac{Y^2}{4} - \frac{1}{2}\ln t - H(Y) + o(1)\right\} \tag{3.129}$$

as $t \to \infty$ with $2v^*(m) + o(1) < Y < \infty$, where

$$H(Y) \sim (r+1) \ln Y - \ln c_\infty - \frac{\sigma Y}{2} \qquad (3.130)$$

as $Y \to \infty$. The function $H(Y)$ remains undetermined. This is a consequence of using the far field asymptotics $x, t \gg 1$ rather than information from the bulk region $x, t = O(1)$ as a basis for the large time asymptotic structure. It is instructive to note, at this point, that the expansion in region **V(b)** is given by

$$u(Y,t) = \exp\{-tv^*(m)[Y - v^*(m)] + \lambda_1 \ln t + \lambda_1 \ln(2v^*(m) - Y)$$
$$+ \lambda_2 + o(1)\} \qquad (3.131)$$

as $t \to \infty$, with $v^*(m) + o(1) < Y < 2v^*(m) - o(1)$ where $\lambda_1$ and $\lambda_2$ are constants to be determined on matching with region **TR** as $Y \to 2v^*(m)^-$.

As $Y \to 2v^*(m)^+$ we move out of region **V(a)** into the transition region, region **TR**. An examination of expansion (3.129) as $Y \to 2v^*(m)^+$ reveals that in this region $Y = 2v^*(m) + O(t^{-1/2})$ as $t \to \infty$. Thus, in region **TR**, we introduce the scaled coordinate $\xi = (Y - 2v^*(m))t^{1/2}$, where $\xi = O(1)$ as $t \to \infty$, and expand as indicated from expansion (3.129), that is,

$$u(\xi, t) = [h_0(\xi)t^\gamma + \psi(\xi, t)] \exp\left\{-[v^*(m)]^2 t - v^*(m)\xi t^{1/2}\right\} \qquad (3.132)$$

as $t \to \infty$, with $h_0(\xi) > 0$ and $\xi = O(1)$. Here the gauge function $\psi(\xi, t) = o(t^\gamma)$ as $t \to \infty$ is to be determined, along with the constant $\gamma$. Substitution of (3.132) into equation (P1) (when written in terms of $\xi$ and $t$) we obtain at leading order

$$h_{0\xi\xi} + \frac{1}{2}\xi h_{0\xi} - \gamma h = 0, \quad -\infty < \xi < \infty. \qquad (3.133)$$

Matching expansion (3.132) (as $\xi \to -\infty$) to expansion (3.131) (as $Y \to 2v^*(m)^-$), to leading order, requires that

$$h_0(\xi) \sim (-\xi)^{\lambda_1} e^{\lambda_2} \quad \text{as} \quad \xi \to -\infty, \qquad (3.134)$$

$$\gamma = \frac{1}{2}\lambda_1. \qquad (3.135)$$

Matching expansion (3.132) (as $\xi \to +\infty$) to expansion (3.129) (as $Y \to 2v^*(m)^+$) requires

$$h_0(\xi) = O\left(e^{-\frac{\xi^2}{4}}\right) \quad \text{as} \quad \xi \to \infty. \qquad (3.136)$$

However, equation (3.133) has no solutions which satisfy both (3.134) and (3.136) when $\gamma < 0$. We conclude that $\gamma \geq 0$. With $\gamma \geq 0$, equation (3.133) with conditions (3.134) and (3.136) has a solution, with

$$h_0(\xi) \sim \hat{A}\xi^{-2\gamma - 1}e^{-\frac{\xi^2}{4}} \quad \text{as} \quad \xi \to \infty, \tag{3.137}$$

with $\hat{A} > 0$ undetermined. On matching expansion (3.132) (as $\xi \to \infty$), using (3.137), to expansion (3.129) (as $Y \to 2v^*(m)^+$) we require, to obtain the least singular behaviour in region $\mathbf{V(a)}$ as $Y \to 2v^*(m)^+$, that

$$\gamma = 0, \tag{3.138}$$

after which

$$H(Y) \sim \ln(Y - 2v^*(m)) - \ln \hat{A}$$

as $Y \to 2v^*(m)^+$. In fact, the solution of equation (3.132) and conditions (3.134) and (3.137), when $\gamma = 0$, may be obtained explicitly as

$$h_0(\xi) = \frac{1}{2}\hat{A}\sqrt{\pi}\,\text{erfc}\left(\frac{1}{2}\xi\right), \tag{3.139}$$

with

$$\lambda_2 = \ln(\sqrt{\pi}\hat{A}),$$

and via (3.138) and (3.135)

$$\lambda_1 = 0.$$

On continuing to next order in region $\mathbf{TR}$ we obtain

$$\psi(\xi, t) = \frac{-[h_0(\xi)]^m}{[v^*(m)]^2(m-1)m}\exp\left\{-(m-1)\left[(v^*(m))^2 t + v^*(m)\xi t^{1/2}\right]\right\} \tag{3.140}$$

as $t \to \infty$ with $\xi = O(1)$ (the details of which are omitted for brevity). Hence, the expansion in region $\mathbf{TR}$ is given via (3.132) and (3.140) as

$$u(\xi, t) = A_0\text{erfc}\left(\frac{1}{2}\xi\right)e^{-[v^*(m)]^2 t - v^*(m)\xi t^{1/2}} + O\left(e^{-m[v^*(m)]^2 t - mv^*(m)\xi t^{1/2}}\right) \tag{3.141}$$

as $t \to \infty$ with $\xi = O(1)$ and where $A_0 = \frac{1}{2}\sqrt{\pi}\hat{A}$ $(> 0)$. As $\xi \to -\infty$ we move out of region $\mathbf{TR}$ into region $\mathbf{V(b)}$, where $v^*(m) + o(1) < Y < 2v^*(m) - o(1)$.

The structure of (3.141) as $(-\xi) \gg 1$ (as we move into region $\mathbf{V(b)}$) indicates that the correction term to (3.131) is $O\left[t^{-1/2}e^{-\frac{Y^2}{4t}}\right]$ as $t \to \infty$. The expansion in region $\mathbf{V(b)}$ is now given (after some calculation) as

$$u(Y, t) = \exp\left\{-tv^*(m)[Y - v^*(m)] + \ln 2A_0\right\} - \exp\left\{-\frac{Y^2}{4}t - \frac{1}{2}\ln t - \hat{H}(Y)\right\}$$
$$+ O\left[\exp\left\{-tmv^*(m)[Y - v^*(m)]\right\}\right] \tag{3.142}$$

as $t \to \infty$ with $v^*(m) + o(1) < Y < 2v^*(m) - o(1)$. The function $\hat{H}(Y)$ remains undetermined at this order but matching to (3.141) as $Y \to 2v^*(m)$ requires that

$$\hat{H}(Y) \sim \ln(2v^*(m) - Y) - \ln\left(\frac{2A_0}{\sqrt{\pi}}\right) \tag{3.143}$$

as $Y \to 2v^*(m)^-$. As $Y \to v^*(m)^+$ we move out of region **V(b)** into the wave front region, region **TW**, where $Y = v^*(m) + O(1/t)$ as $t \to \infty$. In region **TW**, $x \sim s(t)$ and $u$ (when written in terms of the travelling wave coordinate, $z$) has the form (via 3.128)

$$u(z,t) = u_T(z, v^*(m)) + o(1) \tag{3.144}$$

as $t \to \infty$ with $z = O(1)$, where $z = x - s(t)$ and $s(t) = v^*(m)t + \hat{\phi}(t) + \phi_0 + \hat{\psi}(t)$ as $t \to \infty$. Here $1 \ll \hat{\phi}(t) \ll t$, $\phi_0$ is a constant and $\hat{\psi}(t) = o(1)$ as $t \to \infty$ and are as yet undetermined gauge functions (to be fixed on matching with region **V(b)** as $z \to \infty$), whilst $u_T(z, v^*(m))$ represents the minimum speed permanent form travelling wave solution. We recall from Section 3.1 the following asymptotic properties of $u_T(z, v^*(m))$:

$$u_T(z, v^*(m)) \sim \begin{cases} B^* e^{-v^*(m)z} & \text{as } z \to \infty, \\ 1 - A^* e^{\lambda_m(v^*)z} & \text{as } z \to -\infty, \end{cases} \tag{3.145}$$

where $A^*, B^*$ are positive constants which are, in principle, determined. On making the assumption (which we will verify as consistent) that

$$\hat{H}(Y) \sim c_0 \ln(Y - v^*(m)) + \hat{H}_c \quad Y \to v^*(m)^+,$$

for some constants $c_0$ and $\hat{H}_c$, we have on writing (3.142) in terms of $z$, that

$$u(z,t) \sim 2A_0 \exp\left\{-v^*(m)\hat{\phi}(t) - v^*(m)[z + \phi_0]\right\}$$
$$-2A_0 v^*(m)\hat{\psi}(t) \exp\left\{-v^*(m)\hat{\phi}(t) - v^*(m)[z + \phi_0]\right\}$$
$$- \exp\left\{-\frac{[v^*(m)]^2}{4}t - \frac{v^*(m)}{2}\hat{\phi}(t) + \left(c_0 - \frac{1}{2}\right)\ln t - c_0 \ln\left(\hat{\phi}(t)\right)\right.$$
$$\left. + (z + \phi_0) + \hat{\psi}(t)\right) - \frac{v^*(m)}{2}(z + \phi_0) - \hat{H}_c \right\} + \dots \tag{3.146}$$

as $t \to \infty$ with $z \gg 1$. The matching of expansion (3.144) (as $z \to \infty$) with (3.146) up to terms of $O(1)$ requires $\hat{\phi}(t) \equiv 0$ and

$$\phi_0 = -\frac{1}{v^*(m)} \ln\left(\frac{B^*}{2A_0}\right).$$

On rewriting (3.146) we obtain

$$u(z,t) \sim 2A_0 e^{-v^*(m)[z+\phi_0]} - 2A_0 v^*(m)\hat{\psi}(t)e^{-v^*(m)[z+\phi_0]}$$
$$-e^{-\hat{H}_c}z^{-c_0}e^{-\frac{v^*(m)}{2}[z+\phi_0]}t^{c_0-1/2}e^{-\frac{[v^*(m)]^2}{4}t} + \dots \tag{3.147}$$

as $t \to \infty$ with $z \gg 1$. We conclude from (3.147) that in region **TW** we must have

$$u(z,t) = u_T(z; v^*(m)) + O\left(\hat{\psi}(t)\right) \quad \text{as} \quad t \to \infty. \tag{3.148}$$

On substituting (3.148) into equation (P1) (when written in terms of $z$ and $t$) we require to obtain a nontrivial balance at $O\left(\hat{\psi}'(t)\right)$ that $\hat{\psi}'(t) = O\left(\hat{\psi}(t)\right)$ as $t \to \infty$ and we conclude that $\hat{\psi}(t)$ must be exponential in $t$, as $t \to \infty$, and write

$$\hat{\psi}(t) = v_1 t^{\tilde{\gamma}} e^{-\lambda t}[1 + o(1)] \tag{3.149}$$

as $t \to \infty$, where $v_1, \tilde{\gamma}$ and $\lambda(> 0)$ are to be determined. We now continue the expansion in region **TW** as

$$u(z,t) = u_T(z, v^*(m)) + \hat{u}_c(z)\hat{\psi}'(t) + o\left\{\hat{\psi}'(t)\right\} \tag{3.150}$$

as $t \to \infty$ with $z = O(1)$. On solving at $O\left\{\hat{\psi}'(t)\right\}$ for $\hat{u}_c(z)$, we obtain

$$\hat{u}_c(z) \sim \begin{cases} \bar{A}e^{m_- z} + \bar{B}e^{m_+ z} + \frac{v^*(m)B^*}{\lambda}e^{-v^*(m)z} & \text{if } \lambda < \frac{[v^*(m)]^2}{4}, \\ (\bar{A} + z\bar{B})e^{-\frac{v^*(m)z}{2}} + \frac{v^*(m)B^*}{\lambda}e^{-v^*(m)z} & \text{if } \lambda = \frac{[v^*(m)]^2}{4}, \end{cases} \tag{3.151}$$

as $z \to \infty$, where $\bar{A}$ and $\bar{B}$ are constants and $m_\pm = -\frac{v^*(m)}{2} \pm \frac{1}{2}\sqrt{[v^*(m)]^2 - 4\lambda}$. We note that $\lambda > \frac{[v^*(m)]^2}{4}$ would lead to $\hat{u}_c(z)$ being oscillatory and is excluded to allow matching with (3.147). Matching expansion (3.150) (as $z \to \infty$) with expansion (3.147) up to exponentially small terms of $O\left\{t^{\tilde{\gamma}}e^{-\lambda t}\right\}$ as $t \to \infty$, requires that

$$\lambda = \frac{[v^*(m)]^2}{4}, \bar{A} = 0, \tilde{\gamma} = -\frac{3}{2}, v_1 = -\frac{4e^{-\hat{H}_c}}{[v^*(m)]^2 \bar{B}}e^{-\frac{v^*(m)\phi_0}{2}}, c_0 = -1, \tag{3.152}$$

and that

$$\hat{H}(Y) \sim -\ln(Y - v^*(m)) + \hat{H}_c$$

as $Y \to v^*(m)^+$, where $\hat{H}_c$ and $\bar{B}$ remain undetermined at this order. Hence, we have established that the large-$t$ structure in region **TW** is dominated by the evolution of the PTW with asymptotic speed $v = v^*(m)$ (this being the minimum available wave speed). Further, we have established that

$$s(t) = v^*(m)t + \phi_0 + v_1 t^{-\frac{3}{2}}e^{-\frac{[v^*(m)]^2}{4}t} + o\left\{t^{-\frac{3}{2}}e^{-\frac{[v^*(m)]^2}{4}t}\right\} \tag{3.153}$$

as $t \to \infty$, with the asymptotic speed of the PTW being given by

$$\dot{s}(t) \sim v^*(m) - v_1 \frac{[v^*(m)]^2}{4}t^{-\frac{3}{2}}e^{-\frac{[v^*(m)]^2}{4}t} + \dots \tag{3.154}$$

as $t \to \infty$. We note that the correction to the propagation speed is exponential in $t$, as $t \to \infty$, being of $O\left\{t^{-\frac{3}{2}}e^{-\frac{[v^*(m)]^2}{4}t}\right\}$. We further note that the rate of convergence of the solution of [P,m] to the PTW as $t \to \infty$ is exponential in $t$, being of $O\left\{t^{-\frac{3}{2}}e^{-\frac{[v^*(m)]^2}{4}t}\right\}$.

As $z \to -\infty$, we move out of the localized region **TW**. The remainder of the asymptotic structure of [P,m] as $t \to \infty$ in this case is identical to that given in Section 3.2.3 (part (a)) with regions **IV** and **V** (of Section 3.2.3 (part (a))) now being renamed **VI** and **VII**, respectively.

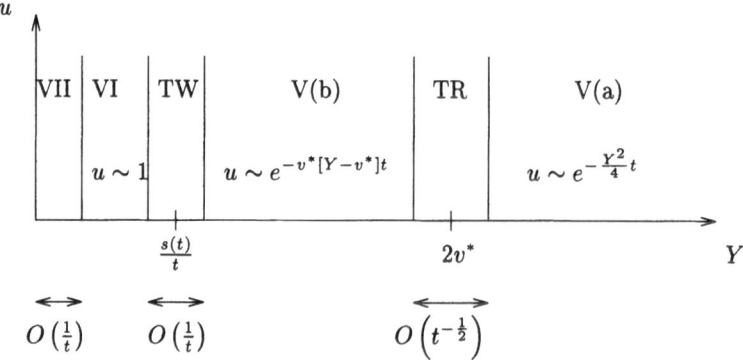

**Fig. 3.5.** Schematic representation of the location and thickness of the asymptotic regions as $t \to \infty$.

This completes the asymptotic structure as $t \to \infty$ in this case. In particular, with $u(x, t)$ the solution to [P,m], we have established (via regions **V(a)-VII**) that

$$u(z + s(t), t) \to u_T(z, v^*) \quad \text{as} \quad t \to \infty, \tag{3.155}$$

uniformly in $z$ (and through terms exponentially small in $t$ as $t \to \infty$) with $s(t)$ as given in (3.153). In the terminology of Volpert *et al* [71] the solution to [P,m], converges to $u_T(z, v^*)$ *uniformly* as $t \to \infty$. A schematic representation of the location and thickness of the asymptotic regions as $t \to \infty$ is given in Figure 3.5.

## 3.3 Numerical Solutions

Finally, we present numerical solutions to initial-boundary value problem [P,m] which support and illustrate the detailed analysis given in Section 3.2. We restrict our attention to the case when $m = 2$ with $v^*(2) = \frac{1}{\sqrt{2}}$.

### 3.3.1 Numerical Solutions when the Initial Data has Exponential Decay Rate as $x \to \infty$

We restrict attention to two values of $\sigma$, namely (a) $\sigma = \frac{1}{2}(< v^*(2))$ and (b) $\sigma = 1(> v^*(2))$ with $u_\infty = e^\sigma$. Equation (P1) was solved using the NAG routine D03PCF (method of lines with finite differences) with the initial data used being given by

$$u_0(x) = \begin{cases} 1, & 0 \le x \le 1, \\ e^{-\sigma(x-1)}, & x > 1. \end{cases} \tag{3.156}$$

We recall from the detailed theory of Section 3.2.3 that in this case

$$u(z + s(t), t) = u_T(z, 1/\sqrt{2}) + O\left(\chi(t)\right) \tag{3.157}$$

as $t \to \infty$ with $z = O(1)$, where $z = x - s(t)$, and

$$s(t) = \frac{1}{\sqrt{2}}t + \phi_0 + O\left(\chi(t)\right), \tag{3.158}$$

where

$$\chi(t) = \begin{cases} e^{-1/2(1/\sqrt{2}-1/2)t}, & \sigma = \frac{1}{2} < v^*(2), \\ t^{-\frac{3}{2}}e^{-\frac{1}{8}t}, & \sigma = 1 > v^*(2), \end{cases} \tag{3.159}$$

as $t \to \infty$ and the asymptotic wave speed of the PTW is given by $\dot{s}(t)$.

Figure 3.6 gives for both cases (a) $\sigma = 1/2$ and (b) $\sigma = 1$ a series of three graphs obtained from numerical integrations:

(i) Shows the development of the solution $u(x,t)$ of [P,2] as $t \to \infty$ to the PTW, as predicted by (3.157).

(ii) Shows a plot of $s(t) - v^*(m)t$ against $t$. The dashed line represents the constant $\phi_0$, with the rate of convergence to $\phi_0$ in both cases being exponential in $t$, although with different rates, as predicted by (3.158).

(iii) Shows a plot of $\ln(s(t) - v^*(m) - \phi_0)$ in case (a) and of $\ln[t^{3/2}(s(t) - v^*(m) - \phi_0)]$ in case (b) against $t$. The dashed line represents the theoretically predicted value of the exponent of the exponential correction, given via (3.159), as $-1/2(1/\sqrt{2} - 1/2)t$ in case (a) and by $-\frac{1}{8}t$ in case (b). Clearly we have good agreement between numerically calculated values and the predicted values for $t \gg 1$ in both cases.

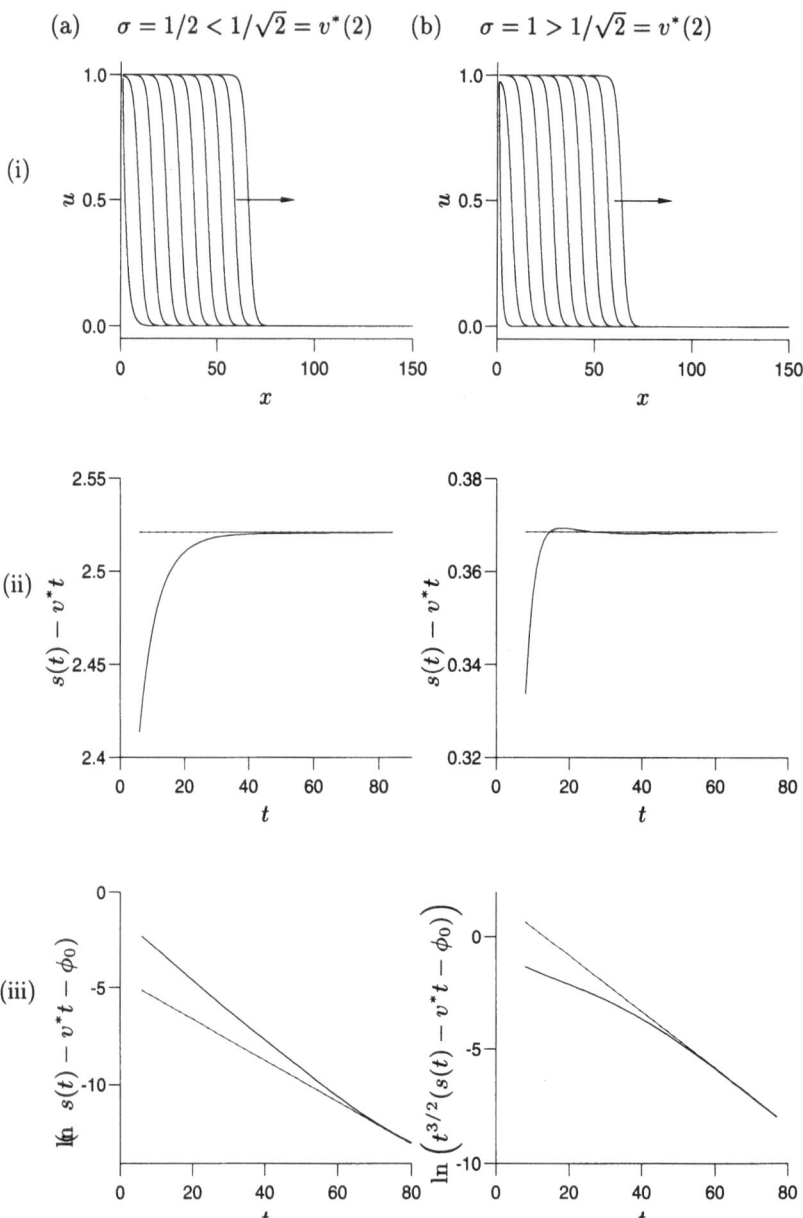

**Fig. 3.6.** Graphs illustrating the development of travelling waves in two cases when $m = 2$, for (a) $\sigma < v^*(2)$ and (b) $\sigma > v^*(2)$. The solid lines show numerically computed solutions, and the dashed lines in (iii) represent the predicted gradients.

### 3.3.2 Numerical Solutions when the Initial Data has Compact Support

We restrict attention to the case when $u_0 = 1$ and $\sigma = 1$. Equation (P1) was solved using the NAG routine D03PCF (method of lines with finite differences) with the initial data used being given by

$$g(x) = (1 - x^2), \quad 0 \le x \le 1.$$

We recall from the detailed theory of Section 3.2.4 that in this case

$$u(z + s(t), t) = u_T(z, 1/\sqrt{2}) + O\left(t^{-\frac{3}{2}} e^{-\frac{1}{8}t}\right) \tag{3.160}$$

as $t \to \infty$ with $z = O(1)$, where $z = x - s(t)$, and

$$s(t) = \frac{1}{\sqrt{2}} t + \phi_0 + O\left(t^{-\frac{3}{2}} e^{-\frac{1}{8}t}\right) \tag{3.161}$$

as $t \to \infty$ and the asymptotic wave speed of the PTW is given by $\dot{s}(t)$.

The results are shown in Figure 3.7 ((i)-(iii)). Figure 3.7(i) shows the developing PTW profile, whilst Figure 3.7(ii) shows the numerically obtained curve of $s(t)$ (the $x$ location where $u = 1/2$) against $t$. As predicted by the theory, $\dot{s}(t) \to v^*(2) \approx 0.7071$ as $t \to \infty$. In Figure 3.7(iii) we plot $\log\left(t^{3/2}|\dot{s}(t) - v^*(2)|\right)$ against $t$ (as generated numerically). We observe that the graph approaches a straight line as $t \to \infty$, with gradient $\left[-\frac{(v^*(2))^2}{4}\right] \approx -0.1250$, in excellent agreement with (3.154).

## 3.4 Summary

In this chapter we have considered, via the method of matched asymptotic expansions, the initial-boundary value problem [P,m] for $m > 1$, with initial data satisfying (g1) [(d1)]; that is, initial data with exponential decay as $x \to \infty$ [compact support] respectively. The results in each case are summarized as follows:

### 3.4.1 Initial Data with Exponential Decay as $x \to \infty$

We have established in Section 3.2.3 that a PTW develops as $t \to \infty$ in [P,m] (with $m > 1$) for any $u_\infty, \sigma > 0$. Further, we have established the wave speed of this PTW (this being the minimum speed available, $v = v^*(m)$, in all cases), its asymptotic correction as $t \to \infty$, together with the rate of convergence of the solution to [P,m] onto the PTW as $t \to \infty$. As we have demonstrated through the detailed theory of Section 3.2, the solution $u(x,t)$ to [P,m] satisfies

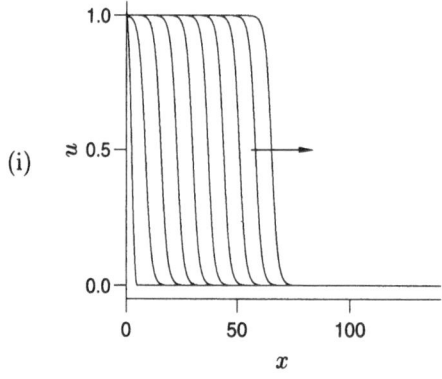

(i)

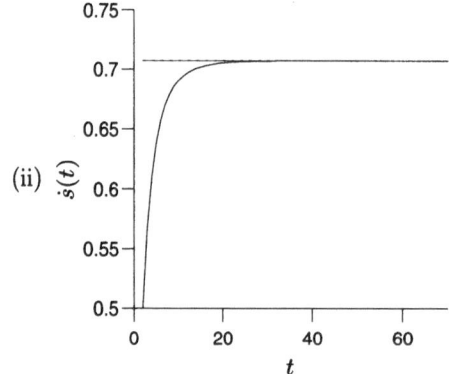

(ii)

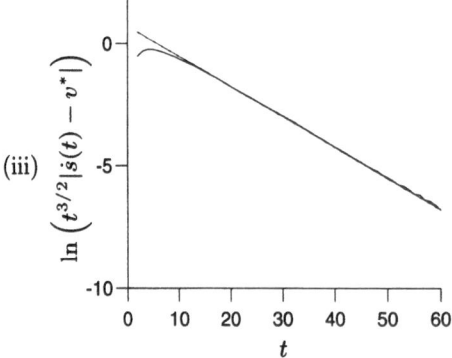

(iii)

**Fig. 3.7.** Graphs illustrating the development of travelling waves when $m = 2$, from compactly supported initial conditions.

$$u(z + s(t), t) = u_T(z, v^*(m)) + O(\delta(t)) \tag{3.162}$$

as $t \to \infty$, uniformly in $-\infty < z < \infty$, where $\dot{s}(t)$ and $\delta(t)$ are displayed in Table 3.1. We observe that, when

(i) $\sigma \geq v^*(m)$ The rate of convergence is exponential in $t$, as $t \to \infty$, being of

$$O\left\{ t^{-\frac{3}{2}} e^{-\frac{[v^*(m)]^2}{4} t} \right\} \text{ as } t \to \infty.$$

(ii) $\sigma < v^*(m)$ The rate of convergence is exponential in $t$, as $t \to \infty$, being of

$$O\left\{ e^{-\sigma[v^*(m)-\sigma]t} \right\} \text{ as } t \to \infty.$$

### 3.4.2 Initial Data with Compact Support

It was established in Section 3.2.4 that a PTW develops as $t \to \infty$ in [P,m] $(m > 1)$, with initial data given by (d1), for any $u_0, \sigma > 0$. The wave speed of this PTW is the minimum speed available, $v = v^*(m)$ and the asymptotic correction to this minimum speed is of $O\left( t^{-\frac{3}{2}} e^{-\frac{[v^*(m)]^2}{4} t} \right)$ as $t \to \infty$. The rate of convergence of the solution to [P,m] onto the PTW is of $O\left( t^{-\frac{3}{2}} e^{-\frac{[v^*(m)]^2}{4} t} \right)$ as $t \to \infty$. We have demonstrated the solution $u(x, t)$ to [P,m] satisfies,

$$u(z + s(t), t) = u_T(z, v^*(m)) + O\left( t^{-\frac{3}{2}} e^{-\frac{[v^*(m)]^2}{4} t} \right)$$

as $t \to \infty$, uniformly in $-\infty < z < \infty$, where $s(t)$ is given by

$$s(t) = v^*(m)t + \phi_0 + v_1 t^{-\frac{3}{2}} e^{-\frac{[v^*(m)]^2}{4} t} + o\left( t^{-\frac{3}{2}} e^{-\frac{[v^*(m)]^2}{4} t} \right)$$

as $t \to \infty$.

**Table 3.1.**

| | $v$ | $\dot{s}(t)$ | $\delta(t)$ |
|---|---|---|---|
| $\sigma \geq v^*(m)$ | $v^*(m)$ | $v^*(m) + O\left\{ t^{-\frac{3}{2}} e^{-\frac{[v^*(m)]^2}{4} t} \right\}$ | $O\left\{ t^{-\frac{3}{2}} e^{-\frac{[v^*(m)]^2}{4} t} \right\}$ |
| $\sigma < v^*(m)$ | $v^*(m)$ | $v^*(m) + O\left\{ e^{-\sigma[v^*(m)-\sigma]t} \right\}$ | $O\left\{ e^{-\sigma[v^*(m)-\sigma]t} \right\}$ |

## 3.5 Consideration of a More General Class of Initial Data with Exponential Decay Rate as $x \to \infty$

In this section we summarize the modifications that are required to the above theory and the results regarding the wave speed of the PTW, its asymptotic correction as $t \to \infty$, together with the rate of convergence of the solution [**P,m**] onto the PTW as $t \to \infty$, when the initial data has the more general form as $x \to \infty$, given by

$$u_0(x) \sim u_\infty x^{-n} e^{-\sigma x} \quad \text{as} \quad x \to \infty, \tag{3.163}$$

where $-\infty < n < \infty$.

It can be readily established (after minor modifications to the theory given in Section 3.2.3) that a PTW develops as $t \to \infty$ in [**P,m**] (with $m > 1$) for any $u_\infty, \sigma > 0$ and $-\infty < n < \infty$.

We first note in this case, following the analysis of Sections 3.2.1 and 3.2.2, that the expansion in region **III** is given by

$$u(Y,t) = \exp\{-t\sigma(Y - \sigma) - n \ln t - n \ln(Y - 2\sigma) + \ln u_\infty + o(1)\} \tag{3.164}$$

as $t \to \infty$ with $2\sigma + o(1) < Y < \infty$.

The main modifications to the theory of Section 3.2.3 are summerized as follows:

(i) $\sigma > v^*(m), n < 0$. The boundary value problem in region **TRA** (located at $Y = 2\sigma$) is now given by

$$h_{\eta\eta} + \frac{1}{2}\eta h_\eta + \frac{n}{2}h = 0, \quad -\infty < \eta < \infty, \tag{3.165}$$

$$h(\eta) \sim u_\infty \eta^{-n} \quad \text{as} \quad \eta \to \infty, \tag{3.166}$$

$$h(\eta) \sim Q\left\{(-\eta)^{n-1} e^{-\frac{\eta^2}{4}}\right\} \quad \text{as} \quad \eta \to -\infty. \tag{3.167}$$

The solution to (3.165)-(3.167) and the complete asymptotic structure of the solution to [**P,m**] as $t \to \infty$ in this case is given in Leach *et al* [36].

(ii) $\sigma > v^*(m), n > 0$. The modified boundary value problem (3.165)-(3.167) in region **TRA** has no solution in this case and we conclude that region **TRA** is located at $Y = 2\sigma + c(t)$ where $c(t) \sim \frac{n}{\sigma t}$ as $t \to \infty$.

(iii) $\sigma = v^*(m), n < 0$. In this case (3.167) is replaced by

$$h(\eta) \sim Q(1) \quad \eta \to -\infty, \tag{3.168}$$

and we note that boundary value problem (3.165), (3.166) and (3.168) has no solution. We conclude that region **TRA** is located at $Y = 2v^* + c(t)$, where $c(t) \sim -\frac{n}{v^* t}$ as $t \to \infty$.

(iv) $\sigma = v^*(m), n > 0$. As (iii) with $c(t) \sim \frac{n}{v^*(m)t}$ as $t \to \infty$.

(v) $0 < \sigma < v^*(m), -\infty < n < \infty$. Region **TR** is in this case located at $Y = v^*(m) + \sigma + \bar{c}(t)$ where $\bar{c}(t) \sim \frac{n}{(v^*(m)-\sigma)} \frac{\ln t}{t}$ as $t \to \infty$.

We note that the thickness of regions **TRA** and **TR** remain unchanged. The remaining structure in each case requires minor or no modification and we conclude that the wave speed of the PTW is the minimum speed available, $v = v^*(m)$, in all cases. Further, the solution $u(x,t)$ to **[P,m]** satisfies

$$u(z + s(t), t) = u_T(z, v^*(m)) + O(\delta(t)) \qquad (3.169)$$

as $t \to \infty$, uniformly in $-\infty < z < \infty$, where $\dot{s}(t)$ and $\delta(t)$ are displayed in Table 3.1.

# $m$th-Order ($m > 1$) Fisher Nonlinearity: Initial Data with Algebraic Decay Rates

In this chapter we extend the analysis of Chapter 3 by considering initial-boundary value problem [**P,m**] for $m > 1$, namely,

$$\left.\begin{array}{ll} u_t = u_{xx} + u^m(1 - u), & x, t > 0, \quad \text{(P1)} \\ u(x, 0) = u_0(x), & x \geq 0, \quad \text{(P2)} \\ u_x(0, t) = 0, & t > 0, \quad \text{(P3)} \\ u(x, t) \to 0 \ \text{as} \ x \to \infty, & t \geq 0, \quad \text{(P4)} \end{array}\right\} \ [\mathbf{P}, \mathbf{m}]$$

when the initial data, $u_0(x)$, is analytic, positive and monotone decreasing function in $x \geq 0$, with algebraic decay (up to exponential corrections) of degree $\alpha \left(\geq \frac{1}{m-1}\right)$ as $x \to \infty$ where

$$u_0(x) \sim \begin{cases} \frac{u_\infty}{x^\alpha} + EST(x) & \text{as } x \to \infty \quad \text{(g1)} \\ \tilde{u}_0 + \sum_{n=1}^{\infty} \tilde{u}_n x^n & \text{as } x \to 0^+ \quad \text{(g2)} \end{cases}$$

for some $\alpha \geq \frac{1}{(m-1)}$, where $u_\infty, \tilde{u}_0 > 0$ and $\tilde{u}_n$ are constants, and $EST(x)$ denotes exponentially small terms in $x$ as $x \to \infty$.

## 4.1 Permanent Form Travelling Waves

The main results concerning the existence and structure of PTWs which may occur in the solution to [**P, m**] ($m > 1$) as $t \to \infty$ are reviewed in Section 3.1 and are not repeated here.

However, we note that the asymptotics of $u_T(z; m, v)$ as $z \to +\infty$, when $v > v^*(m)$ can be continued, and will be of interest in what follows. In fact we have, for $v > v^*(m)$,

(i) $\underline{1 < m < 2}$

$$u_T(z; m, v) \sim \left[\frac{v}{(m-1)}\right]^{\frac{1}{(m-1)}} \frac{1}{z^{\frac{1}{(m-1)}}}$$

$$-\frac{\left[\frac{m}{v(m-1)^2}\right]\left[\frac{v}{m-1}\right]^{\frac{1}{(m-1)}} \ln z}{z^{\frac{1}{(m-1)}+1}} + o\left(z^{-\left(1+\frac{1}{(m-1)}\right)}\right), \tag{4.1}$$

(ii) $\underline{m = 2}$

$$u_T(z; m, v) \sim \frac{v}{z} - \frac{(2 - v^2)\ln z}{z^2} + o\left(z^{-2}\right), \tag{4.2}$$

(iii) $\underline{m > 2}$

$$u_T(z; m, v) \sim \left[\frac{v}{(m-1)}\right]^{\frac{1}{(m-1)}} \frac{1}{z^{\frac{1}{(m-1)}}}$$

$$+ \left[\frac{v}{(m-1)}\right]^{\frac{2}{(m-1)}} \frac{1}{(m-2)} \frac{1}{z^{\frac{2}{(m-1)}}} + o\left(z^{-\frac{2}{(m-1)}}\right) \tag{4.3}$$

as $z \to \infty$. Note the change in structure between the cases $1 < m < 2$, $m = 2$ and $m > 2$ ; in particular, the change in order of the correction terms as $m$ passes through $m = 2$, and the presence of the logarithmic term for $1 < m \le 2$. These features will recur in what follows.

## 4.2 Asymptotic Solution as $t \to 0$ for $0 \le x < \infty$

We first consider region **I**, where $x = O(1)$ as $t \to 0$ and expand the solution to [**P, m**] as

$$u(x, t) = u_0(x) + tu_1(x) + O\left(t^2\right) \tag{4.4}$$

as $t \to 0$. On substitution into equation (P1) and applying initial condition (P2) we readily obtain

$$u(x, t) = u_0(x) + t\left[u_0''(x) + u_0^m(x)\left(1 - u_0(x)\right)\right] + O\left(t^2\right) \tag{4.5}$$

as $t \to 0$. Now, for $x \gg 1$, expansion (4.5), with (g1), takes the form

$$u(x, t) \sim \frac{u_\infty}{x^\alpha} + t\left[u_\infty \frac{\alpha(\alpha + 1)}{x^{\alpha+2}} + \frac{u_\infty^m}{x^{\alpha m}} + \dots\right] + \dots \tag{4.6}$$

as $t \to 0$, where $\alpha \geq \frac{1}{(m-1)}$, and we conclude that (4.5) remains uniform for $x \gg 1$ as $t \to 0$. However, for $x \ll 1$, we note that expansion (4.5) with (g2) will not in general satisfy the boundary condition (P3) at $x = 0$. Therefore we require an inner region, region $\mathbf{I}_0$, when $x = O\left(t^{\frac{1}{2}}\right)$ as $t \to 0$, over which this condition is satisfied. The details of this region are given in Section 3.2.1 and are repeated here.

This completes the asymptotic structure as $t \to 0$, with expansions (4.5) (in region $\mathbf{I}$) and (3.7) (in region $\mathbf{I}_0$) providing a uniform approximation in $x \geq 0$ to the solution of $[\mathbf{P}, \mathbf{m}]$ as $t \to 0$.

## 4.3 Asymptotic Solution as $x \to \infty$ for $t = O(1)$

We now investigate the structure of the solution to $[\mathbf{P}, \mathbf{m}]$ as $x \to \infty$ for $t = O(1)$. We shall need to consider the cases $\alpha = \frac{1}{(m-1)}$ and $\alpha > \frac{1}{(m-1)}$ separately.

### 4.3.1 $\alpha = \frac{1}{m-1}$

The form of (4.6) indicates that in this case $u = O\left(x^{-\frac{1}{(m-1)}}\right)$ as $x \to \infty$ with $t = O(1)$, which suggests that in this region, region $\mathbf{II}$, we expand as

$$u(x,t) = \frac{f_0(t)}{x^{\frac{1}{(m-1)}}} + \frac{f_1(t)}{x^r} + \frac{f_2(t)}{x^s} + o\left(\frac{1}{x^s}\right) \tag{4.7}$$

as $x \to \infty$ with $t = O(1)$. A balancing of terms then requires that

$$r = 1 + \frac{1}{(m-1)},$$

$$s = \begin{cases} 2 + \frac{1}{(m-1)}, & 1 < m < 2, \\ 3, & m = 2, \\ 1 + \frac{2}{(m-1)}, & m > 2. \end{cases} \tag{4.8}$$

On substituting (4.7) into equation (P1) and solving at each order in turn, we find (after matching with (4.6) as $t \to 0$) that

$$f_0(t) = u_\infty, \quad f_1(t) = u_\infty^m t,$$

$$f_2(t) = \begin{cases} \frac{1}{2} m u_\infty^{(2m-1)} t^2 + \frac{m}{(m-1)^2} u_\infty t, & 1 < m < 2, \\ u_\infty[2 - u_\infty^2] t + u_\infty^3 t^2, & m = 2, \\ -u_\infty^{(m+1)} t, & m > 2. \end{cases} \tag{4.9}$$

Thus, the structure of (4.7) depends upon whether $1 < m < 2, m = 2$ or $m > 2$. However, in each case, we observe that expansion (4.7) remains uniform for $t \gg 1$ *provided* that $x \gg t$. However, expansion (4.7) fails to provide an asymptotic approximation for $t \gg 1$ when $x \leq O(t)$. Therefore, to complete the asymptotic structure of the solution to [**P, m**], when $\alpha = \frac{1}{(m-1)}$ for $t \gg 1$, we must introduce a further region when $x = O(t)$ as $t \to \infty$, with, from (4.7), $u = O\left(t^{-\frac{1}{(m-1)}}\right)$ as $t \to \infty$. We will return to this in the next section.

## 4.3.2 $\alpha > \frac{1}{m-1}$

The form of (4.6) indicates that in this case $u = O\left(x^{-\alpha}\right)$ as $x \to \infty$ with $t = O(1)$, which now suggests that in region **II** we expand as

$$u(x,t) = \frac{g_0(t)}{x^\alpha} + \frac{g_1(t)}{x^q} + o\left(\frac{1}{x^q}\right) \tag{4.10}$$

as $x \to \infty$, with $t = O(1)$. Here

$$q = \begin{cases} m\alpha, & \text{for } \frac{1}{(m-1)} < \alpha \leq \frac{2}{(m-1)}, \\ \\ \alpha + 2, & \text{when } \alpha > \frac{2}{(m-1)}. \end{cases} \tag{4.11}$$

On substituting from (4.10) into equation (P1) and solving, we find (after matching with (4.5) as $t \to 0$) that

$$g_0(t) = u_\infty,$$

$$g_1(t) = \begin{cases} u_\infty^m t, & \frac{1}{(m-1)} < \alpha < \frac{2}{(m-1)}, \\ \\ \left[u_\infty^m + \frac{2u_\infty(m+1)}{(m-1)^2}\right] t, & \alpha = \frac{2}{(m-1)}, \\ \\ u_\infty \alpha(\alpha + 1)t, & \alpha > \frac{2}{(m-1)}. \end{cases} \tag{4.12}$$

Consideration of higher order terms in (4.10) reveals that expansion (4.10) again remains uniform for $t \gg 1$, *provided* $x \gg t$, but fails to provide an asymptotic approximation when $x \leq O(t)$, with

$$u = O\left(t^{-\alpha}\right). \tag{4.13}$$

Thus, to complete the asymptotic structure when $t \gg 1$ in this case, we will require a further region when $x = O(t)$ and $u = O\left(t^{-\alpha}\right)$ as $t \to \infty$.

# 4.4 Asymptotic Solution as $t \to \infty$ when $\alpha = \frac{1}{m-1}$

We now investigate the structure of the solution to [**P, m**] as $t \to \infty$ when $\alpha = \frac{1}{(m-1)}$. In this case we have seen in Section 4.3.1 that expansion (4.7) continues to remain uniform for $t \gg 1$ when $x \gg t$. Thus we begin the structure as $t \to \infty$ with a region where $x = O(t)$ as $t \to \infty$, and then, via (4.7), $u = O\left(t^{-\frac{1}{(m-1)}}\right)$ as $t \to \infty$. We will refer to this region as region **III**. To examine region **III** we introduce the scaled coordinate $y = \frac{x}{t}$, where $y = O(1)$ as $t \to \infty$. Consideration of expansion (4.7) indicates that there are three distinct cases to consider, these being $1 < m < 2$, $m = 2$ and $m > 2$. We consider these in turn.

### 4.4.1 $1 < m < 2$

In this case, expansion (4.7) of region **II** suggests that we expand in the form

$$u(y,t) = \bar{u}_0(y)\, t^{-\frac{1}{(m-1)}} + \bar{u}_1(y)\, t^{-\frac{1}{(m-1)}-1} + o\left(t^{-\frac{1}{(m-1)}-1}\right) \qquad (4.14)$$

as $t \to \infty$ with $y = \frac{x}{t} = O(1)$. After substituting (4.14) into equation (P1) (when written in terms of $y$ and $t$), the leading order problem for $\bar{u}_0(y)$ becomes

$$y\, \bar{u}_{0y} + \bar{u}_0^m + \frac{1}{(m-1)}\, \bar{u}_0 = 0, \quad y > 0, \qquad (4.15)$$

$$\bar{u}_0(y) \sim u_\infty\, y^{-\frac{1}{(m-1)}}, \quad \text{as} \quad y \to \infty \qquad (4.16)$$

with condition (4.16) arising from matching expansion (4.14) ($y \gg 1$) with the far field expansion (4.7) ($x = O(t)$). The solution to (4.15), (4.16) is readily obtained as

$$\bar{u}_0(y) = \frac{u_\infty}{(y - y_b)^{\frac{1}{(m-1)}}}, \quad y > y_b, \qquad (4.17)$$

where $y_b = u_\infty^{(m-1)}(m-1)$. Hence, via (4.17), $\bar{u}_0(y)$ develops a singularity as $y \to y_b^+$, and thus expansion (4.14) becomes nonuniform when $y = y_b + o(1)$ as $t \to \infty$. On proceeding to next order, we obtain the following problem for $\bar{u}_1(y)$, namely

$$y\, \bar{u}_{1y} + \bar{u}_1\left(m\, \bar{u}_0^{(m-1)} + \frac{m}{(m-1)}\right) = -\bar{u}_{0yy}, \quad y > y_b, \qquad (4.18)$$

$$\bar{u}_1(y) \sim \frac{m}{(m-1)^2}\, u_\infty\, y^{-\frac{1}{(m-1)}-2} \quad \text{as} \quad y \to \infty \qquad (4.19)$$

with condition (4.19) arising from matching expansion (4.14) ($y \gg 1$) with the far field expansion (4.7) ($x = O(t)$). The solution to (4.18), (4.19) is given by

$$\bar{u}_1(y) = -\frac{m}{(m-1)^2}\frac{u_\infty}{y_b}\frac{\ln(1-\frac{y_b}{y})}{(y-y_b)^{\frac{m}{(m-1)}}}, \quad y > y_b. \tag{4.20}$$

Finally, we have, via (4.14), (4.17) and (4.20), that the expansion in region **III** has the form

$$u(y,t) = \frac{u_\infty\, t^{-\frac{1}{(m-1)}}}{(y-y_b)^{\frac{1}{(m-1)}}} - \frac{m}{(m-1)^2}\frac{u_\infty}{y_b}\frac{\ln(1-\frac{y_b}{y})}{(y-y_b)^{\frac{m}{(m-1)}}}t^{-\frac{1}{(m-1)}-1}$$
$$+ o\left(t^{-\frac{1}{(m-1)}-1}\right) \tag{4.21}$$

as $t \to \infty$ with $y_b < y < \infty$. An examination of expansion (4.21) reveals that it becomes nonuniform when, on balancing when the first two terms become comparable,

$$y \sim y_b + c(t), \tag{4.22}$$

where

$$c(t) \sim \frac{m}{(m-1)^2 y_b}\frac{\ln t}{t} - \frac{m}{(m-1)^2 y_b}\frac{\ln(\ln t)}{t} + \ldots + o\left(\frac{1}{t}\right) \tag{4.23}$$

as $t \to \infty$, and with $u = O\left((\ln t)^{-\frac{1}{(m-1)}}\right)$. Thus we continue the asymptotic structure with a transition region in which $\xi = O(1)$ as $t \to \infty$, where

$$y = y_b + c(t) + t^{-\gamma}\xi. \tag{4.24}$$

We denote this region by region **TR**, and expand as

$$u(\xi,t) = \frac{F(\xi)}{(\ln t)^{\frac{1}{(m-1)}}} + o\left(\frac{1}{(\ln t)^{\frac{1}{(m-1)}}}\right) \tag{4.25}$$

as $t \to \infty$ with $\xi = O(1)$. On substituting form (4.24) and (4.25) into equation (P1), we find that a non-trivial leading order balance requires

$$\gamma = 1, \tag{4.26}$$

after which the leading order problem becomes

$$F_{\xi\xi} + y_b F_\xi = 0, \quad -\infty < \xi < \infty. \tag{4.27}$$

Equation (4.27) is to be solved subject to matching with region **III** as $\xi \to \infty$, which gives, via (4.21) and (4.24),

$$F(\xi) \to u_\infty\left[\frac{(m-1)^2}{m}y_b\right]^{\frac{1}{(m-1)}} \quad \text{as} \quad \xi \to \infty. \tag{4.28}$$

The solution to (4.27) and (4.28) is readily obtained as

$$F(\xi) = u_\infty \left[ \frac{(m-1)^2}{m} y_b \right]^{\frac{1}{(m-1)}} + B e^{-y_b \xi} \quad -\infty < \xi < \infty, \qquad (4.29)$$

with $B(\geq 0)$ being an arbitrary constant. Thus, in region **TR** we have

$$u(\xi, t) = \frac{(F_\infty + B e^{-y_b \xi})}{(\ln t)^{\frac{1}{(m-1)}}} + o \left( \frac{1}{(\ln t)^{\frac{1}{(m-1)}}} \right) \qquad (4.30)$$

as $t \to \infty$ with $\xi = O(1)$, and the constant $F_\infty$ given by

$$F_\infty = u_\infty \left[ \frac{(m-1)^2}{m} y_b \right]^{\frac{1}{(m-1)}}. \qquad (4.31)$$

An examination of the scaling (4.24), together with (4.23), shows that we re-emerge from this transition region **TR** when $(-\xi) \gg 1$ and, in particular, when

$$1 \ll (-\xi) \leq O(\ln t). \qquad (4.32)$$

Thus, we suppose that we leave the transition region **TR** when $\xi = -\theta(t) + O(1)$, where $1 \ll \theta(t) \leq O(\ln t)$, and $\theta(t)$ is to be determined. We refer to this next region as region **TW**. We introduce the new coordinate $z$ by

$$\xi = -\theta(t) + z \qquad (4.33)$$

so that $z = O(1)$ as $t \to \infty$ in region **TW**. In addition, via (4.30), we suppose that $u = O(1)$ as $t \to \infty$ in region **TW**. Therefore, we expand as

$$u(z, t) = u_c(z) + o(1) \quad \text{as} \quad t \to \infty \qquad (4.34)$$

with $z = O(1)$. After substituting from (4.33), (4.34) into the full equation (P1), the leading order equation is

$$u_c'' + y_b u_c' + u_c^m (1 - u_c) = 0, \quad -\infty < z < \infty. \qquad (4.35)$$

Moreover, we require, via (1.22), that

$$u_c(z) > 0 \quad \text{for all} \quad -\infty < z < \infty, \qquad (4.36)$$

and

$$u_c(z) \quad \text{remains bounded as} \quad z \to -\infty. \qquad (4.37)$$

The remaining condition is obtained by matching expansion (4.34) (for $z \gg 1$) to expansion (4.30) (for $\xi \sim -\theta(t) + O(1)$). Matching these expansions to $O(1)$ requires

$$u_c(z) \to 0 \quad \text{as} \quad z \to +\infty. \qquad (4.38)$$

The leading order problem is now complete and is given by the nonlinear boundary value problem (4.35)-(4.38). However, a phase plane analysis of equation (4.35) (using the Poincaré-Bendixson theorem) allows boundary condition (4.37) to be replaced by

$$u_c(z) \to 1 \quad \text{as} \quad z \to -\infty \tag{4.39}$$

(all other possibilities leading to unbounded behaviour as $z \to -\infty$). Thus, the complete boundary value problem for $u_c(z)$ may be stated as

$$\left.\begin{array}{ll} u_c'' + y_b u_c' + u_c^m(1 - u_c) = 0, & -\infty < z < \infty, \\ u_c(z) \to 0 & \text{as} \quad z \to +\infty, \\ u_c(z) \to 1 & \text{as} \quad z \to -\infty, \\ u_c(z) \geq 0 & \text{for all} \quad -\infty < z < \infty. \end{array}\right\} \text{[TW]}$$

We observe that [TW] is precisely the boundary value problem (3.2) of Chapter 3 for PTWs with $v = y_b$, and its solutions are classified in Theorem 3.1. There are a number of cases to consider, namely $0 < y_b < v^*(m)$, $y_b = v^*(m)$ and $y_b > v^*(m)$.

## (a) $y_b > v^*(m)$

In this case, via Theorem 3.1, the solution to [TW] is given by

$$u_c(z) = u_T(z; m, y_b), \quad -\infty < z < \infty, \tag{4.40}$$

and

$$u_c(z) \sim \left[\frac{y_b}{(m-1)z}\right]^{\frac{1}{(m-1)}} \quad \text{as} \quad z \to \infty. \tag{4.41}$$

We can now match to terms of $O\left((\ln t)^{-\frac{1}{(m-1)}}\right)$ with expansion (4.30). From expansion (4.34), with (4.40) and (4.1) we have as we move into region TR

$$u \sim \left[\frac{y_b}{(m-1)\theta(t)}\right]^{\frac{1}{m-1}} - \frac{\left[\frac{m}{y_b(m-1)^2}\right]\left[\frac{y_b}{(m-1)}\right]^{\frac{1}{(m-1)}} \ln[\theta(t)]}{[\theta(t)]^{\frac{1}{m-1}+1}} + \cdots, \tag{4.42}$$

whilst from expansion (4.30), as we move into region TW, we have

$$u \sim \frac{F_\infty}{[\ln t]^{\frac{1}{(m-1)}}} + \frac{B\, e^{y_b \theta(t)}}{[\ln t]^{\frac{1}{(m-1)}}} e^{-y_b z}. \tag{4.43}$$

The generalized Van Dyke Matching Principle (Van Dyke [70]) requires agreement between (4.42) and (4.43). Recalling that

$$1 \ll \theta(t) \leq O(\ln t),$$

we require $B = 0$ in (4.43) and then

$$\theta(t) = \frac{m}{(m-1)^2 y_b} \ln t - \frac{m}{(m-1)^2 y_b} \ln(\ln t) + o\left(\ln(\ln t)\right) \qquad (4.44)$$

as $t \to \infty$, and the matching is complete. Thus, in this case the large-$t$ structure in region **TW** is dominated by the evolution of the PTW with asymptotic speed given by $v = y_b = u_\infty^{(m-1)}(m-1)$. In terms of the original coordinates (via (4.44), (4.33), (4.24) and (4.23)) the location of the region **TW** is

$$x = y_b t + t\left[c(t) - t^{-1}\theta(t)\right] + z \qquad (4.45)$$

as $t \to \infty$ with $z = O(1)$. However, via (4.23) and (4.44) (after continuation through logarithmic terms of all orders) we observe that

$$\left[c(t) - t^{-1}\theta(t)\right] \equiv \psi(t) = o\left(\frac{1}{t}\right) \qquad (4.46)$$

as $t \to \infty$, through cancellations. Thus, the PTW has location $x = s(t)$, where

$$s(t) = y_b t + t\,\psi(t) + \ldots \qquad \text{as} \qquad t \to \infty, \qquad (4.47)$$

with, via (4.46), $t\,\psi(t) = o(1)$ as $t \to \infty$. It then follows from (4.47) and (4.46) that the asymptotic speed of the PTW is

$$\dot{s}(t) \sim y_b + o\left(\frac{1}{t}\right) \qquad \text{as} \qquad t \to \infty, \qquad (4.48)$$

and that the correction to expansion (4.34) in region **TW** is of $O\left(\dot{s}(t) - y_b\right) = o\left(\frac{1}{t}\right)$ as $t \to \infty$.

Now, for $(-z) \gg 1$ we move out of the localized region **TW**, in particular, when $(-z) = O(t)$ and $y = O(1)$. The remainder of the asymptotic structure of $[\mathbf{P},\mathbf{m}]$ as $t \to \infty$ in this case follows, after minor modification, regions **IV** and **V** (with $s(t)$ given by (4.47) and $v^*(m)$ and $Y$ replaced by $y_b$ and $y$ respectively) given in Section 3.2.3(a).

This completes the asymptotic structure in this case. In particular, with $u(x, t)$ the solution of $[\mathbf{P},\mathbf{m}]$, we have established (via regions **III** - **V**) that

$$u(z + s(t), t) \to u_T(z; m, y_b) \qquad \text{as} \qquad t \to \infty, \qquad (4.49)$$

uniformly in $z$ with $s(t)$ as given by (4.47). A schematic representation of the location and thickness of the asymptotic regions as $t \to \infty$ is given in Figure 4.1.

**(b)** $y_b = v^*(m)$

In this case, via Theorem 3.1, the solution to $[\mathbf{TW}]$ is given by

$$u_c(z) = u_T\left(z; m, v^*(m)\right), \qquad -\infty < z < \infty, \qquad (4.50)$$

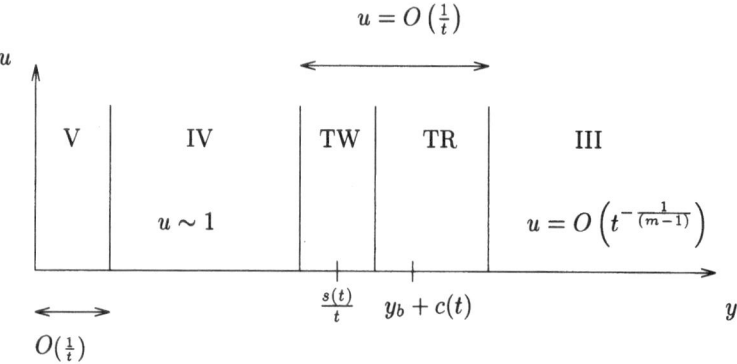

**Fig. 4.1.** Schematic representation of the location and thickness of the asymptotic regions as $t \to \infty$ when $\alpha = \frac{1}{(m-1)}$, $1 < m < 2$ and $y_b > v^*(m)$. Note that $u = O\left((\ln t)^{-\frac{1}{(m-1)}}\right)$ in region **TR**.

and

$$u_c(z) \sim B^* e^{-v^*(m)z} \quad \text{as} \quad z \to \infty, \tag{4.51}$$

with $B^* > 0$ fixed. We can now match to terms of $O\left([\ln t]^{-\frac{1}{(m-1)}}\right)$ with expansion (4.30). From expansion (4.34), with (4.50) and (4.51), we have, as we move into region **TR**

$$u \sim B^* e^{-v^*(m)z}, \tag{4.52}$$

whilst from expansion (4.30), as we move into region **TW**, we have

$$u \sim \frac{B}{(\log t)^{\frac{1}{(m-1)}}} e^{v^*(m)\,\theta(t)} \, e^{-v^*(m)z}. \tag{4.53}$$

The generalized Van Dyke Matching Principle requires agreement between (4.52) and (4.53), which gives

$$B = B^*, \tag{4.54}$$

after which the gauge function $\theta(t)$ is given by

$$\theta(t) = \frac{1}{(m-1)\,v^*(m)} \, \ln(\ln t), \tag{4.55}$$

and matching is complete. Therefore, in this case the large-$t$ structure in region **TW** is again dominated by the evolution of the PTW with asymptotic wave speed $v = y_b = v^*(m)$. In terms of the original coordinates, the PTW has location $x = s(t)$, where, now,

$$s(t) \sim v^*(m)t + \frac{m}{(m-1)^2 v^*(m)} \, \ln t - \frac{(2m-1)}{(m-1)^2 v^*(m)} \, \ln(\ln t) + O(1) \tag{4.56}$$

as $t \to \infty$; hence the asymptotic speed of the PTW is

$$\dot{s}(t) \sim v^*(m) + \frac{m}{(m-1)^2 v^*(m)} \frac{1}{t} - \frac{(2m-1)}{(m-1)^2 v^*(m)} \frac{1}{t \ln t} + o\left(\frac{1}{t \ln t}\right) \quad (4.57)$$

as $t \to \infty$. Notice the change in the correction term between (4.48) and (4.57). This has been driven by the change in structure of the PTW for $z \gg 1$; the PTW has algebraic decay when $y_b > v^*(m)$ but exponential decay when $y_b = v^*(m)$. In addition, in this case, the correction to expansion (4.34) in region **TW** is of $O\left(\dot{s}(t) - v^*(m)\right) = O\left(\frac{1}{t}\right)$ as $t \to \infty$.

The remainder of the structure in this case follows that in (a) (with $y_b = v^*(m)$ and $s(t)$ given by (4.56)) and is not repeated.

## (c) $0 < y_b < v^*(m)$

In this case the leading order problem [**TW**] has no solution, via Theorem 3.1. We must return to region **III** and conclude that expansion (4.21) must fail before $y = y_b^+$. Therefore, in this case we assume that expansion (4.21) fails as $y \to y_T^+$ where now

$$y_T \geq v^*(m) \ (> y_b). \quad (4.58)$$

When $y = y_T + o(1)$ we introduce a transition region **TR'**, and write

$$y = y_T + \hat{c}(t) + t^{-\gamma}\xi \quad (4.59)$$

with $\xi = O(1)$ as $t \to \infty$ and $\gamma > 0$ with $\hat{c}(t) = o(1)$ as $t \to \infty$. From expansion (4.21), via (4.59), (4.58), we have that

$$u = O\left(t^{-\frac{1}{(m-1)}}\right) \quad (4.60)$$

as $t \to \infty$, in region **TR'**. Thus in region **TR'** we write

$$u(\xi, t) = H(\xi) t^{-\frac{1}{(m-1)}} + o\left(t^{-\frac{1}{(m-1)}}\right) \quad (4.61)$$

as $t \to \infty$ with $\xi = O(1)$. On substituting from (4.59), (4.61) into equation (P1), we find that a non-trivial leading order balance requires

$$\gamma = 1, \quad (4.62)$$

after which the leading order problem is

$$H_{\xi\xi} + y_T H_\xi = 0, \quad -\infty < \xi < \infty, \quad (4.63)$$

$$H(\xi) \to \frac{u_\infty}{(y_T - y_b)^{\frac{1}{(m-1)}}} \quad \text{as} \quad \xi \to \infty, \quad (4.64)$$

after matching expansion (4.61) (as $\xi \to \infty$) with expansion (4.21) (as $y \to y_T^+$). The solution to this problem is

$$H(\xi) = \frac{u_\infty}{(y_T - y_b)^{\frac{1}{(m-1)}}} + B\,e^{-y_T\xi} \tag{4.65}$$

with $B$ as yet undetermined. Therefore, in region **TR′** we have

$$u(\xi, t) = \left[\frac{u_\infty}{(y_T - y_b)^{\frac{1}{(m-1)}}} + B\,e^{-y_T\xi}\right] t^{-\frac{1}{(m-1)}} + o\left(t^{-\frac{1}{(m-1)}}\right) \tag{4.66}$$

as $t \to \infty$, with $\xi = O(1)$. An examination of the scaling (4.59) shows that we re-emerge from the transition region **TR′** when $(-\xi) \gg 1$, and in particular, when

$$1 \ll (-\xi) \le O(t\,\hat{c}(t)). \tag{4.67}$$

We therefore suppose that we leave the transition region **TR′** when $\xi = -\phi(t) + O(1)$, where

$$1 \ll \phi(t) \le O(t\,\hat{c}(t)) = o(t), \tag{4.68}$$

and $\phi(t)$ is to be determined. We refer to this next region as region **TW′**. We introduce the new coordinate $z$ by

$$\xi = -\phi(t) + z, \tag{4.69}$$

so that $z = O(1)$ as $t \to \infty$ in region **TW′**. In addition, via (4.66), we suppose that $u = O(1)$ as $t \to \infty$ in region **TW′**, and therefore expand as

$$u(z, t) = u_c(z) + o(1) \quad \text{as} \quad t \to \infty \tag{4.70}$$

with $z = O(1)$. The leading order problem in region **TW′** is now given by problem **[TW]**, except that now $z$ is defined by (4.69) and $y_b$ is replaced by $y_T$. With $y_T \ge v^*(m)$, **[TW]** has the solution

$$u_c(z) = u_T(z; m, y_T) \tag{4.71}$$

with, from Theorem 3.1,

$$u_c(z) \sim \begin{cases} B^*e^{-v^*(m)z}, & y_T = v^*(m), \\[2mm] \left[\dfrac{y_T}{(m-1)z}\right]^{\frac{1}{(m-1)}}, & y_T > v^*(m). \end{cases} \tag{4.72}$$

It remains to match expansion (4.70) ($z \gg 1$) to expansion (4.66) ($\xi \sim -\phi(t)$). First suppose that $y_T > v^*(m)$. Then, from (4.72) we have

$$u \sim \left[\frac{y_T}{(m-1)\,\phi(t)}\right]^{\frac{1}{(m-1)}} \tag{4.73}$$

as we move into region **TR′**. Conversely, from (4.66) we have

$$u \sim \frac{u_\infty}{(y_T - y_b)^{\frac{1}{(m-1)}}} t^{-\frac{1}{(m-1)}} + B\, e^{y_T \phi(t)} e^{-y_T z} t^{-\frac{1}{(m-1)}} \tag{4.74}$$

as we move into region **TW'**. Matching requires that (4.73) and (4.74) agree. This is only possible with $B = 0$ and $\phi(t) = O(t)$ as $t \to \infty$. However, this violates the order requirement (4.68). We conclude that matching requires

$$y_T = v^*(m), \tag{4.75}$$

after which (4.73) is replaced, via (4.72), by

$$u \sim B^*\, e^{-y_T z}. \tag{4.76}$$

This will now match with (4.74) (up to exponential terms) provided

$$B = B^*, \tag{4.77}$$

and

$$\phi(t) = \frac{1}{(m-1)\, v^*(m)} \ln t, \tag{4.78}$$

after using (4.75). It now follows from (4.68) that

$$O\left[\frac{\ln t}{t}\right] \le \hat{c}(t) \ll 1. \tag{4.79}$$

However, the results of the previous cases (a) and (b) together with the comparison theorem require, from (4.79), that $\hat{c}(t) = O\left[\frac{\ln t}{t}\right]$, and so we write (without loss of generality)

$$\hat{c}(t) = \gamma_0 \frac{\ln t}{t} \tag{4.80}$$

for some constant $\gamma_0 \neq 0$. In fact, case (b) together with the comparison theorem requires

$$\gamma_0 < \frac{(2m-1)}{(m-1)^2 v^*(m)}, \tag{4.81}$$

whilst the results of Chapters 2 and 3 concerning initial data with compact support, together with the comparison theorem, require

$$\gamma_0 \ge \frac{1}{(m-1)\, v^*(m)}. \tag{4.82}$$

Unfortunately, we are unable to determine $\gamma_0$ exactly, without knowledge of terms in expansion (4.21) (in region **III**) beyond all algebraic orders (that is, exponentially small terms in $t$ as $t \to \infty$). Thus, in this case, the large-$t$ structure in region **TW'** is dominated by the emergence of the PTW with asymptotic speed $v = v^*(m)$. In terms of the original coordinates, the PTW has location, $x = s(t)$, where now

$$s(t) \sim v^*(m)\, t + \left[\gamma_0 - \frac{1}{(m-1)\, v^*(m)}\right] \ln t + O(1) \qquad (4.83)$$

as $t \to \infty$, and so the asymptotic speed of the PTW is

$$\dot{s}(t) \sim v^*(m) + \left[\gamma_0 - \frac{1}{(m-1)\, v^*(m)}\right]\frac{1}{t} + o\left(\frac{1}{t}\right) \qquad (4.84)$$

as $t \to \infty$. Note that the correction term in (4.84) has changed from the cases (a) and (b). This is due to the relocation of the transition region **TR′**, which was necessary in the present case. Again the correction to expansion (4.34) in region **TW′** is of $O\left(\dot{s}(t) - v^*(m)\right) \le O\left(\frac{1}{t}\right)$ as $t \to \infty$.

The remainder of the structure of **[P,m]** as $t \to \infty$ in this case again follows that in Section 3.2.3(a) (with $s(t)$ given by (4.83) and $y_b = v^*(m)$) and is not repeated here. A schematic representation of the location and thickness of the asymptotic regions as $t \to \infty$ is given in Figure 4.2.

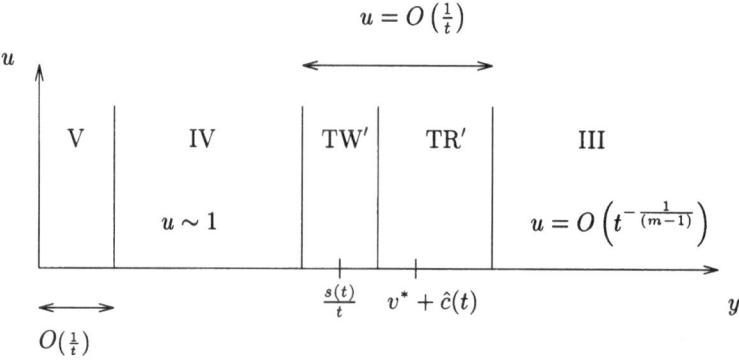

**Fig. 4.2.** Schematic representation of the location and thickness of the asymptotic regions as $t \to \infty$ when $\alpha = \frac{1}{(m-1)}$, $1 < m < 2$ and $0 < y_b < v^*$. Note that $u = O\left(t^{-\frac{1}{(m-1)}}\right)$ in region **TR′**.

### 4.4.2 $m = 2$

In this case, expansion (4.7) of region **II** suggests that we expand in region **III** in the form

$$u(y,t) = \bar{u}_0(y)\, t^{-1} + \bar{u}_1(y)\, t^{-2} + o\left(t^{-2}\right) \qquad (4.85)$$

as $t \to \infty$ with $y = \frac{x}{t} = O(1)$. After substituting (4.85) into equation (P1) (when written in terms of $y$ and $t$) the leading order problem for $\bar{u}_0(y)$ is given by (4.15), (4.16) (of Section 4.4.1) on setting $m = 2$, with solution

$$\bar{u}_0(y) = \frac{u_\infty}{[y - y_b]}, \quad y > y_b, \tag{4.86}$$

where $y_b = u_\infty$. Again $\bar{u}_0(y)$ develops a singularity as $y \to y_b^+$, and hence expansion (4.85) becomes non-uniform when $y = y_b + o(1)$ as $t \to \infty$. On proceeding to next order, we obtain the following problem for $\bar{u}_1(y)$, namely

$$y\bar{u}_{1y} + 2\bar{u}_1 (1 + \bar{u}_0) = -\bar{u}_{0yy} + \bar{u}_0^3, \quad y > y_b, \tag{4.87}$$

$$\bar{u}_1(y) \sim u_\infty[2 - u_\infty^2] y^{-3} \quad \text{as} \quad y \to \infty, \tag{4.88}$$

with condition (4.88) arising from matching expansion (4.85) $(y \gg 1)$ with the far field expansion (4.7) $(x = O(t))$. The solution to (4.87), (4.88) is given by

$$\bar{u}_1(y) = [y_b^2 - 2] \frac{\ln\left(1 - \frac{y_b}{y}\right)}{(y - y_b)^2}, \quad y > y_b, \tag{4.89}$$

recalling that $y_b = u_\infty$ in this case. Finally, we have, via (4.85), (4.86) and (4.89), that the expansion in this region, region **III**, has the form

$$u(y,t) = \frac{u_\infty}{(y - y_b)} t^{-1} + (y_b^2 - 2) \frac{\ln\left(1 - \frac{y_b}{y}\right)}{(y - y_b)^2} t^{-2} + o\left(t^{-2}\right) \quad \text{as} \quad t \to \infty \tag{4.90}$$

with $y > y_b$. An examination of expansion (4.90) reveals that it becomes nonuniform when

$$y = y_b + c(t) + O\left(t^{-1}\right), \tag{4.91}$$

where now

$$c(t) = \frac{|2 - y_b^2|}{y_b} \frac{\ln t}{t} - \frac{|2 - y_b^2|}{y_b} \frac{\ln(\ln t)}{t} + \ldots + o\left(\frac{1}{t}\right) \tag{4.92}$$

as $t \to \infty$ (having used $y_b = u_\infty$), with $u = O\left[\frac{1}{\ln t}\right]$ when $y_b \neq \sqrt{2}$ but $u = O(1)$ when $y_b = \sqrt{2}$.

## (a) $y_b \neq \sqrt{2}$

In this case, continuation of the asymptotic structure from region **III** follows precisely that given in Section 4.4.1 for $1 < m < 2$, with each of the individual cases (a), (b) and (c) being required, with the corresponding region **TR**, region **TW** and regions **IV/V**. Note that in this case $v^*(2) = \frac{1}{\sqrt{2}}(< \sqrt{2})$. Without giving the details in full, we conclude that in each case a PTW develops in region **TW**, which has propagation speed $y_b$ for $y_b > \frac{1}{\sqrt{2}}$ but propagation speed $\frac{1}{\sqrt{2}}$ for $0 < y_b \leq \frac{1}{\sqrt{2}}$. In terms of the original coordinates, the asymptotic location of the PTW is $x = s(t)$, where

$$s(t) \sim \begin{cases} y_b t + \left[ t \, o \left( \frac{1}{t} \right) \right], & y_b > \frac{1}{\sqrt{2}}, \\[2mm] \frac{1}{\sqrt{2}} t + \frac{3}{\sqrt{2}} \ln t + O \left( \ln(\ln t) \right), & y_b = \frac{1}{\sqrt{2}}, \\[2mm] \frac{1}{\sqrt{2}} t + \left( \gamma_0 - \sqrt{2} \right) \ln t + O(1), & 0 < y_b < \frac{1}{\sqrt{2}}, \end{cases} \tag{4.93}$$

as $t \to \infty$, and where $\gamma_0$ is undetermined, but satisfies the inequality

$$\sqrt{2} \le \gamma_0 < \frac{5}{\sqrt{2}}.$$

The corresponding asymptotic PTW propagation speeds are

$$\dot{s}(t) \sim \begin{cases} y_b + o \left( \frac{1}{t} \right), & y_b > \frac{1}{\sqrt{2}}, \\[2mm] \frac{1}{\sqrt{2}} + \frac{3}{\sqrt{2}} \frac{1}{t} + o \left( \frac{1}{t} \right), & y_b = \frac{1}{\sqrt{2}}, \\[2mm] \frac{1}{\sqrt{2}} + \left( \gamma_0 - \sqrt{2} \right) \frac{1}{t} + o \left( \frac{1}{t} \right), & 0 < y_b < \frac{1}{\sqrt{2}}. \end{cases} \tag{4.94}$$

Again the correction term in expansion (4.34) is of $O \left( \dot{s}(t) - y_b \right) = o \left( \frac{1}{t} \right)$ for $y_b > \frac{1}{\sqrt{2}}$, but of $O \left( \dot{s}(t) - \frac{1}{\sqrt{2}} \right) \le O \left( \frac{1}{t} \right)$ for $0 < y_b \le \frac{1}{\sqrt{2}}$.

## (b) $y_b = \sqrt{2}$

In this case we see, via (4.92), that $c(t) = o \left( \frac{1}{t} \right)$. We conclude that region **TR** is not present in this case, and we move straight into region **TW** when, via (4.91), $y = y_b + O \left( t^{-1} \right)$ and $u = O(1)$. The problem at leading order in region **TW** is as in Section 4.4.1 and the PTW develops in this region with propagation speed $y_b = \sqrt{2}$. The subsequent regions **IV/V** are as in Section 4.4.1 and the details are not repeated. In terms of the original coordinates the PTW location is $x = s(t)$, where now

$$s(t) \sim \sqrt{2} \, t + \left[ t \, o \left( \frac{1}{t} \right) \right] \quad \text{as} \quad t \to \infty \tag{4.95}$$

with asymptotic propagation speed

$$\dot{s}(t) \sim \sqrt{2} + o \left( \frac{1}{t} \right) \quad \text{as } t \to \infty. \tag{4.96}$$

This change in structure when $y_b = \sqrt{2}$ is driven by the change in structure of $u_T(z; 2, v)$ when $v = \sqrt{2}$, as given in (4.2). In this case, the correction to expansion (4.34) is of $O \left( \dot{s}(t) - \sqrt{2} \right) = o \left( \frac{1}{t} \right)$ as $t \to \infty$.

A schematic representation of the location and thickness of the asymptotic regions as $t \to \infty$ is given in Figure 4.3.

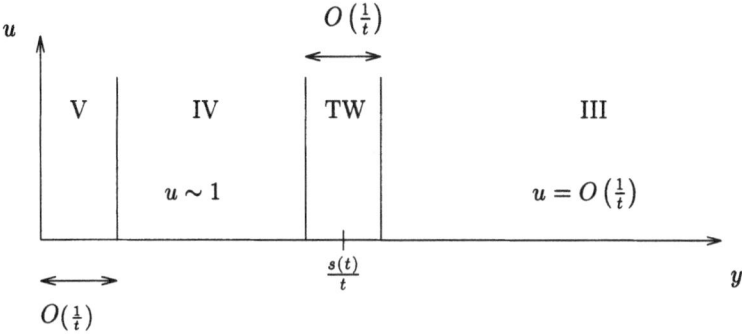

**Fig. 4.3.** Schematic representation of the location and thickness of the asymptotic regions as $t \to \infty$ when $\alpha = \frac{1}{(m-1)} = 1$, $m = 2$ and $y_b = \sqrt{2}$.

### 4.4.3 $m > 2$

In this case expansion (4.7) suggests that we now expand in region **III** in the form

$$u(y,t) = \bar{u}_0(y)t^{-\frac{1}{(m-1)}} + \bar{u}_1(y)t^{-\frac{2}{(m-1)}} + o\left(t^{-\frac{2}{(m-1)}}\right) \qquad (4.97)$$

as $t \to \infty$ with $y = O(1)$. After substituting (4.97) into equation (P1) (when written in terms of $y$ and $t$) the leading order problem for $\bar{u}_0(y)$ is given by (4.15), (4.16) (of Section 4.4.1) with solution

$$\bar{u}_0(y) = \frac{u_\infty}{(y - y_b)^{\frac{1}{(m-1)}}}, \qquad y > y_b, \qquad (4.98)$$

where $y_b = u_\infty^{(m-1)}(m-1)$. Again $\bar{u}_0(y)$ develops a singularity as $y \to y_b^+$, and hence expansion (4.97) becomes non-uniform when $y = y_b + o(1)$ as $t \to \infty$. On proceeding to next order, we obtain the following problem for $\bar{u}_1(y)$, namely

$$y\,\bar{u}_{1y} + \bar{u}_1\left(\frac{2}{(m-1)} + m\,\bar{u}_0^{(m-1)}\right) = \bar{u}_0^{(m+1)}, \qquad y > y_b, \qquad (4.99)$$

$$\bar{u}_1(y) \sim -u_\infty^{(m+1)}y^{-\left(1+\frac{2}{(m+1)}\right)} \qquad \text{as} \quad y \to \infty, \qquad (4.100)$$

with condition (4.100) arising from matching expansion (4.97) ($y \gg 1$) with the far field expansion (4.7) [$x = O(t)$]. The solution to (4.99), (4.100) is given by

$$\bar{u}_1(y) = -\frac{u_\infty^2}{(m-2)}\frac{y^{\frac{(m-2)}{(m-1)}}}{(y - y_b)^{\frac{m}{m-1}}} + \frac{u_\infty^2}{(m-2)}\frac{1}{(y - y_b)^{\frac{2}{(m-1)}}}, \qquad y > y_b. \quad (4.101)$$

Finally, we have, via (4.97), (4.98) and (4.101), that the expansion in region **III** is given by

$$u(y,t) = \frac{u_\infty}{(y-y_b)^{\frac{1}{(m-1)}}} t^{-\frac{1}{(m-1)}}$$

$$+ \left[ -\frac{u_\infty^2}{(m-2)} \frac{y^{\frac{(m-2)}{(m-1)}}}{(y-y_b)^{\frac{m}{(m-1)}}} + \frac{u_\infty^2}{(m-2)} \frac{1}{(y-y_b)^{\frac{2}{(m-1)}}} \right] t^{-\frac{2}{(m-1)}} + o\left( t^{-\frac{2}{(m-1)}} \right)$$

$$(4.102)$$

as $t \to \infty$ with $y_b < y < \infty$. Expansion (4.102) becomes nonuniform when $y = y_b + O\left( t^{-\frac{1}{(m-1)}} \right)$, with $u = O\left( t^{-\frac{(m-2)}{(m-1)^2}} \right)$ as $t \to \infty$. This nonuniformity indicates a transition region, region **TR**, now of thickness $O\left( t^{-\frac{1}{(m-1)}} \right)$. We therefore continue the asymptotic structure by writing, in region **TR**,

$$y = y_b + \frac{\xi}{t^{\frac{1}{(m-1)}}} \tag{4.103}$$

with $\xi = O(1)$ as $t \to \infty$, whilst expanding

$$u(\xi, t) = F(\xi) t^{-\frac{(m-2)}{(m-1)^2}} + o\left( t^{-\frac{(m-2)}{(m-1)^2}} \right) \tag{4.104}$$

as $t \to \infty$ with $\xi = O(1)$. On substituting from (4.103), (4.104) into equation (P1), we obtain the leading order problem as

$$y_b F_\xi - F^m = 0, \quad -\infty < \xi < \infty, \tag{4.105}$$

$$F(\xi) \sim \frac{u_\infty}{\xi^{\frac{1}{(m-1)}}} - \frac{\left[ \frac{u_\infty^2}{(m-2)} \right] y_b^{\frac{(m-2)}{(m-1)}}}{\xi^{\frac{1}{(m-1)}+1}} \quad \text{as} \quad \xi \to \infty, \tag{4.106}$$

after matching expansion (4.104) ($\xi \gg 1$) with expansion (4.102) ($y \to y_b^+$). The solution to (4.105), (4.106) is readily obtained as

$$F(\xi) = \frac{u_\infty}{(\xi - \xi_c)^{\frac{1}{(m-1)}}}, \quad \xi > \xi_c, \tag{4.107}$$

where

$$\xi_c = -\frac{(m-1)}{(m-2)} u_\infty y_b^{\frac{(m-2)}{(m-1)}}. \tag{4.108}$$

We immediately observe that $F(\xi)$ develops a singularity as $\xi \to \xi_c^+$, and thus expansion (4.104) becomes nonuniform when $\xi = \xi_c + o(1)$ as $t \to \infty$, and a further transition region will be required. A consideration of the ratio of neglected to retained terms in equation (P1) in this region indicates that expansion (4.104) becomes nonuniform when

$$\xi = \xi_c + O\left( t^{-\frac{(m-2)}{(m-1)}} \right) \tag{4.109}$$

as $t \to \infty$, with via (4.104) that $u = O(1)$. Hence, as $\xi \to \xi_c^+$ in region **TR**, we move into region **TW**. In region **TW** we introduce the coordinate $z$ so that

$$\xi = \xi_c + t^{-\frac{(m-2)}{(m-1)}} z \tag{4.110}$$

with $z = O(1)$ as $t \to \infty$. Note that $y = y_b + \xi_c t^{-\frac{1}{(m-1)}} + t^{-1} z$ in region **TW**. We expand as

$$u(z,t) = u_c(z) + o(1) \quad \text{as} \quad t \to \infty \tag{4.111}$$

with $z = O(1)$. The leading order problem in region **TW** is precisely (with our new definition of $z$ in (4.110)) that in region **TW** of Section 4.4.1, namely problem [**TW**]. Thus there are three cases to again consider, that is, $0 < y_b < v^*(m)$, $y_b = v^*(m)$ and $y_b > v^*(m)$.

### (a) $y_b > v^*(m)$

In this case, via Theorem 3.1, the solution to [**TW**] is given by

$$u_c(z) = u_T(z; m, y_b), \quad -\infty < z < \infty, \tag{4.112}$$

and has

$$u_c(z) \sim \left[ \frac{y_b}{(m-1)z} \right]^{\frac{1}{(m-1)}} \quad \text{as} \quad z \to \infty. \tag{4.113}$$

Thus expansion (4.111) in region **TW** ($z \to \infty$) matches directly with expansion (4.104) in region **TR** ($\xi \to \xi_c$), and the solution at leading order is complete. Therefore, in this case the large-$t$ structure in region **TW** is dominated by the evolution of a PTW with asymptotic speed given by $v = y_b = u_\infty^{(m-1)}(m-1)$. In terms of the original coordinates, the PTW has location $x = s(t)$, where

$$s(t) \sim y_b t + \xi_c t^{\frac{(m-2)}{(m-1)}} + O(1) \tag{4.114}$$

as $t \to \infty$, and so the asympotic speed of the PTW is

$$\dot{s}(t) \sim y_b + \frac{(m-2)}{(m-1)} \xi_c t^{-\frac{1}{(m-1)}} + o\left(t^{-\frac{1}{(m-1)}}\right) \tag{4.115}$$

as $t \to \infty$. Notice now that the correction to the asymptotic wave speed is of $O\left(t^{-\frac{1}{(m-1)}}\right)$. The correction term in expansion (4.111) is of $O\left(\dot{s}(t) - y_b\right) = O\left(t^{-\frac{1}{(m-1)}}\right)$ as $t \to \infty$.

Finally, for $(-z) \gg 1$, we move out of region **TW**, and to complete the asymptotic structure as $t \to \infty$ in this case, we require two further regions; namely regions **IV** and **V**, details of which follow those given in Section 3.2.3(a) (with $v^*$ replaced by $y_b$ and $s(t)$ given by (4.114)).

This completes the asymptotic structure in this case. In particular, with $u(x,t)$ the solution of [**P,m**], we have established (via regions **III** - **V**) that

$$u(z + s(t), t) \to u_T(z; m, v) \quad \text{as} \quad t \to \infty, \qquad (4.116)$$

uniformly in $z$ with $s(t)$ as given by (4.114). A schematic representation of the location and thickness of the asymptotic regions as $t \to \infty$ is given in Figure 4.4.

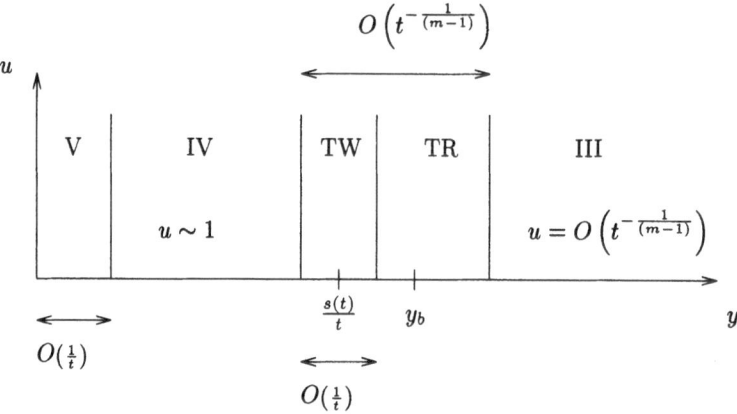

**Fig. 4.4.** Schematic representation of the location and thickness of the asymptotic regions as $t \to \infty$ when $\alpha = \frac{1}{(m-1)}$, $m > 2$ and $y_b > v^*$. Note that $u = O\left(t^{-\frac{(m-2)}{(m-1)^2}}\right)$ in region **TR**.

**(b) $y_b = v^*(m)$**

In this case, via Theorem 3.1, the solution to [**TW**] is given by

$$u_c(z) = u_T(z; m, v^*(m)), \quad -\infty < z < \infty.$$

However, this has

$$u_c(z) \sim B^* e^{-v^*(m)z} \quad \text{as} \quad z \to \infty, \qquad (4.117)$$

via Theorem 3.1, and matching of expansion (4.111) of region **TW** to expansion (4.104) in region **TR** is not possible. We conclude that in this case expansion (4.102) in region **III** must become nonuniform in a displaced transition region based at $y = y_b$. Thus we modify region **TR** so that now in region **TR**

$$y = y_b + \tilde{c}(t) + t^{-\theta}\xi \qquad (4.118)$$

with $\xi = O(1)$ as $t \to \infty$, $\theta > 0$ and $\tilde{c}(t) = o(1)$ as $t \to \infty$. To obtain a non-trivial balance at leading order, and to enable matching with region TW, we require

$$\theta = 1,$$
$$\tilde{c}(t) \sim \gamma_0 \frac{\ln t}{t} + O\left(\frac{\ln(\ln t)}{t}\right)$$

with $\gamma_0$ undetermined. The details of this case now follow those of Section 4.4.1(b). In region **TR** we have

$$u(\xi, t) = \left[\frac{u_\infty}{\gamma_0^{\frac{1}{(m-1)}}} + B^* e^{-y_b \xi}\right] \frac{1}{(\ln t)^{\frac{1}{(m-1)}}} + o\left[\frac{1}{(\ln t)^{\frac{1}{(m-1)}}}\right]$$

as $t \to \infty$ with $\xi = O(1)$. For $(-\xi) \gg 1$, we move into region **TW**, where $z = O(1)$ as $t \to \infty$ with

$$\xi = -\theta(t) + z,$$

and

$$\theta(t) \sim \frac{1}{(m-1)y_b} \ln(\ln t)$$

as $t \to \infty$. Within region **TW**

$$u(z, t) = u_T\left(z; m, v^*(m)\right) + o(1)$$

as $t \to \infty$ with $z = O(1)$. In terms of the original coordinates, the PTW has location $x = s(t)$, where

$$s(t) \sim v^*(m)\, t + \gamma_0 \ln t + O\left[\ln(\ln t)\right] \qquad (4.119)$$

as $t \to \infty$, and so the asymptotic speed of the PTW is

$$\dot{s}(t) \sim v^*(m) + \gamma_0\, t^{-1} + O\left[\frac{1}{(t \ln t)}\right] \qquad (4.120)$$

as $t \to \infty$. The correction to expansion (4.111) is now of $O(\dot{s}(t) - v^*(m)) \leq O\left(t^{-1}\right)$ as $t \to \infty$. Note that although $\gamma_0$ is undetermined to the order which we have taken our asymptotic expansions, comparison arguments require that

$$\gamma_0 \geq 0.$$

Finally, for $(-z) \gg 1$, we move out of region **TW**, and to complete the asymptotic structure we again require regions **IV** and **V**. These follow those given in Section 3.2.3(a) (with $s(t)$ given by (4.119)), and are not repeated here.

This completes the asymptotic structure in this case. In particular, with $u(x, t)$ the solution of **[P,m]**, we have established (via regions **III - V**) that

$$u(z + s(t), t) \to u_T(z; m, v^*(m)) \quad \text{as} \quad t \to \infty, \qquad (4.121)$$

uniformly in $z$ with $s(t)$ as given by (4.119). A schematic representation of the location and thickness of the asymptotic regions as $t \to \infty$ is given in Figure 4.5.

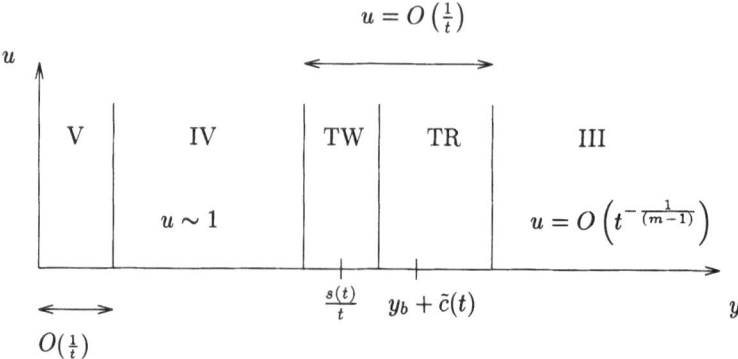

**Fig. 4.5.** Schematic representation of the location and thickness of the asymptotic regions as $t \to \infty$ when $\alpha = \frac{1}{(m-1)}$, $m > 2$ and $y_b = v^*$. Note that $u = O\left((\ln t)^{-\frac{1}{(m-1)}}\right)$ in region TR.

## (c) $0 < y_b < v^*(m)$

In this case, via Theorem 3.1, the leading order problem [**TW**] has no solution. We must return to region **III** and conclude that expansion (4.102) fails before $y = y_b$. The details in this case follow precisely those given in Section 4.4.1(c), with region **TR'** and region **TW'** following in exactly the same form. We conclude that region **TW'** is dominated by the emergence of the PTW with asymptotic speed $v^*(m)$, and that, in terms of the original coordinates, the PTW has location $x = s(t)$, where

$$s(t) \sim v^*(m)\, t + \left(\gamma_0 - \frac{1}{(m-1)v^*}\right)\ln t + O(1) \tag{4.122}$$

as $t \to \infty$ and $\gamma_0 \geq \frac{1}{(m-1)v^*}$ is undetermined at this order. The asymptotic speed of the PTW is

$$\dot{s}(t) \sim v^*(m) + \left(\gamma_0 - \frac{1}{(m-1)v^*}\right)\frac{1}{t} + o\left(\frac{1}{t}\right) \tag{4.123}$$

as $t \to \infty$. The correction to expansion (4.111) is of $O\left(\dot{s}(t) - v^*(m)\right) \leq O\left(\frac{1}{t}\right)$ as $t \to \infty$.

## 4.5 Asymptotic Solution as $t \to \infty$ when $\alpha > \frac{1}{m-1}$

As in the case when $\alpha = \frac{1}{(m-1)}$ (see Section 4.4), to examine region **III** we introduce the scaled coordinate $y = \frac{x}{t}$, where $y = O(1)$ as $t \to \infty$, and, following (4.10), look for an expansion of the form

$$u(y,t) = F(y)t^{-\alpha} + o\left(t^{-\alpha}\right) \quad \text{as} \quad t \to \infty \tag{4.124}$$

with $y = O(1)$. On substituting (4.124) into equation (P1) (when written in terms of $y$ and $t$), the leading order problem becomes

$$y\,F_y + \alpha F = 0, \quad y > 0, \tag{4.125}$$

$$F(y) \sim u_\infty y^{-\alpha} \quad \text{as} \quad y \to \infty \tag{4.126}$$

with condition (4.126) arising from matching expansion (4.124) ($y \gg 1$) with (4.10) [$x = O(t)$]. The solution to (4.125), (4.126) is readily obtained as

$$F(y) = u_\infty y^{-\alpha}, \quad 0 < y < \infty. \tag{4.127}$$

Thus in region **III** we have

$$u(y,t) = \frac{u_\infty}{y^\alpha} t^{-\alpha} + o\left(t^{-\alpha}\right) \quad \text{as } t \to \infty \tag{4.128}$$

with $y = O(1)$. The form of expansion (4.128) suggests that the structure for $t \gg 1$ in this case will follow that given in Section 4.4.1(c). Accordingly we assume that approximation (4.128) fails as $y \to y_T(> 0)$ (with $y_T$ to be determined). When $y = y_T + o(1)$ we introduce a transition layer, region **TR**, and write

$$y = y_T + c(t) + t^{-\gamma}\xi \tag{4.129}$$

with $\xi = O(1)$ as $t \to \infty$, $\gamma > 0$ and $c(t) = o(1)$ as $t \to \infty$ (to be determined). From equation (4.128) we have that

$$u = O\left(t^{-\alpha}\right) \quad \text{as} \quad t \to \infty \tag{4.130}$$

in region **TR**. Therefore, in region **TR** we write

$$u(\xi,t) = \hat{H}(\xi)\,t^{-\alpha} + o\left(t^{-\alpha}\right) \quad \text{as} \quad t \to \infty \tag{4.131}$$

with $\xi = O(1)$. On substituting from (4.130), (4.131) into the full equation (P1), we find that a non-trivial leading order balance requires

$$\gamma = 1, \tag{4.132}$$

after which the leading order problem is

$$\hat{H}_{\xi\xi} + y_T \hat{H}_\xi = 0, \quad -\infty < \xi < \infty, \tag{4.133}$$

$$\hat{H}(\xi) \to \frac{u_\infty}{y_T^\alpha} \quad \text{as} \quad \xi \to \infty, \tag{4.134}$$

after matching (4.131) to (4.128). The solution to (4.133), (4.134) is given by

$$\hat{H}(\xi) = \frac{u_\infty}{y_T^\alpha} + \tilde{B}\,e^{-y_T\xi} \tag{4.135}$$

with $\tilde{B}$ as yet undetermined. Thus, in region **TR** we have

$$u(\xi, t) = \left[ \frac{u_\infty}{y_T^\alpha} + \tilde{B} e^{-y_T \xi} \right] t^{-\alpha} + o\left( t^{-\alpha} \right) \qquad (4.136)$$

as $t \to \infty$, with $\xi = O(1)$. An examination of the scaling (4.129) indicates that we re-emerge from the transition region **TR** when $(-\xi) \gg 1$, and specifically when

$$1 \ll (-\xi) \le O\left( c(t)t \right). \qquad (4.137)$$

We thus suppose that we leave the transition region **TR** when $\xi = -\hat{\phi}(t) + O(1)$ where

$$1 \ll \hat{\phi}(t) \le O\left( c(t)t \right), \qquad (4.138)$$

and $\hat{\phi}(t)$ is to be determined. As before, we refer to the next region as region **TW**. We introduce the new coordinate $z$ by

$$\xi = -\hat{\phi}(t) + z, \qquad (4.139)$$

so that $z = O(1)$ as $t \to \infty$ in region **TW**, and we suppose that $u = O(1)$ as $t \to \infty$ in region **TW**. Thus we expand as

$$u(z, t) = u_c(z) + o(1) \quad \text{as} \quad t \to \infty, \qquad (4.140)$$

with $z = O(1)$. The leading order problem in region **TW** is now given by problem [**TW**] (see Section 4.4.1), with $y_b$ replaced by $y_T$ and $z$ given by (4.139). For this to have a solution we require $y_T \ge v^*(m)$, and then the solution is (via Theorem 3.1)

$$u_c(z) = u_T(z; m, y_T), \qquad (4.141)$$

which has

$$u_c(z) \sim \begin{cases} B^* e^{-v^*(m)z}, & y_T = v^*(m), \\ \left[ \frac{y_T}{(m-1)z} \right]^{\frac{1}{(m-1)}}, & y_T > v^*(m) \end{cases} \qquad (4.142)$$

as $z \to \infty$. It remains to match expansion (4.140) $(z \gg 1)$ to expansion (4.131) $(\xi \sim -\hat{\phi}(t))$. Following Section 4.4.1(c) directly, we find that this matching can only be achieved when

$$\left. \begin{array}{l} y_T = v^*(m), \\ \tilde{B} = B^*, \\ \hat{\phi}(t) = \frac{\alpha}{v^*(m)} \ln t. \end{array} \right\} \qquad (4.143)$$

Thus, from (4.143) and (4.138), we have that

$$c(t) \ge O\left[ \frac{\ln t}{t} \right].$$

However, comparison arguments require that $c(t) \leq O\left[\frac{\ln t}{t}\right]$, and so we have

$$c(t) = O\left[\frac{\ln t}{t}\right].$$

Hence, without loss of generality we write

$$c(t) \sim \frac{\gamma_0 \ln t}{t} \quad \text{as} \quad t \to \infty$$

with $\gamma_0$ undetermined. Therefore in this case, the large-$t$ structure in region **TW** is dominated by the emergence of the PTW with asymptotic speed $v = v^*(m)$. In terms of the original coordinates, the PTW has location $x = s(t)$, where

$$s(t) = v^*(m)t + \left(\gamma_0 - \frac{\alpha}{v^*}\right)\ln t + o(\ln t) \qquad (4.144)$$

as $t \to \infty$, and so the asymptotic speed of the PTW is given by

$$\dot{s}(t) \sim v^*(m) + \left(\gamma_0 - \frac{\alpha}{v^*}\right)\frac{1}{t} + o\left(\frac{1}{t}\right) \qquad (4.145)$$

as $t \to \infty$. The correction to expansion (4.140) is now of $O\left(\dot{s}(t) - v^*(m)\right) \leq O\left(t^{-1}\right)$ as $t \to \infty$. We note that, although $\gamma_0$ has remained undetermined at this order, comparison arguments require

$$\gamma_0 \geq \frac{\alpha}{v^*(m)}.$$

To complete the asymptotic structure for $t \gg 1$, we require again the regions **IV** and **V**. These follow those given in Section 3.2.3(a) (with $s(t)$ given by (4.144)), and are not repeated here. This completes the asymptotic structure in this case. In particular, with $u(x,t)$ the solution to **[P,m]**, we have established (via regions **III** - **V**) that

$$u(z + s(t), t) \to u_T(z; m, v^*) \quad \text{as } t \to \infty,$$

uniformly in $z$ with $s(t)$ as given by (4.144). A schematic representation of the location and thickness of the asymptotic regions as $t \to \infty$ is given in Figure 4.6

## 4.6 Numerical Solutions

The results obtained in Sections 4.4 and 4.5, via matched asymptotic expansions, are summarized in Table 4.1. In this section we perform representative numerical solutions of **[P,m]** for comparison. For this purpose we take

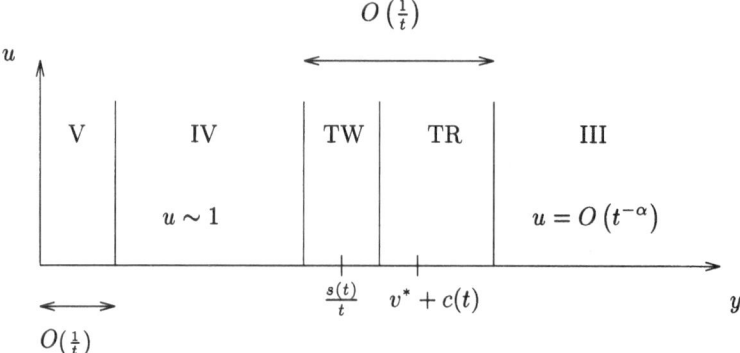

**Fig. 4.6.** Schematic representation of the location and thickness of the asymptotic regions as $t \to \infty$ in the case when $\alpha > \frac{1}{(m-1)}$. Note that $u = O\left(t^{-\alpha}\right)$ in region **TR**.

$$u_0(x) = \frac{u_\infty}{(1+x)^\alpha}, \quad x \geq 0. \tag{4.146}$$

The numerical solutions were performed using the NAG routine DO3PCF which incorporates the method of lines with finite differences. A large spatial domain, $0 \leq x \leq L$, was required to ensure that the imposed Dirichlet boundary condition, $u = 0$ at $x = L$, had negligible effect on the numerical solution over the $t$ intervals for which the computations were performed. We note in what follows that

$$\ell = \left[\frac{v^*(m)}{(m-1)}\right]^{\frac{1}{(m-1)}}.$$

### $1 < m < 2$

In this case we selected $m = \frac{3}{2}$, which gives $v^* = 0.983...$, $\ell = 3.865...$ .

### (a) $\alpha > \frac{1}{(m-1)}$

We take $\alpha = 3$ and $u_\infty = 1$. The results are given in Figures 4.7(a) and 4.8(a). The PTW of wave speed $v^*$ is seen to develop rapidly, and the correction to $\dot{s}(t)$ for $t \gg 1$ is certainly of $o\left(t^{-1}\right)$, in line with the theory, and appears to be exponentially small in $t$ as $t \to \infty$, from Figure 4.7(a).

### (b) $\alpha = \frac{1}{(m-1)}$

We take $\alpha = 2$.

(i) $u_\infty < \ell$

We set $u_\infty = 2$. The results are given in Figures 4.7(b) and 4.8(b). Again we observe a rapid approach to the PTW with wave speed $v^*$, and the correction to $\dot{s}(t)$ is certainly of order $o\left(t^{-1}\right)$, which is in line with the theory.

(ii) $u_\infty = \ell$

We set $u_\infty = 3.865...$ . The results are given in Figures 4.7(c) and 4.8(c). In accord with the theory, there is a rapid approach to the PTW with wave speed $v = v^*$, whilst the correction to $\dot{s}(t)$ as $t \to \infty$ is larger than in the previous cases, being of $O\left(\frac{1}{t}\right)$. Note that $\dot{s}(t)$ approaches $v^*$ as $t \to \infty$ from above in accord with the theory in this case.

(iii) $u_\infty > \ell$

We set $u_\infty = 5.0$ . The results are given in Figures 4.7(d) and 4.8(d). In this case we observe approach to a PTW with waves speed $v = u_\infty^{m-1}(m-1) = 1.118...$, in accord with the theory.

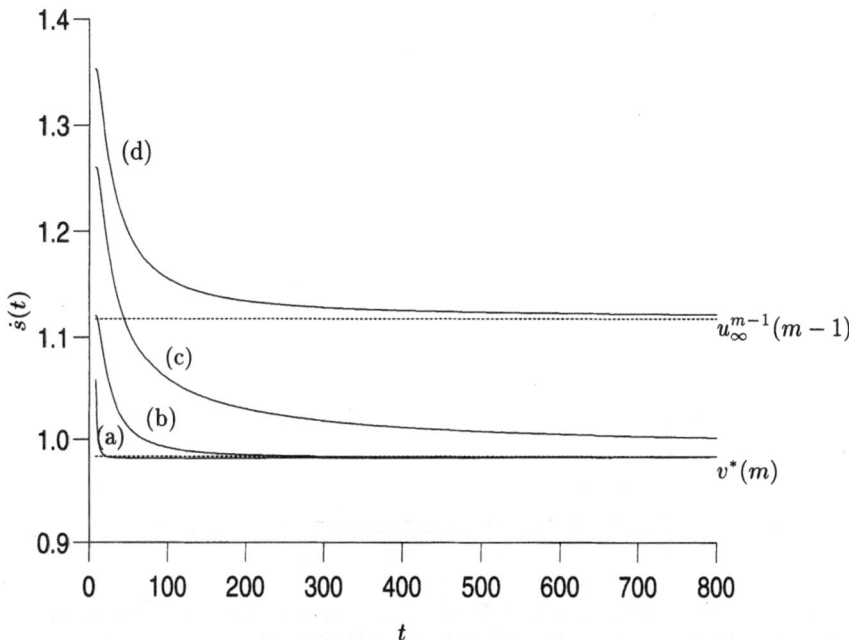

Fig. 4.7. Numerical solution of [P,m], $\dot{s}(t)$ versus $t$ for $m = 3/2$ with (a) $\alpha = 3$, $u_\infty = 1$; (b) $\alpha = 2$, $u_\infty = 2$; (c) $\alpha = 2$, $u_\infty = 3.865$; (d) $\alpha = 2$, $u_\infty = 5$.

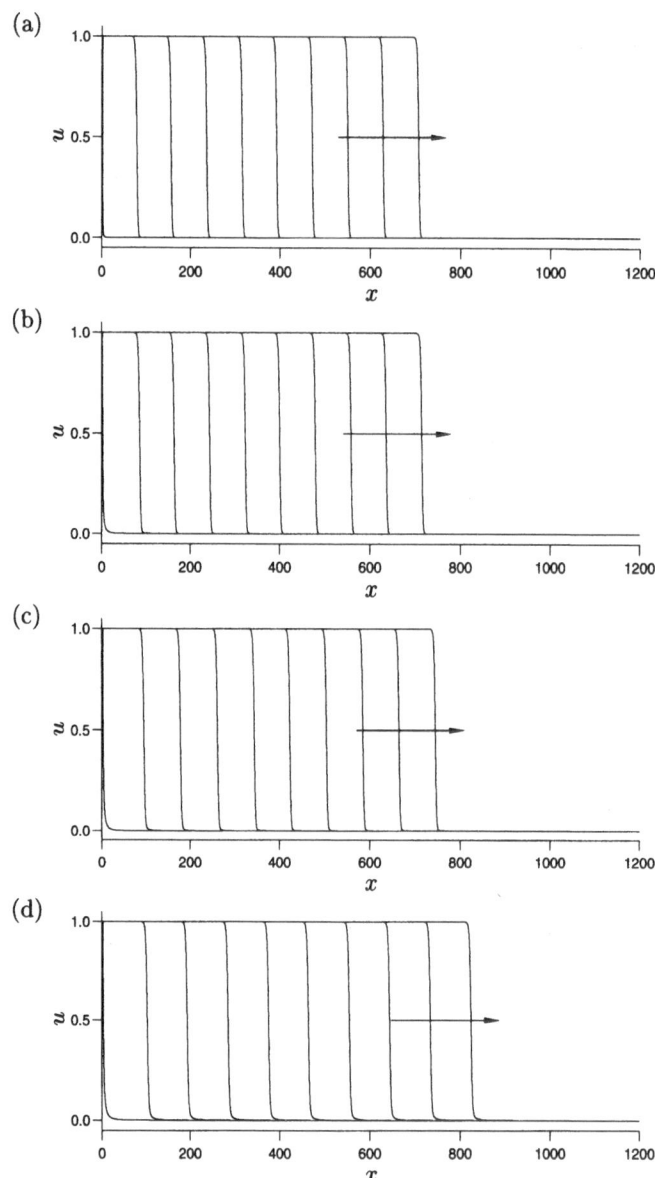

**Fig. 4.8.** Numerical solution of [**P,m**], $u(x,t)$ versus $x$ for $m = 3/2$ with (a) $\alpha = 3$, $u_\infty = 1$; (b) $\alpha = 2$, $u_\infty = 2$; (c) $\alpha = 2$, $u_\infty = 3.865$; (d) $\alpha = 2$, $u_\infty = 5$.

## $m = 2$

In this case we have $v^* = \frac{1}{\sqrt{2}}$ and $\ell = \frac{1}{\sqrt{2}}$.

(a) $\underline{\alpha > \frac{1}{(m-1)}}$

We take $\alpha = 2$ and $u_\infty = 1$. As predicted by the theory, we observe a rapid approach to the PTW with wave speed $v^*$, with the correction to $\dot{s}(t)$ as $t \to \infty$ being exponentially small in $t$. The results are illustrated in Figures 4.9(a) and 4.10(a).

(b) $\underline{\alpha = \frac{1}{(m-1)} = 1}$

(i) $u_\infty < \ell$
We put $u_\infty = 0.5$. The results are shown in Figures 4.9(b) and 4.10(b). In agreement with the theory, we see the approach to the PTW with wave speed $v^*$. The correction to $\dot{s}(t)$ as $t \to \infty$ appears to be $o\left(\frac{1}{t}\right)$.

(ii) $u_\infty = \ell = \frac{1}{\sqrt{2}}$
The results are shown in Figures 4.9(c) and 4.10(c). In accord with the theory we see the approach to the PTW with wave speed $v^*$. Moreover, the correction to $\dot{s}(t)$ appears now to be $O\left(\frac{1}{t}\right)$, and $\dot{s}(t) \to v^*$ from above as $t \to \infty$.

(iii) $u_\infty > \ell$
We consider the case when $u_\infty = 1$. The results are illustrated in Figures 4.9(d) and 4.10(d). In agreement with the theory, the solution rapidly approaches the PTW with speed $v = u_\infty = 1$. The correction to $\dot{s}(t)$ as $t \to \infty$ appears to be $o\left(\frac{1}{t}\right)$, again in accord with the theory.

## $m > 2$

We now take $m = 3$ and have $v^* = 0.463...$, $\ell = 0.481...$ .

(a) $\underline{\alpha > \frac{1}{(m-1)}}$

We take $\alpha = 1.5$ and $u_\infty = 1$. The results are shown in Figures 4.11(a) and 4.12(a). A PTW develops with speed $v^* = 0.463...$, and the correction to $\dot{s}(t)$ as $t \to \infty$ appears to be exponential in $t$. This is in line with the theory.

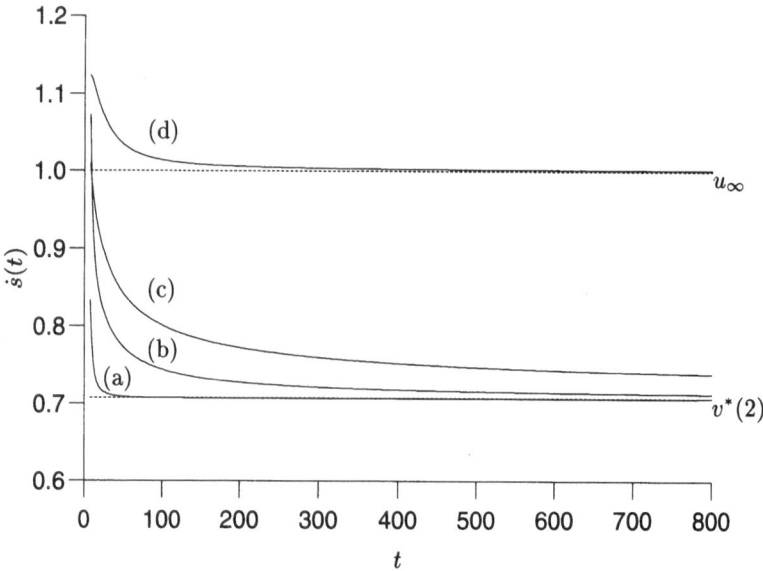

**Fig. 4.9.** Numerical solution of [**P,m**], $\dot{s}(t)$ versus $t$ for $m = 2$ with (a) $\alpha = 2$, $u_\infty = 1$; (b) $\alpha = 1$, $u_\infty = 1/2$; (c) $\alpha = 1$, $u_\infty = 1/\sqrt{2}$; (d) $\alpha = 1$, $u_\infty = 1$.

(b) $\alpha = \frac{1}{(m-1)}$

We set $\alpha = 0.5$.

(i) $u_\infty < \ell$

The integrations are performed with $u_\infty = 0.3$, and the results are shown in Figures 4.11(b) and 4.12(b). Again the PTW develops with speed $v^*$ in accord with the theory. The correction to $\dot{s}(t)$ as $t \to \infty$ appears to be $o\left(\frac{1}{t}\right)$.

(ii) $u_\infty = \ell$

The integrations are performed with $u_\infty = 0.481... = \ell$, and the results are given in Figures 4.11(c) and 4.12(c). In agreement with the theory, the PTW with speed $v^*$ develops. The correction to $\dot{s}(t)$ as $t \to \infty$ now appears to be $O\left(\frac{1}{t}\right)$.

(iii) $u_\infty > \ell$

We take $u_\infty = 0.6$. The results are shown in Figures 4.11(d) and 4.12(d). A PTW develops with speed $v = u_\infty^{m-1}(m - 1) = 0.72...$, and $\dot{s}(t)$ approaches $v$ from below as $t \to \infty$. This is in agreement with the theory.

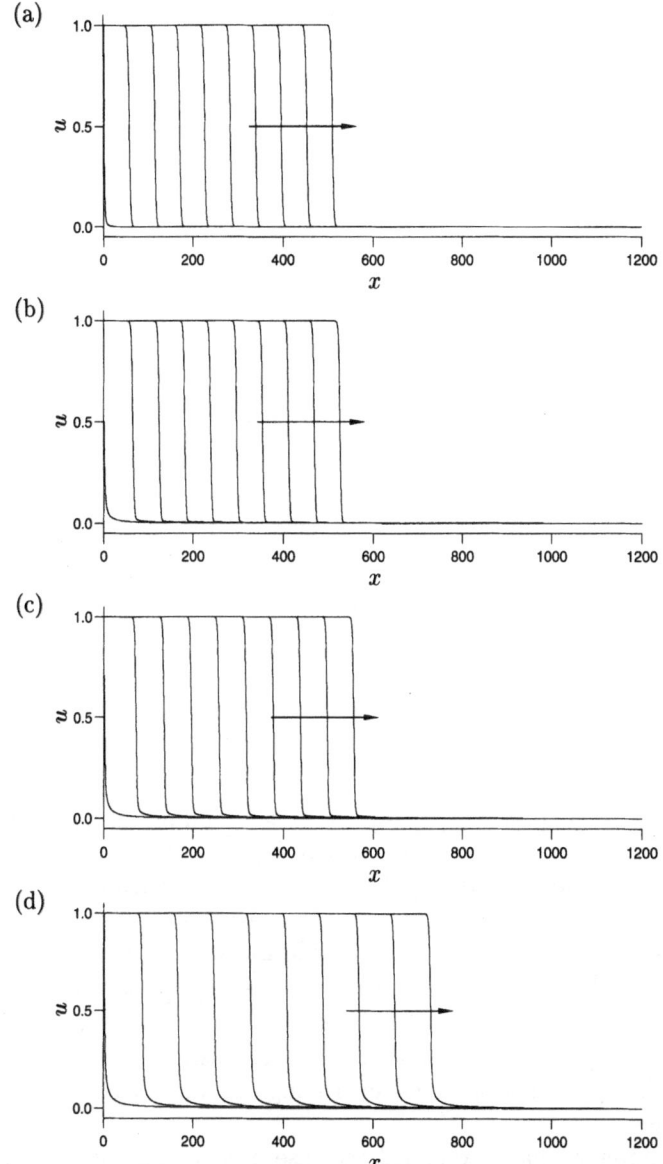

**Fig. 4.10.** Numerical solution of **[P,m]**, $u(x,t)$ versus $x$ for $m = 2$ with (a) $\alpha = 2$, $u_\infty = 1$; (b) $\alpha = 1$, $u_\infty = 1/2$; (c) $\alpha = 1$, $u_\infty = 1/\sqrt{2}$; (d) $\alpha = 1$, $u_\infty = 1$.

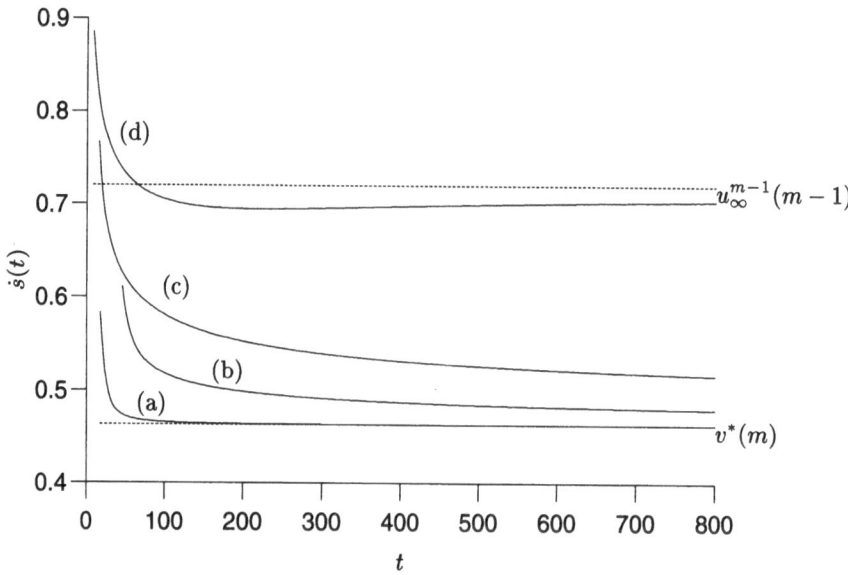

**Fig. 4.11.** Numerical solution of [**P**,**m**], $\dot{s}(t)$ versus $t$ for $m = 3$ with (a) $\alpha = 1.5$, $u_\infty = 1$; (b) $\alpha = 0.5$, $u_\infty = 0.3$; (c) $\alpha = 0.5$, $u_\infty = 0.481$; (d) $\alpha = 0.5$, $u_\infty = 0.6$.

## 4.7 Summary

In this chapter, we have extended the analysis of Chapter 3 by considering the initial-boundary value problem [**P, m**] for $m > 1$, and initial data satisfying (g1); that is initial data with algebraic decay (up to exponential corrections) of degree $\alpha$ as $x \to \infty$, with $\alpha \geq \frac{1}{(m-1)}$. The situation when $0 < \alpha < \frac{1}{(m-1)}$ has been considered using matched asymptotic expansions by Needham and Barnes [56] and numerically by Sherratt and Marchant [64]. It was established in Needham and Barnes [56] that, for $0 < \alpha < \frac{1}{(m-1)}$, a travelling wave of permanent form (PTW) does not develop in the large-$t$ structure of the solution to [**P, m**] ($m > 1$), but that an accelerating phase wave (PHW) emerges, of the type first observed by Billingham and Needham [8] for a related problem.

In this chapter we have examined the case of algebraic initial data with degree $\alpha \geq \frac{1}{(m-1)}$, again via the method of matched asymptotic expansions, which has again allowed the full asymptotic structure of [**P, m**] to be determined as $t \to \infty$. Primarily, we have seen that, in contrast to the case $0 < \alpha < \frac{1}{(m-1)}$, the large-$t$ structure always involves the development of an PTW in the solution to [**P, m**]. However, the structure in the critical case $\alpha = \frac{1}{(m-1)}$ is considerably more sensitive to the form of the initial data than that when $\alpha > \frac{1}{(m-1)}$. This should not be unexpected, as we have shown that $\alpha = \frac{1}{(m-1)}$ forms a bifurcation point for the formation of PTW ($\alpha \geq \frac{1}{(m-1)}$) or PHW ($0 < \alpha < \frac{1}{(m-1)}$) as $t \to \infty$ in the solution to [**P, m**]. Not only have we

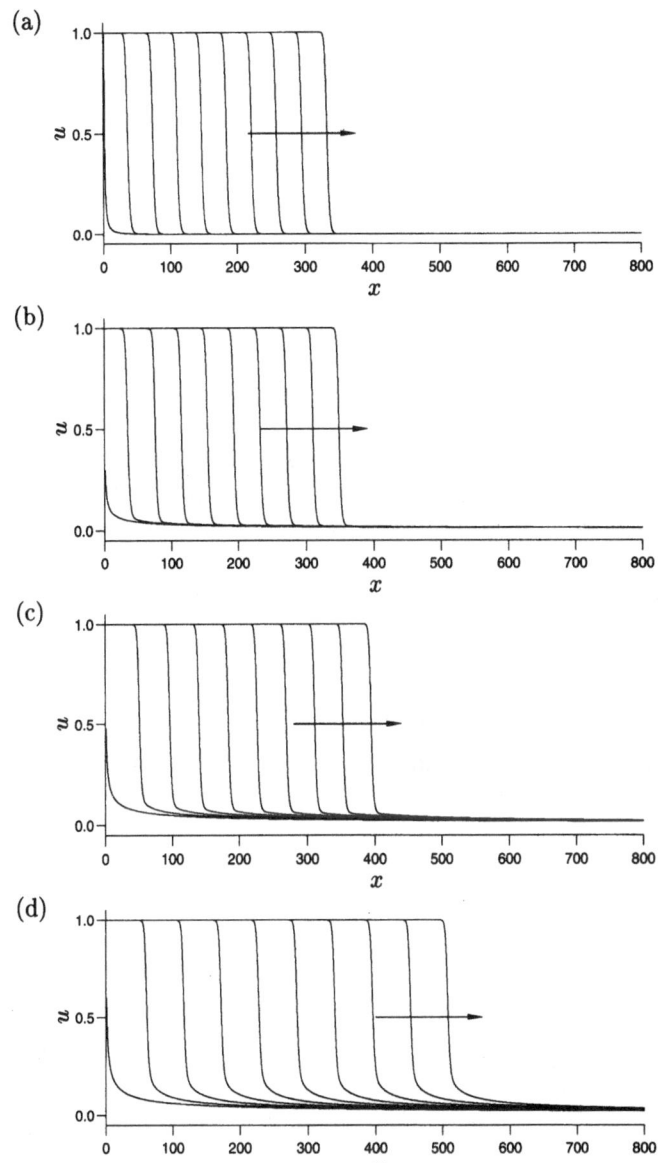

**Fig. 4.12.** Numerical solution of [**P,m**], $u(x,t)$ versus $x$ for $m = 3$ with (a) $\alpha = 1.5$, $u_\infty = 1$; (b) $\alpha = 0.5$, $u_\infty = 0.3$; (c) $\alpha = 0.5$, $u_\infty = 0.481$; (d) $\alpha = 0.5$, $u_\infty = 0.6$.

established that an PTW develops as $t \to \infty$ in [**P, m**] with $\alpha \geq \frac{1}{(m-1)}$, but we have also obtained the speed of this PTW and its asymptotic correction as $t \to \infty$, together with the rate of convergence of the solution to [**P, m**] onto the PTW as $t \to \infty$. The rate of convergence depends crucially upon the decay rate of the initial data (g1) and how this compares to the asymptotic form of the emerging PTW for $z \gg 1$, as given in (4.1)-(4.3). This indicates that we should not only expect bifurcations on the rate of convergence between $\alpha = \frac{1}{(m-1)}$ and $\alpha > \frac{1}{(m-1)}$, but also, when $\alpha = \frac{1}{(m-1)}$, between the cases $1 < m < 2$, $m = 2$ and $m > 2$. This is borne out by the matched asymptotic theory. As we have demonstrated the solution $u(x, t)$ to [**P, m**] satisfies

$$u(z + s(t), t) \sim u_T(z; m, v) + O(\delta(t)) \tag{4.147}$$

as $t \to \infty$, uniformly in $-\infty < z < \infty$, where $s(t)$, $v$ and $\delta(t)$ are displayed in Table 4.1. We observe that the convergence rate to the PTW in the critical case is very sensitive both to the value of $m$ and the parameter $u_\infty$, being either $O\left(t^{-\frac{1}{(m-1)}}\right)$, $O\left(t^{-1}\right)$ or $o\left(t^{-1}\right)$. In the cases when the convergence rate has been demonstrated to be of $o\left(t^{-1}\right)$, the asymptotic structure we have developed still leads us to expect that the convergence rate is algebraic, but of degree larger than unity. To confirm this, the present theory needs modification, so that terms beyond all orders of the type $O\left[\frac{(\ln(\ln(...(\ln t))))}{t}\right]$ in both $c(t)$ (4.23) and $\theta(t)$ (4.44) can be accessed. The authors are considering this at present.

**Table 4.1.** Note that $\ell = \left(\frac{v^*(m)}{(m-1)}\right)^{\frac{1}{(m-1)}}$, $\xi_c < 0$ and $\lambda_i(t) \leq O\left(\frac{1}{t}\right)$ as $t \to \infty$

| | | $v$ | $\dot{s}(t)$ | $\delta(t)$ |
|---|---|---|---|---|
| $\alpha = \frac{1}{(m-1)}$ | $u_\infty > \ell$ | $u_\infty^{(m-1)}(m-1)$ | $u_\infty^{(m-1)}(m-1) + o\left(\frac{1}{t}\right)$ | $o\left(\frac{1}{t}\right)$ |
| $1 < m < 2$ | $u_\infty = \ell$ | $v^*(m)$ | $v^*(m) + \frac{m}{(m-1)^2 v^*(m)}\frac{1}{t} + o\left(\frac{1}{t}\right)$ | $O\left(\frac{1}{t}\right)$ |
| | $u_\infty < \ell$ | $v^*(m)$ | $v^*(m) + \lambda_1(t)$ | $O(\lambda_1(t))$ |
| $\alpha = \frac{1}{(m-1)}$ | $u_\infty > \frac{1}{\sqrt{2}}$ | $u_\infty$ | $u_\infty + o\left(\frac{1}{t}\right)$ | $o\left(\frac{1}{t}\right)$ |
| $m = 2$ | $u_\infty = \frac{1}{\sqrt{2}}$ | $\frac{1}{\sqrt{2}}$ | $\frac{1}{\sqrt{2}} + \frac{3}{\sqrt{2}}\frac{1}{t} + o\left(\frac{1}{t}\right)$ | $O\left(\frac{1}{t}\right)$ |
| | $u_\infty < \frac{1}{\sqrt{2}}$ | $\frac{1}{\sqrt{2}}$ | $\frac{1}{\sqrt{2}} + \lambda_2(t)$ | $O(\lambda_2(t))$ |
| $\alpha = \frac{1}{(m-1)}$ | $u_\infty > \ell$ | $u_\infty^{(m-1)}(m-1)$ | $u_\infty^{(m-1)}(m-1) + \frac{(m-2)}{(m-1)}\xi_c t^{-\frac{1}{(m-1)}}$ | $o\left(t^{-\frac{1}{(m-1)}}\right)$ |
| $m > 2$ | $u_\infty = \ell$ | $v^*(m)$ | $v^*(m) + \lambda_3(t)$ | $O(\lambda_3(t))$ |
| | $u_\infty < \ell$ | $v^*(m)$ | $v^*(m) + \lambda_4(t)$ | $O(\lambda_4(t))$ |
| $\alpha > \frac{1}{(m-1)}$ | $u_\infty > 0$ | $v^*(m)$ | $v^*(m) + \lambda_5(t)$ | $O(\lambda_5(t))$ |

# Extension to Systems of Fisher-Kolmogorov Equations. Example: A Simple Model for an Ionic Autocatalytic System

In this chapter we extend the approach developed in the previous chapters to determine the large-$t$ structure of scalar reaction-diffusion equations of the Fisher-Kolmogorov tpye by considering a system of Fisher-Kolmogorov equations. This system arises as a simple model for an isothermal, autocatalytic, chemical reaction scheme. The scheme is based on the autocatalytic step

$$A + nB \to (n + 1)B, \quad \text{rate} = kab^n. \tag{5.1}$$

Here $a$ and $b$ are the concentrations of the reactant, A, and the autocatalyst, B, respectively, $k$ is the reaction rate constant and $n$ the general order of autocatalysis, $(n \geq 1)$.

The above autocatalytic step (5.1) occurs in several models of real chemical systems. These include the Belousov-Zhabotinskii reaction [15], the iodate-arsenous acid reaction [61] and hydroxylamite-nitrate reaction [19]. Similar autocatalytic reaction-rate laws also arise in enzyme reactions such as glycosis [62] and in gas-phase, radical chain-branching, oxidation reactions, examples of which are the $CO + O_2$ and $H_2 + O_2$ systems [46].

Observations show that chemical systems for which autocatalysis forms a key step can support propagating chemical wavefronts, when the reaction mixture is unstirred (see for example, Zaikin and Zhabotinskii [75] and Hanna *et al* [25]). These wavefronts, arise via a combination of reaction and molecular diffusion. Physically, the typical situation which leads to the development of travelling waves is that which arises when a quantity of the autocatalyst, B, is introduced locally into an expanse of the reactant, A, which is initially at uniform concentration. The developing reaction is often observed to generate wavefronts, which propagate outward from the initial reaction zone. This phenomenon has been discussed by Billingham and Needham [5], [6] and [7].

In this chapter we aim to address the problem when the reactant, A, and the autocatalyst, B, are not only mobile through diffusion, but are also ionic species, and their mobility may thus be further enhanced by the application of an external electric field. The effects of an applied electric field on ionic autocatalytic chemical systems have recently be examined in [67], [50], [51] and [17]. For simplicity, we focus attention on the case of the quadratic autocatalytic step (5.1) with $n = 1$, and we restrict attention to the case of one-dimensional slab geometry, with the coordinate $\bar{x}$ measuring distance, and assume that the external electric field acts along the $\bar{x}$-axis. The equations that govern the reaction and diffusion of the species A and B under the quadratic autocatalytic reaction scheme and the electric field are then

$$\frac{\partial a}{\partial \bar{t}} + \bar{c}_\alpha \frac{\partial a}{\partial \bar{x}} = D_A \frac{\partial^2 a}{\partial \bar{x}^2} - kab, \quad \frac{\partial b}{\partial \bar{t}} + \bar{c}_\beta \frac{\partial b}{\partial \bar{x}} = D_B \frac{\partial^2 b}{\partial \bar{x}^2} + kab, \quad (5.2)$$

where $-\infty < \bar{x} < \infty$ and $\bar{t} > 0$. Here $D_A$ and $D_B$ are the constant diffusion rates of the reactant, A, and the autocatalyst, B, respectively, $\bar{t}$ is time, and $\bar{c}_\alpha$ and $\bar{c}_\beta$ are the constant convection speeds induced on the ions of A and B, respectively, by the external electric field. The initial conditions to be considered are

$$a(\bar{x}, 0) = a_0, \quad b(\bar{x}, 0) = \begin{cases} b_0 g(\bar{x}), & |\bar{x}| \leq \sigma, \\ 0, & |\bar{x}| > \sigma. \end{cases} \quad (5.3)$$

Here $g(\bar{x})$ is a given, even, analytic, non-negative function of $\bar{x}$, which is positive for $|\bar{x}| < \sigma$, has $g(\pm\sigma) = 0$ and $\max_{\bar{x} \in [-\sigma, \sigma]} g(\bar{x}) = 1$ where

$$g(\bar{x}) \sim \begin{cases} g_\sigma(\sigma - \bar{x})^r & \text{as} \quad \bar{x} \to \sigma^-, \\ g_\sigma(\sigma + \bar{x})^r & \text{as} \quad \bar{x} \to -\sigma^+, \end{cases} \quad (5.4)$$

where $r \in \mathbb{N}$ and $g_\sigma > 0$ is a constant. Here $a_0$ and $b_0$ are the positive, constant, initial concentration of the reactant A, and the maximum initial concentration of the autocatalyst B, respectively. In addition, we have the following boundary conditions:

$$\left. \begin{array}{c} a(\bar{x}, \bar{t}) \to a_0 \\ b(\bar{x}, \bar{t}) \to 0 \end{array} \right\} \quad as \ |\bar{x}| \to \infty, \ \bar{t} \geq 0. \quad (5.5)$$

It is now convenient to introduce dimensionless variables as

$$\alpha = \frac{a}{a_0}, \quad \beta = \frac{b}{a_0}, \quad t = ka_0\bar{t}, \quad \hat{x} = \left\{\frac{ka_0}{D_A}\right\}^{\frac{1}{2}} \bar{x}, \quad (5.6)$$

in terms of which equations (5.2), together with initial and boundary conditions (5.3) and (5.5), become

$$\frac{\partial \alpha}{\partial t} + c_\alpha \frac{\partial \alpha}{\partial \hat{x}} = \frac{\partial^2 \alpha}{\partial \hat{x}^2} - \alpha\beta, \quad \frac{\partial \beta}{\partial t} + c_\beta \frac{\partial \beta}{\partial \hat{x}} = D \frac{\partial^2 \beta}{\partial \hat{x}^2} + \alpha\beta, \quad (5.7)$$

for $-\infty < \hat{x} < \infty$ and $t > 0$, where $D = \frac{D_B}{D_A}$, $c_\alpha = \left\{\frac{1}{ka_0 D_A}\right\}^{\frac{1}{2}} \bar{c}_\alpha$, $c_\beta = \left\{\frac{1}{ka_0 D_A}\right\}^{\frac{1}{2}} \bar{c}_\beta$ and

$$\alpha(\hat{x},0) = 1 \quad -\infty < \hat{x} < \infty, \quad \beta(\hat{x},0) = \begin{cases} \beta_0 g(\hat{x}), & |\hat{x}| \leq \sigma, \\ 0, & |\hat{x}| > \sigma, \end{cases} \tag{5.8}$$

$$\left. \begin{array}{l} \alpha(\hat{x},t) \to 1 \\ \beta(\hat{x},t) \to 0 \end{array} \right\} \quad as \ |\hat{x}| \to \infty, \ t \geq 0. \tag{5.9}$$

The dimensionless parameter $\beta_0 = \frac{b_0}{a_0}$ provides a measure of the maximum concentration of the initial input of the autocatalyst, while the dimensionless parameter $D$ measures the rate of diffusion of the autocatalyst B, relative to that of the reactant A. In chemical systems involving reactants of similar molecular mass, it is a reasonable approximation to assume that the diffusion rates $D_A$ and $D_B$ are equal, which gives $D = 1$. This simplifying assumption has been used in what follows. Equations (5.7) can now be simplified by introducing the translating spatial coordinate

$$x = \hat{x} - c_\alpha t, \tag{5.10}$$

in terms of which equations (5.7) reduce to

$$\alpha_t = \alpha_{xx} - \alpha\beta, \quad \beta_t + c\beta_x = \beta_{xx} + \alpha\beta, \tag{5.11}$$

where $-\infty < x < \infty$ and $t > 0$. Here $c = c_\beta - c_\alpha$, the difference in convection speeds of the ions of B to those of A. Initial and boundary conditions (5.8) and (5.9) are in terms of (5.10) are rewritten as

$$\alpha(x,0) = 1, \quad -\infty < x < \infty \quad \beta(x,0) = \begin{cases} \beta_0 g(x), & |x| \leq \sigma, \\ 0, & |x| > \sigma, \end{cases} \tag{5.12}$$

$$\left. \begin{array}{l} \alpha(x,t) \to 1 \\ \beta(x,t) \to 0 \end{array} \right\} \quad as \ |x| \to \infty, \ t \geq 0. \tag{5.13}$$

In this chapter we investigate the initial value problem (5.11)-(5.13) (which we shall from now on refer to as **IVP**), with particular emphasis on how the evolution is affected by the parameter $c$, measuring the effect of the external electric field. We anticipate that the evolution may involve the propagation of permanent form travelling waves away from the initial reaction zone, which is the case when $c = 0$ (see Merkin and Needham [44]), when travelling waves of speed 2 propagate both to the left and right. However, when $c \neq 0$, the symmetry of equations (5.11) about $x = 0$ is broken, and we must expect that the propagation of waves to the left and right will differ in nature. Thus, an important preliminary to the study of the initial value problem **IVP** is an investigation of the permanent form travelling wave solutions of equations (5.11), which may be generated from **IVP**. We consider left and right travelling waves separately. We make the following definitions

**Definition 5.1.** *A right permanent form travelling wave solution (RPTW) of equations (5.11) is a non-trivial, non-negative solution which depends only on*

*the single variable $z = x - \gamma(t)$, where $\gamma(t)$ is the position of the wavefront, and satisfies the conditions $\alpha \to 1, \beta \to 0$ as $z \to \infty$, and $\alpha \to \alpha_{-\infty}, \beta \to \beta_{-\infty}$ as $z \to -\infty$, where $\alpha_{-\infty}$ and $\beta_{-\infty}$ are the uniform, non-negative concentrations behind the wavefront.*

**Definition 5.2.** *A left permanent form travelling wave solution (LPTW) of equations (5.11) is a non-trivial, non-negative solution which depends only on the single variable $z = x - \gamma(t)$, where $\gamma(t)$ is the position of the wavefront, and satisfies the conditions $\alpha \to \alpha_{\infty}, \beta \to \beta_{\infty}$ as $z \to \infty$, and $\alpha \to 1, \beta \to 0$ as $z \to -\infty$, where $\alpha_{\infty}$ and $\beta_{\infty}$ are the uniform, non-negative concentrations behind the wavefront.*

The equations which govern both the right and left permanent form travelling waves (RPTWs, LPTWs) are obtained by looking for a solution of equations (5.11) in the form $\alpha \equiv \alpha(z)$ and $\beta \equiv \beta(z)$, which become

$$\alpha_{zz} + v\alpha_z - \alpha\beta = 0, \quad \beta_{zz} + v^*\beta_z + \alpha\beta = 0, \quad -\infty < z < \infty, \qquad (5.14)$$

where $v(t) = \frac{d\gamma}{dt}$ and $(v - c) = v^*$. However, since $\alpha$ and $\beta$ are functions of $z$ alone, equations (5.14) show that the wavefront propagation speed $v$ must be a constant. The boundary conditions to be considered with (5.14) are

$$\alpha(z), \beta(z) \ge 0 \quad -\infty < z < \infty, \qquad (5.15)$$

$$\left.\begin{array}{ll} \alpha(z) \to 1, & \beta(z) \to 0 \quad \text{as } z \to \infty \\ \alpha(z) \to \alpha_{-\infty}, & \beta(z) \to \beta_{-\infty} \text{ as } z \to -\infty \end{array}\right\} \text{ RPTW} \qquad (5.16)$$

$$\left.\begin{array}{ll} \alpha(z) \to \alpha_{\infty}, & \beta(z) \to \beta_{\infty} \text{ as } z \to \infty \\ \alpha(z) \to 1, & \beta(z) \to 0 \quad \text{as } z \to -\infty \end{array}\right\} \text{ LPTW.} \qquad (5.17)$$

We can now regard (5.14) with (5.15) and either of (5.16) or (5.17) as a nonlinear eigenvalue problem for the propagation speed $v$ as a function of the parameter $c$. We will determine at which points in the $(c, v)$ plane that a solution to either (5.14),(5.15) and (5.16) or (5.14),(5.15) and (5.17) exists.

## 5.1 General Properties of Travelling Wave Solutions

In this section we prove some elementary results, which concern both RPTWs and LPTWs. First we observe that we can reduce our task by only considering RPTWs. This arises due to the symmetry properties of the boundary value problems for RPTWs and LPTWs respectively. We denote the boundary value problems for RPTWs and LPTWs by RP[c, v] and LP[c, v], respectively. Thus we have:

**Theorem 5.3. (symmetry)** $\alpha = F(z; v, c), \beta = G(z; v, c), -\infty < z < \infty$ *is a solution to RP[c,v] if and only if $\alpha = F(-z; v, c), \beta = G(-z; v, c), -\infty < z < \infty$ is a solution to LP[-c,-v].*

*Proof.* Via direct substitution.                                              □

Interpreted in the $(c, v)$ plane, Theorem 5.3 establishes that an RPTW exists at point $(c, v)$ if and only if an LPTW exists at point $(-c, -v)$. Henceforth we need only consider RPTWs. Corresponding LPTWs are located via Theorem 5.3. We have the following propositions:

**Proposition 5.4.** *Let* $\alpha(z)$, $\beta(z)$ *be an RPTW, then* $\alpha(z) \not\equiv 1$ *for all* $-\infty < z < \infty$.

*Proof.* Let $\alpha(z), \beta(z)$ be an RPTW and suppose that $\alpha(z) \equiv 1$. Hence $\alpha_z = \alpha_{zz} = 0$, and from equation (5.14(a)), we obtain $\alpha(z)\beta(z) = 0$. Since $\alpha(z) \equiv 1$, we must have $\beta(z) \equiv 0$. However, this is the trivial solution, and is not therefore an RPTW, so $\alpha(z) \not\equiv 1$ in an RPTW.                    □

**Proposition 5.5.** *Let* $\alpha(z), \beta(z)$ *be an RPTW, then* $\beta(z) \not\equiv 0$ *for all* $-\infty < z < \infty$.

*Proof.* Let $\alpha(z), \beta(z)$ be an RPTW and suppose that $\beta(z) \equiv 0$. Firstly suppose that $v \neq 0$. From (5.14(a)), we then have

$$\alpha_{zz} + v\alpha_z = 0. \tag{5.18}$$

Hence

$$\alpha = A + Be^{-vz}, \tag{5.19}$$

for some constants A,B. Consider the case where $v > 0$. After applying the conditions as $z \to \pm\infty$, we obtain $A = 1, B = 0$ and hence $\alpha(z) \equiv 1$, which contradicts Proposition 5.4. Now consider the case where $v < 0$. After applying the conditions as $z \to \pm\infty$, we obtain $A = 1, B = 0$ and again $\alpha(z) \equiv 1$, which contradicts Proposition 5.4. Finally, suppose that $v = 0$. From equation (5.14(a)) we now have

$$\alpha_{zz} = 0, \tag{5.20}$$

and hence $\alpha(z) = A + Bz$. Applying conditions as $z \to \pm\infty$ again gives $A = 1, B = 0$ and so $\alpha(z) \equiv 1$, again contradicting Proposition 5.4. We conclude that an RPTW cannot have $\beta(z) \equiv 0$.                          □

**Proposition 5.6.** *Let* $\alpha(z), \beta(z)$ *be an RPTW, then* $\alpha(z), \beta(z) > 0$ *for all* $-\infty < z < \infty$.

*Proof.* Let $\alpha(z), \beta(z)$ be an RPTW and suppose that there exists a $z_0$ such that $\alpha(z_0) = 0$. Then, since $\alpha(z)$ is non-negative, we have that $\alpha_z(z_0) = 0$. Moreover, for given $\beta(z)$, equation (5.14(a)) can be regarded as a second order, linear, ordinary differential equation for $\alpha(z)$, which has no singular points for any $-\infty < z < \infty$. Thus any initial value problem for $\alpha(z)$ has a unique solution in $-\infty < z < \infty$ (see, for example, Burkhill [11]). Equation (5.14(a)) together with the above homogeneous conditions at $z_0$ form an initial value problem for $\alpha(z)$, which has the unique solution $\alpha(z) \equiv 0$ for $-\infty < z < \infty$. However, we must have $\alpha(z) \to 1$ as $z \to \infty$ for an RPTW. Hence, no such $z_0$ exists and $\alpha(z) > 0$ for all $-\infty < z < \infty$. Following similar arguments, we readily establish the equivalent result for $\beta(z)$. $\qquad\square$

**Proposition 5.7.** *Let* $\alpha(z), \beta(z)$ *be an RPTW, then* $\alpha_{-\infty} = 0$ *and/or* $\beta_{-\infty} = 0$.

*Proof.* From equation (5.14(a)), and letting $z \to -\infty$ we obtain $\alpha_{-\infty}\beta_{-\infty} = 0$. Hence $\alpha_{-\infty} = 0$ and/or $\beta_{-\infty} = 0$.

*Remark 5.8.* Similarly, for a LPTW, $\alpha_{\infty} = 0$ and/or $\beta_{\infty} = 0$, via Theorem 5.3.

**Proposition 5.9.** *Let* $\alpha(z), \beta(z)$ *be an RPTW, then* $\beta_{-\infty} > 0$ *and* $\alpha_{-\infty} = 0$.

*Proof.* After integrating equation (5.14(b)) with respect to $z$ on the range $-\infty < z < \infty$ we obtain,

$$\int_{-\infty}^{\infty} \alpha\beta dz = - \{\beta_z + v^*\beta\}_{-\infty}^{\infty} = v^*\beta_{-\infty}. \qquad (5.21)$$

Thus via Proposition 5.6, we must have $v^*\beta_{-\infty} > 0$. Hence $\beta_{-\infty} \neq 0$ and since $\beta_{-\infty} \geq 0$ we obtain $\beta_{-\infty} > 0$. It then follows via Proposition 5.7 that $\alpha_{-\infty} = 0$. $\qquad\square$

*Remark 5.10.* Similarly, for an LPTW, $\beta_{\infty} > 0$ and $\alpha_{\infty} = 0$ via Theorem 5.3.

**Proposition 5.11.** *Let* $\alpha(z), \beta(z)$ *be an RPTW, then* $v^* > 0$.

*Proof.* From Proposition 5.9 we have $v^*\beta_{-\infty} > 0$ and $\beta_{-\infty} > 0$. Hence $v^* > 0$. $\qquad\square$

*Remark 5.12.* Conversely, for an LPTW, $v^* < 0$, via Theorem 5.3.

**Proposition 5.13.** *Let* $\alpha(z), \beta(z)$ *be an RPTW, then* $v > 0$.

*Proof.* Integrate equation (5.14(a)) with respect to $z$ on the range $-\infty < z < \infty$ to obtain

$$\int_{-\infty}^{\infty} \alpha\beta dz = \{\alpha_z + v\alpha\}_{-\infty}^{\infty} = v(1 - \alpha_{-\infty}). \qquad (5.22)$$

Via Proposition 5.6, we obtain that $v(1 - \alpha_{-\infty}) > 0$, and since $\alpha_{-\infty} = 0$ (Proposition 5.9) we obtain $v > 0$. $\qquad\square$

*Remark 5.14.* Conversely, for an LPTW, $v < 0$, via Theorem 5.3.

**Proposition 5.15.** *Let* $\alpha(z), \beta(z)$ *be an RPTW, then* $\beta_{-\infty} = \frac{v}{v^*}$.

*Proof.* On addition, equations (5.14(a)) and (5.14(b)) may be integrated once to give

$$(\alpha + \beta)_z + (v\alpha + v^*\beta) = \text{constant}, \quad -\infty < z < \infty. \tag{5.23}$$

After application of the conditions on $\alpha(z)$ and $\beta(z)$ as $z \to \infty$, we obtain

$$(\alpha + \beta)_z + (v\alpha + v^*\beta) = v, \quad -\infty < z < \infty. \tag{5.24}$$

Now let $z \to -\infty$ using $\alpha_{-\infty} = 0$ (Proposition 5.9) we obtain

$$v^*\beta_{-\infty} = v. \tag{5.25}$$

Since $v^* > 0$ via Proposition 5.11, $\beta_{-\infty} = \frac{v}{v^*}$. □

**Remark 5.16.** Similarly, for an LPTW, $\beta_\infty = \frac{v}{v^*}$, via Theorem 5.3.

**Proposition 5.17.** *Let $\alpha(z), \beta(z)$ be an RPTW, then $\alpha(z)$ is strictly monotone increasing in $z$ while $\beta(z)$ is strictly monotone decreasing in $z$, with $0 < \alpha < 1$ and $0 < \beta < \frac{v}{v^*}$ for $-\infty < z < \infty$.*

*Proof.* Using an integrating factor on equation (5.14(a)) gives

$$\frac{d}{dz}\{e^{vz}\alpha_z(z)\} = e^{vz}\alpha\beta, \quad -\infty < z < \infty. \tag{5.26}$$

Using Proposition 5.6 and integrating with respect to $s$ over the range $-\infty < s < z$ gives

$$\alpha_z e^{vz} = \int_{-\infty}^{z} e^{vs}\alpha(s)\beta(s)ds > 0, \quad -\infty < z < \infty. \tag{5.27}$$

Since $e^{vz} > 0$, then $\alpha_z > 0$ for all $-\infty < z < \infty$. A similar argument establishes the analogous result for $\beta(z)$

**Remark 5.18.** Conversely, for a LPTW, $\alpha(z)$ is strictly monotone decreasing in $z$ while $\beta(z)$ is strictly monotone increasing in $z$, with $0 < \alpha < 1$ and $0 < \beta < \frac{v}{v^*}$ for $-\infty < z < \infty$.

**Proposition 5.19.** *Let $\alpha(z), \beta(z)$ be an RPTW, then $v > \max[0, c]$.*

*Proof.* From Propositions 5.11 and 5.13 we have $v > 0$ and $v > c$. □

**Remark 5.20.** Conversely, for a LPTW, $v < \min[0, c]$, via Theorem 5.3.

**Proposition 5.21.** *Let $\alpha(z), \beta(z)$ be an RPTW, then $(\alpha + \beta)(z) \gtrless 1$ for $-\infty < z < \infty$ when $c \gtrless 0$.*

*Proof.* Rearrangement of equation (5.24) gives

$$(\alpha + \beta)_z + v(\alpha + \beta) = v + (v - v^*)\beta = v + c\beta, \quad -\infty < z < \infty. \quad (5.28)$$

Using an integrating factor on the above equation results in

$$(\alpha + \beta)(z) = e^{-vz} \int_{-\infty}^{z} e^{vs}[v + c\beta]ds, \quad -\infty < z < \infty. \quad (5.29)$$

Considering each case in turn:
(i) $c > 0$

$$(\alpha + \beta)(z) > e^{-vz} \int_{-\infty}^{z} e^{vs} v ds \equiv 1. \quad (5.30)$$

(ii) $c < 0$

$$(\alpha + \beta)(z) < e^{-vz} \int_{-\infty}^{z} e^{vs} v ds \equiv 1. \quad (5.31)$$

(iii) $c = 0$

$$(\alpha + \beta)(z) = e^{-vz} \int_{-\infty}^{z} e^{vs} v ds \equiv 1. \quad (5.32)$$

Hence $(\alpha + \beta)(z) \gtrless 1$ for $-\infty < z < \infty$ when $c \gtrless 0$, as required.    □

**Proposition 5.22.** *Let* $\alpha(z), \beta(z)$ *be an RPTW, then* $(\alpha+\beta)(z) \gtrless 1 + \frac{c}{(v-c)}$ *for* $-\infty < z < \infty$ *when* $c \lessgtr 0$.

*Proof.* Via (5.29) and Propositions 5.15 and 5.17

$$\begin{array}{l} (i)\ c > 0\ (\alpha + \beta)(z) < e^{-vz} \int_{-\infty}^{z} e^{vs}[v + cv/v^*]ds = 1 + \frac{c}{(v-c)}, \\ (ii)\ c = 0\ (\alpha + \beta)(z) = e^{-vz} \int_{-\infty}^{z} e^{vs} v ds = 1, \\ (iii)\ c < 0\ (\alpha + \beta)(z) > e^{-vz} \int_{-\infty}^{z} e^{vs}[v + cv/v^*]ds = 1 + \frac{c}{(v-c)}. \end{array} \quad (5.33)$$

Hence $(\alpha + \beta)(z) \gtrless 1 + \frac{c}{(v-c)}$ for $-\infty < z < \infty$ when $c \lessgtr 0$, as required.    □

Hence we have, for an RPTW,
$c > 0$

$$1 < (\alpha + \beta)(z) < 1 + \frac{c}{(v - c)} \quad \text{for}\ -\infty < z < \infty, \quad (5.34)$$

$c < 0$

$$1 + \frac{c}{(v - c)} < (\alpha + \beta)(z) < 1 \quad \text{for}\ -\infty < z < \infty, \quad (5.35)$$

$c = 0$

$$(\alpha + \beta)(z) \equiv 1 \quad \text{for}\ -\infty < z < \infty. \quad (5.36)$$

*Remark 5.23.* Similarly for a LPTW we have

$c > 0$

$$1 + \frac{c}{(v - c)} < (\alpha + \beta)(z) < 1 \text{ for } -\infty < z < \infty, \qquad (5.37)$$

$c < 0$

$$1 < (\alpha + \beta)(z) < 1 + \frac{c}{(v - c)} \text{ for } -\infty < z < \infty, \qquad (5.38)$$

$c = 0$

$$(\alpha + \beta)(z) \equiv 1 \text{ for } -\infty < z < \infty. \qquad (5.39)$$

### 5.1.1 Summary

Summarizing the contents of this section, for an RPTW we require a solution of

$$\alpha_{zz} + v\alpha_z - \alpha\beta = 0, \quad \beta_{zz} + v^*\beta_z + \alpha\beta = 0, \quad -\infty < z < \infty, \qquad (5.40)$$

subject to the conditions

$$\begin{aligned}
\alpha(z) &\to 0, \ \beta(z) \to \tfrac{v}{v^*} \text{ as } z \to -\infty, \\
\alpha(z) &\to 1, \ \beta(z) \to 0 \quad \text{as } z \to \infty,
\end{aligned} \qquad (5.41)$$

where $v > 0, v^* > 0$. Any RPTW will be strictly monotone increasing in $\alpha$ and strictly monotone decreasing in $\beta$, with $0 < \alpha < 1$ and $0 < \beta < \frac{v}{v^*}$ for $-\infty < z < \infty$. It will also have

$$1 + c/(v - c) \stackrel{>}{\underset{<}{=}} (\alpha + \beta)(z) \stackrel{>}{\underset{<}{=}} 1 \quad \text{for} \quad -\infty < z < \infty \text{ when } c \stackrel{>}{\underset{<}{=}} 0, \quad (5.42)$$

and

$$v > \max[0, c]. \qquad (5.43)$$

Similar conditions hold for an LPTW via Theorem 5.3.

In the next section we consider, at each point in the $(c, v)$ plane, the possible existence and uniqueness of an RPTW and/or an LPTW. Clearly we can restrict our attention to the case of RPTWs, with the corresponding results for LPTWs following via the symmetry Theorem 5.3.

## 5.2 The Existence of Travelling Wave Solutions

We now consider the existence of solutions to $RP[c, v]$ and $LP[c, v]$. As mentioned earlier we need only consider $RP[c, v]$, with the corresponding results for $LP[c, v]$ following via Theorem 5.3.

## 5.2.1 Equivalent Dynamical System

To proceed it is convenient to write equations (5.14) as the equivalent three dimensional dynamical system

$$
\begin{aligned}
\alpha_z &= v - v\alpha - v^*\beta - w, \\
\beta_z &= w \qquad\qquad\qquad -\infty < z < \infty, \\
w_z &= -\alpha\beta - v^*w,
\end{aligned}
\tag{5.44}
$$

which we obtain from (5.14) after performing a first integral of (5.14(a)) + (5.14(b)) and applying conditions (5.16). We will analyze this dynamical system in the $(\alpha, \beta, w)$ phase space. For any fixed pair of parameters $(c, v)$ an RPTW solution of $RP[c, v]$ corresponds to the existence of a directed phase path of the dynamical system (5.44) connecting (via Proposition 5.15) the point $(\alpha, \beta, w) = (0, v/v^*0)$ to the point $(\alpha, \beta, w) = (1, 0, 0)$, which remains in the region defined by $\alpha(z), \beta(z) \geq 0$. From Propositions 5.11 and 5.13 we need only consider those points in the $(c, v)$ plane where $v > 0$ and $v^* > 0$. That is

$$
v \begin{cases} > 0, c < 0, \\ > c, c \geq 0. \end{cases}
\tag{5.45}
$$

RPTWs cannot exist at other points in the $(c, v)$ plane.

Under conditions (5.45) it is readily established that (5.44) has just two finite equilibrium points at $\underline{e}_1 = (0, v/v^*, 0)$ and $\underline{e}_2 = (1, 0, 0)$. Thus for any RPTW, we require a directed phase path in $(\alpha, \beta, w)$ space, which connects $\underline{e}_1$ to $\underline{e}_2$ and remains in the region $\alpha, \beta \geq 0$. Let us now fix the parameters $(c, v)$ in the region satisfied by inequalities (5.45). We begin by examining the local behaviour in the neighbourhood of the two finite equilibrium points. Linearization of equation (5.44) about the point $\underline{e}_1$ shows that it is a simple equilibrium point with a two-dimensional stable manifold and a one-dimensional unstable manifold. The eigenvalues and associated eigenvectors are

$$
\begin{aligned}
\lambda_1 &= -v^*, & e_{\lambda_1} &= (0, -1, -\lambda_1)^T, \\
\lambda_2 &= -\tfrac{1}{2}\left\{\sqrt{v^2 + \tfrac{4v}{v^*}} + v\right\}, & e_{\lambda_2} &= (\lambda_2\tfrac{v^*}{v}(v^* + \lambda_2), -1, -\lambda_2)^T, \\
\lambda_3 &= \tfrac{1}{2}\left\{\sqrt{v^2 + \tfrac{4v}{v^*}} - v\right\}, & e_{\lambda_3} &= (\lambda_3\tfrac{v^*}{v}(v^* + \lambda_3), -1, -\lambda_3)^T.
\end{aligned}
\tag{5.46}
$$

Therefore the only integral path which satisfies condition (5.41(a)) and has $\alpha, \beta \geq 0$, as $z \to -\infty$ is the unstable manifold of the point $\underline{e}_1$ in $\alpha, \beta \geq 0$, which we label $S^-$. Linearization of equation (5.44) about the other equilibrium point $\underline{e}_2$ shows that it is a simple, stable equilibrium point with eigenvalues and associated eigenvectors given by

$$
\begin{aligned}
\mu_1 &= -v^*, & e_{\mu_1} &= (1, 0, 0)^T, \\
\mu_2 &= -\tfrac{1}{2}\left\{\sqrt{(v^*)^2 - 4} + v^*\right\}, & e_{\mu_2} &= ((v^* + \mu_2), -(v + \mu_2), -\mu_2(v + \mu_2))^T, \\
\mu_3 &= \tfrac{1}{2}\left\{\sqrt{(v^*)^2 - 4} - v^*\right\}, & e_{\mu_3} &= ((v^* + \mu_3), -(v + \mu_3), -\mu_3(v + \mu_3))^T.
\end{aligned}
\tag{5.47}
$$

Thus, when $0 < v^* < 2$ the equilibrium point $\underline{e}_2$ is a stable spiral, whilst for $v^* \geq 2$, the equilibrium point $\underline{e}_2$ is a stable node.

The existence of an RPTW requires the unstable manifold at $\underline{e}_1$, which we have labelled as $\mathcal{S}^-$, to connect with the stable equilibrium point $\underline{e}_2$, whilst remaining in the region $\alpha, \beta \geq 0$. This is impossible when $\underline{e}_2$ is a spiral (all phase paths entering $\underline{e}_2$ oscillate about $\beta = 0$). Thus we immediately have the following necessary condition for the existence of an RPTW:

**Proposition 5.24.** *A necessary condition for the existence of an RPTW is $v > 0$ and $v^* \geq 2$.*

This proposition establishes a necessary condition for the existence of an RPTW. The following proposition establishes the sufficiency of this condition:

**Proposition 5.25.** *A unique RPTW exists for each $v > 0$ and $v^* \geq 2$.*

*Proof.* Fix $(c, v)$ so that $v > 0$ and $v^* \geq 2$. Next define the region $R$ by

$$R = \left\{ (\alpha, \beta, w) : 0 \leq \alpha \leq 1, 0 \leq \beta \leq \frac{v}{v^*}, w \geq -\beta \right\}, \qquad (5.48)$$

as shown in Figure 5.1. An examination of equations (5.44) for $v > 0$ and

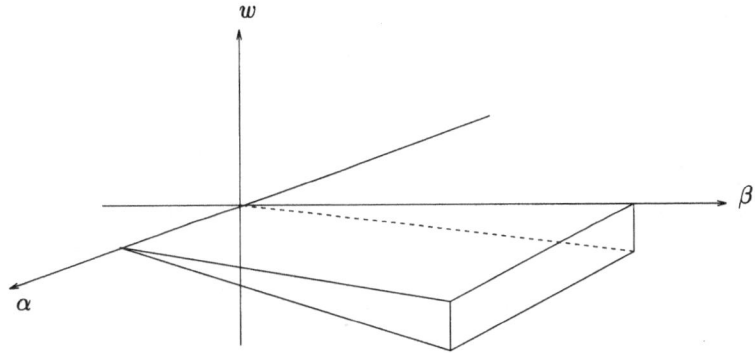

**Fig. 5.1.** Region $R$.

$v^* \geq 2$ shows that the region $R$ is a positively invariant region. Now we observe, via (5.46), that the unstable manifold $\mathcal{S}^-$ at $\underline{e}_1$ enters $R$ for any $v > 0$ and $v^* \geq 2$, and thus, as $R$ is positively invariant, it cannot subsequently leave $R$ with increasing $z$. Moreover, denoting the integral path $\mathcal{S}^-$ by $(\alpha_s(z), \beta_s(z), w_s(z))$ we see from (5.44(b)) that $\beta_s(z)$ is monotone decreasing with $z$, and bounded below by zero (it must remain in $R$). Thus $\beta_s$ has a finite limit as $z \to \infty$, say $\beta_s(z) \to \beta_\infty$ as $z \to \infty$, with

$$0 \leq \beta_\infty < \frac{v}{v^*}. \qquad (5.49)$$

It then follows from (5.44(b)) that $w_s(z) \to 0$ as $z \to \infty$, and from (5.44(c)) that $\alpha_s(z)\beta_s(z) \to 0$ as $z \to \infty$. There are two possibilities. Suppose $\alpha_s(z) \to 0$ as $z \to \infty$, then (5.44(a)) requires $\beta_\infty = \frac{v}{v^*}$, which contradicts (5.49). We conclude that $\beta_s(z) \to 0$ as $z \to \infty$, and $\beta_\infty = 0$. Equation (5.44(a)) then requires, as $z \to \infty$,

$$(\alpha_s)_z \sim v(1 - \alpha_s), \quad \alpha_s \in \mathbb{R} \tag{5.50}$$

and so, $\alpha_s(z) \to 1$ as $z \to \infty$. Hence $(\alpha_s(z), \beta_s(z), w_s(z))$ leaves $\underline{e}_1$ and remains in $R$, with $(\alpha_s(z), \beta_s(z), w_s(z)) \to (1, 0, 0) = \underline{e}_2$ as $z \to \infty$. Thus, for each $v > 0, v^* \geq 2$ we have established that $S^-$ connects $\underline{e}_1$ to $\underline{e}_2$, whilst remaining in $R$, and this is the only connection. The proof is complete. □

Propositions 5.24 and 5.25 may be combined to give the following.

**Theorem 5.26.** (i) $c > -2$: A unique RPTW exists for each $v \geq v_m(c) = 2 + c$, no RPTW exists for $v < v_m(c)$, (ii) $c \leq -2$: A unique RPTW exists for each $v > 0$, no RPTW exists for $v \leq 0$.

**Theorem 5.27.** (i) $c < 2$: A unique LPTW exists for each $v \leq v_m(c) = -2 + c$, no LPTW exists for $v > v_m(c)$, (ii) $c \geq 2$: A unique LPTW exists for each $v < 0$, no LPTW exists for $v \geq 0$.

In the $(c, v)$ plane, the regions of existence of LPTWs and RPTWs are illustrated in Figure 5.2. Note the symmetry of the figure, in accordance with Theorem 5.3.

*Remark 5.28.* When $-2 < c < 2$ there exists an RPTW of minimum speed (with speed $v_+ = 2 + c > 0$) together with an LPTW of maximum speed (with speed $v_- = -2 + c < 0$). However, for $c \geq 2$ there continues to be an RPTW of minimum speed ($v_+ = 2 + c > 0$) but there is no longer a maximum speed LPTW (LPTW requiring $v < 0$). At $v = 0$, although there is still a connection in the phase space, the connection exists entirely of equilibrium points. Similarly, for $c \leq -2$, there continues to be an LPTW of maximum speed ($v_- = -2 + c < 0$) but there is no longer a minimum speed RPTW (RPTW requiring $v > 0$).

## 5.3 The Initial Value Problem (IVP)

In this section we consider properties of the full initial-value problem (**IVP**). In particular, we examine whether or not the large-$t$ evolution of the solution to **IVP** involves the propagation of RPTW and/or LPTW waves. First we observe the following symmetry result:

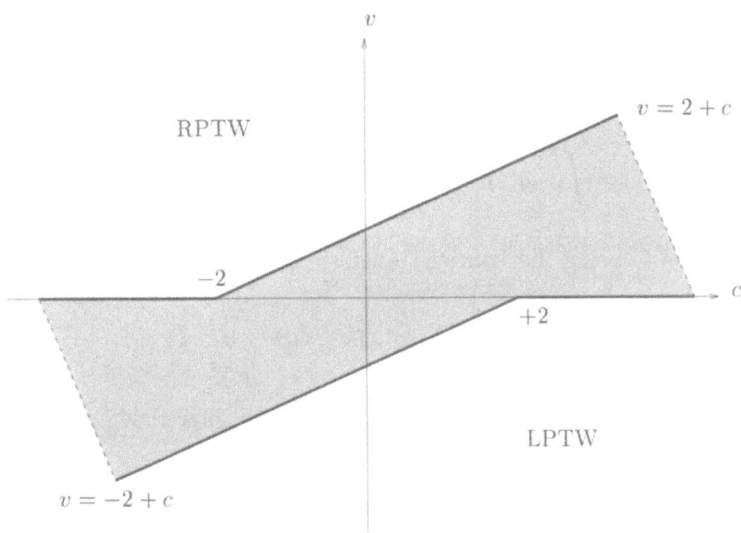

**Fig. 5.2.** Region of existence of LPTWs and RPTWs.

**Theorem 5.29.** *(Symmetry)*

$$\alpha = F(x,t), \quad \beta = G(x,t), \quad -\infty < x < \infty, \ t \ge 0, \tag{5.51}$$

*solves* **IVP** *with* $c = \bar{c}$ *if and only if*

$$\alpha = F(-x,t), \quad \beta = G(-x,t), \quad -\infty < x < \infty, \ t \ge 0, \tag{5.52}$$

*solves* **IVP** *with* $c = -\bar{c}$.

*Proof.* Via direct substitution. □

Thus we need only consider **IVP** with $c \ge 0$. A simple reversal of the $x$-axis leads to the corresponding solution when $c < 0$. Let $\alpha, \beta : \bar{D}_T \to \mathbb{R}$ be a solution to **IVP** on $\bar{D}_T$, where

$$D_T = \{(x,t) : -\infty < x < \infty, \ 0 < t \le T\}, \tag{5.53}$$

and $\bar{D}_T$ is the closure of $D_T$. It is straightforward to establish, via the maximum principle and the comparison theorem, that

$$\begin{aligned}
0 &\le \alpha(x,t) \le 1 \quad \text{for all } (x,t) \in \bar{D}_T, \\
0 &\le \beta(x,t) \le \beta_0 e^t \text{ for all } (x,t) \in \bar{D}_T,
\end{aligned} \tag{5.54}$$

for any $T > 0$, (see for example Fife, [14]). The *a priori* bounds (5.54) then guarantee the global existence of a unique solution to **IVP** (see for example Smoller [66]). A further application of the comparison theorem establishes that

$$0 \leq \beta(x,t) \leq \beta_0 e^t D(x - ct, t), \tag{5.55}$$

in $\bar{D}_T$ for any $T > 0$. Here $D(X,t)$ is the solution of the corresponding pure diffusion problem, and is given by

$$D(X,t) = \int_{-\infty}^{\infty} \hat{g}(s) e^{isX - s^2 t} ds, \quad -\infty < X < \infty, \ t \geq 0, \tag{5.56}$$

with $\hat{g} : \mathbb{C} \to \mathbb{C}$ being the Fourier transform of $g : [-\sigma, \sigma] \to \mathbb{R}$, namely,

$$\hat{g}(s) = \frac{1}{2\pi} \int_{-\sigma}^{\sigma} g(\lambda) e^{-i\lambda s} d\lambda, \quad s \in \mathbb{C}. \tag{5.57}$$

On using the method of steepest descent, we may show that

$$D(X,t) \sim \frac{\sqrt{\pi}}{t^{\frac{1}{2}}} \hat{g}(i\hat{v}/2) e^{-t\hat{v}^2/4}, \quad as \ t \to \infty, \tag{5.58}$$

uniformly for $-\infty < \hat{v} < \infty$, where $\hat{v} = \frac{X}{t}$. The estimate (5.58) together with (5.57) then establishes that for t sufficiently large

$$0 \leq \beta(x,t) \leq \frac{2\beta_0\sqrt{\pi}}{\hat{v}t^{\frac{1}{2}}} \sinh \hat{v} \ e^{[4-\hat{v}^2]t/4}, \quad -\infty < \hat{v} < \infty, \tag{5.59}$$

where now $\hat{v} = (x - ct)/t$. We may now conclude from above that, for any $\delta > 0$,

$$\beta(x,t) \to 0 \quad as \ t \to \infty, \tag{5.60}$$

(exponentially in t) uniformly on $\hat{v} \in [2 + \delta, \infty)$ and on $\hat{v} \in (-\infty, -2 - \delta]$. Thus

$$\beta((\hat{v} + c)t, t) \to 0 \ as \ t \to \infty, \tag{5.61}$$

(exponentially in $t$) uniformly on $\hat{v} \in (-\infty, -2 - \delta] \bigcup [2 + \delta, \infty)$. This result allows us the following observation: it is possible for an RPTW to develop in **IVP** for any $c \geq 0$. However it is only possible for an LPTW to develop in **IVP** when $0 \leq c < 2$. Moreover, when an RPTW develops it must have propagation speed $v_R = 2 + c$ $(c \geq 0)$, whilst when an LPTW develops it must have propagation speed $v_L = -2 + c$ $(0 \leq c < 2)$.

## 5.4 Asymptotic Solution to IVP as $t \to \infty$

In this section we develop the asymptotic structure to **IVP** (with $c \geq 0$) as $t \to \infty$ with particular attention to travelling wave formation. We must first begin by examining the asymptotic structure of the solution to **IVP** as $t \to 0$.

### 5.4.1 Asymptotic Solution as $t \to 0$

We first consider region **I**, in which $|x| \leq \sigma - O(1)$ and $\alpha(x,t), \beta(x,t) = O(1)$ as $t \to 0$. Since $\alpha(x,0), \beta(x,0) > 0$ are analytic in region **I**, we expand $\alpha(x,t)$ and $\beta(x,t)$ as regular power series in $t$. After substitution into equations (5.11(a)) and (5.11(b)), equating powers of $t$ to zero, and applying initial conditions (5.12), we obtain

$$\alpha(\dot{x},t) = 1 - t[\beta_0 g(x)] + O(t^2), \tag{5.62}$$

$$\beta(x,t) = \beta_0 g(x) + t[\beta_0 g''(x) + \beta_0 g(x) - c\beta_0 g'(x)] + O(t^2), \tag{5.63}$$

as $t \to 0$ with $|x| \leq \sigma - O(1)$. Now as $x \to \sigma^-$, expansions (5.62) and (5.63) become

$$\alpha(x,t) \sim 1 - t[\beta_0 g_\sigma(\sigma - x)^r + \ldots] + \ldots, \tag{5.64}$$

$$\beta(x,t) \sim [\beta_0 g_\sigma(\sigma - x)^r + \cdots] + t[\beta_0 g_\sigma r(r-1)(\sigma - x)^{r-2} + \ldots] + \ldots \tag{5.65}$$

as $t \to 0$, and a nonuniformity develops first in expansion (5.63) when $x = \sigma - O(t^{1/2})$, when we observe, via (5.64) and (5.65), that $\alpha(x,t) = 1 - O\left(t^{1+\frac{r}{2}}\right)$ and $\beta(x,t) = O\left(t^{\frac{r}{2}}\right)$ respectively. We must therefore introduce a further region, which we refer to as region **II(a)**, in which $x = \sigma \pm O(t^{1/2})$ as $t \to 0$. As $x \to -\sigma^+$, expansions (5.62) and (5.63) become

$$\alpha \sim 1 - t[\beta_0 g_\sigma(\sigma + x)^r + \ldots] + \ldots, \tag{5.66}$$

$$\beta \sim [\beta_0 g_\sigma(\sigma + x)^r + \cdots] + t[\beta_0 g_\sigma r(r-1)(\sigma + x)^{r-2} + \ldots] + \ldots \tag{5.67}$$

as $t \to 0$, and a nonuniformity develops first in expansion (5.63) when $x = -\sigma + O(t^{1/2})$, when we observe, via (5.66) and (5.67), that $\alpha(x,t) = 1 - O\left(t^{1+\frac{r}{2}}\right)$ and $\beta(x,t) = O\left(t^{\frac{r}{2}}\right)$ respectively. We must therefore introduce a further region, which we refer to as region **II(b)**, in which $x = -\sigma \pm O(t^{1/2})$ as $t \to 0$. We first consider the asymptotic structure of the solution of **IVP** as $t \to 0$ for $x \geq 0$. To examine region **II(a)**, we introduce the scaled coordinate $\eta = (x - \sigma)t^{-\frac{1}{2}}$ and look for asymptotic expansions of the form

$$\alpha(\eta,t) = 1 - t^{1+\frac{r}{2}}\hat{\alpha}_1(\eta) + o\left(t^{1+\frac{r}{2}}\right), \tag{5.68}$$

$$\beta(\eta,t) = t^{\frac{r}{2}}\hat{\beta}_1(\eta) + o\left(t^{\frac{r}{2}}\right) \tag{5.69}$$

as $t \to 0$ with $\eta = O(1)$. On substitution of expansions (5.68) and (5.69) into equations (5.11) (when written in terms of $\eta$ and $t$) we obtain at leading order

$$\hat{\alpha}_{1\eta\eta} + \frac{\eta}{2}\hat{\alpha}_{1\eta} - \left(1 + \frac{r}{2}\right)\hat{\alpha}_1 = -\hat{\beta}_1, \tag{5.70}$$

$$\hat{\beta}_{1\eta\eta} + \frac{\eta}{2}\hat{\beta}_{1\eta} - \frac{r}{2}\hat{\beta}_1 = 0, \tag{5.71}$$

where $-\infty < \eta < \infty$. Equations (5.70) and (5.71) are to be solved subject to matching with region **I** as $\eta \to -\infty$, and the initial condition (5.12) as $t \to 0$, that is,

$$\hat{\alpha}_1(\eta) \sim \beta_0 g_\sigma(-\eta)^r \quad \text{as} \quad \eta \to -\infty, \tag{5.72}$$

$$\hat{\beta}_1(\eta) \sim \beta_0 g_\sigma(-\eta)^r \quad \text{as} \quad \eta \to -\infty, \tag{5.73}$$

and

$$\hat{\alpha}_1(\eta), \hat{\beta}_1(\eta) \text{ remain bounded as } \eta \to \infty. \tag{5.74}$$

The solution to the boundary value problem (5.71), (5.73) and (5.74) is unique and readily obtained as

$$\hat{\beta}_1(\eta) = \begin{cases} \dfrac{\beta_0 g_\sigma r!}{(r/2)! k_1} A(\eta) \displaystyle\int_\eta^\infty \dfrac{e^{-\frac{s^2}{4}}}{A^2(s)} ds, & r \text{ even,} \\[4ex] \dfrac{\beta_0 g_\sigma r!}{(r/2-1/2)! k_2} A(\eta) \left[ \dfrac{1}{\eta} - \displaystyle\int_\eta^\infty \left\{ \dfrac{1}{s^2} - \dfrac{e^{-\frac{s^2}{4}}}{A^2(s)} \right\} ds \right], & r \text{ odd,} \end{cases} \tag{5.75}$$

where $A(\eta)$, $k_1$ and $k_2$ are as given by (2.16), (2.17) and (2.18) of Section 2.2.1 respectively.

The unique solution to the boundary value problem (5.70),(5.72) and (5.74), can be written down in terms of $\hat{\beta}_1(\eta)$ as

$$\hat{\alpha}_1(\eta) = e^{-\frac{\eta^2}{4}} \left[ k_3 u_1(\eta) + k_4 u_2(\eta) + u_1(\eta) \int_0^\eta u_2(s)\hat{\beta}_1(s)ds - u_2(\eta) \int_0^\eta u_1(s)\hat{\beta}_1(s)ds \right], \tag{5.76}$$

where

$$u_1(\eta) = 1 + \sum_{p=1}^\infty \frac{(R+\frac{1}{2})\cdots(R+\frac{(2p-1)}{2})}{[2p]!} \eta^{2p}, \tag{5.77}$$

$$u_2(\eta) = u_1(\eta) \int_{-\infty}^\eta \frac{e^{s^2/4}}{u_1^2(s)} ds, \tag{5.78}$$

and

$$k_3 = \int_{-\infty}^0 u_2(s)\hat{\beta}_1(s)ds, \tag{5.79}$$

$$k_4 = \frac{1}{k_5} \left[ \int_0^\infty (k_5 u_1(s) - u_2(s))\hat{\beta}_1(s)ds - k_3 \right], \tag{5.80}$$

$$k_5 = \int_{-\infty}^\infty \frac{e^{s^2/4}}{u_1^2(s)} ds, \tag{5.81}$$

with $R = 1 + \frac{r}{2}$. We note from (5.76) and (5.75) that $\hat{\alpha}_1(\eta)$ and $\hat{\beta}_1(\eta)$ are positive monotone decreasing for all $-\infty < \eta < \infty$, and, in particular, we observe that

$$\hat{\alpha}_1(\eta), \hat{\beta}_1(\eta) \sim c_\infty \eta^{-(r+1)} e^{-\eta^2/4} \text{ as } \eta \to \infty, \tag{5.82}$$

where

$$c_\infty = \begin{cases} \dfrac{2\beta_0 g_\sigma [r!]^2}{[(r/2)!]^2 k_1}, & r \text{ even,} \\[3mm] \dfrac{2\beta_0 g_\sigma [r!]^2}{[(r/2 - 1/2)!]^2 k_2}, & r \text{ odd.} \end{cases} \tag{5.83}$$

An examination of higher order terms in expansions (5.68) and (5.69) of region **II(a)** indicates that a nonuniformity will develop in expansion (5.69) when $\eta = O(t^{-1/2})$ [that is, $x = \sigma + O(1)$] as $t \to 0$. We therefore anticipate that the structure will contain a further asymptotic region, region **III(a)**. The existence of this region may be confirmed in the following way: a typical term retained from equation (5.11(b)) at leading order in region **II(a)** is $t^{\frac{r}{2}-1}\hat{\beta}_1$, whereas a typical neglected term is $t^{\frac{r}{2}-\frac{1}{2}}\hat{\beta}_{1\eta}$. The ratio of neglected to retained terms is then

$$R(\eta, t) = \frac{t^{1/2}\hat{\beta}_{1\eta}}{\hat{\beta}_1}, \tag{5.84}$$

which is of $O(t^{1/2})$ for $\eta = O(1)$ as $t \to 0$. However, when $\eta \gg 1, \hat{\beta}_1(\eta) \sim O\left(\eta^{-(r+1)}e^{-\eta^2/4}\right)$, and so

$$R(\eta, t) \sim \eta t^{\frac{1}{2}} + \dots \tag{5.85}$$

as $t \to 0$ and $R(\eta, t)$ becomes of $O(1)$ when $\eta$ is sufficiently large ($\eta = O(t^{-1/2})$), confirming the onset of a nonuniformity in expansion (5.69) as $\eta \to \infty$. We then have, from (5.68),(5.69) and (5.82), that for $\eta \gg 1$, as we move into region **III(a)**,

$$\alpha(\eta, t) \sim 1 - t^{1+\frac{r}{2}}c_\infty \eta^{-(r+1)}e^{-\eta^2/4} + \dots, \tag{5.86}$$

$$\beta(\eta, t) \sim t^{\frac{r}{2}}c_\infty \eta^{-(r+1)}e^{-\eta^2/4} + \dots \tag{5.87}$$

as $t \to 0$. In region **III(a)**, where $x = \sigma + O(1), \alpha = 1 - o(1)$ and $\beta = o(1)$, expansions (5.86) and (5.87) suggest that we look for expansions of the form,

$$\alpha(x, t) = 1 - \psi(x, t), \tag{5.88}$$

$$\beta(x, t) = \phi(x, t), \tag{5.89}$$

as $t \to 0$ where $0 < \psi(x,t), \phi(x,t) = e^{-O(\frac{1}{t})}$ as $t \to 0$ with $x = \sigma + O(1)$. On substitution of expansions (5.88) and (5.89) into equations (5.11(a)) and (5.11(b)) we obtain a balance at leading order, given by

$$\psi_t = \psi_{xx} + \phi, \tag{5.90}$$

$$\phi_t + c\phi_x = \phi_{xx} + \phi, \tag{5.91}$$

as $t \to 0$ with $x = \sigma + O(1)$. We first consider the uncoupled equation (5.91). The structure of the solution for $\beta$ in region **II(a)** as $\eta \to \infty$ (given by (5.87)) suggests that we look for a solution of (5.91) of the form

$$\phi(x,t) = e^{-\frac{F(x,t)}{t}} \quad \text{as } t \to 0, \tag{5.92}$$

with

$$F(x,t) = F_0(x) + F_1(x)t\ln t + F_2(x)t + F_3(x)t^2 + o(t^2), \tag{5.93}$$

where $x = \sigma + O(1)$ and $F(x,t) > 0$ for all $x > \sigma$. On substituting (5.92) and (5.93) into equation (5.91) and solving at each order in turn, we find (after matching with (5.87) as $x \to \sigma^+$) that

$$\begin{aligned}
\beta(x,t) = \exp\Bigg[&-\frac{(x-\sigma)^2}{4t} + \left(r+\frac{1}{2}\right)\ln t + \left(\frac{c}{2}(x-\sigma) + \ln c_\infty\right. \\
&- (r+1)\ln(x-\sigma)\Bigg) + t\left(\left(1 - \frac{c^2}{4}\right)\frac{x}{(x-\sigma)} - \frac{(r+1)(r+2)}{(x-\sigma)^2}\right. \\
&\left.-\frac{E}{(x-\sigma)}\right) + o(t)\Bigg]
\end{aligned} \tag{5.94}$$

as $t \to 0$ with $x = \sigma + O(1)$. The constant $E$ will be fixed by matching expansion (5.94) (as $x \to \sigma^+$) to expansion (5.69) (as $\eta \to \infty$), when expansion (5.69) is taken to next order. The details of which we omit for brevity. Equation (5.90) has to be solved subject to the matching condition

$$\psi(x,t) \sim t^{r+\frac{3}{2}}\frac{c_\infty}{(x-\sigma)^{r+1}}e^{-\frac{(x-\sigma)^2}{4t}} \quad \text{as } x \to \sigma^+. \tag{5.95}$$

Now, (5.94) and (5.95) suggest looking for a solution of (5.90) of the form

$$\psi(x,t) = e^{-\frac{G(x,t)}{t}} \quad \text{as } t \to 0, \tag{5.96}$$

with

$$G(x,t) = G_0(x) + G_1(x)t\ln t + G_2(x)t + G_3(x)t^2 + o(t^2), \tag{5.97}$$

as $t \to 0$ with $x = \sigma + O(1)$. On substituting (5.96) and (5.97) into equation (5.90) and solving at each order subject to the matching condition (5.95), we obtain

$$\begin{aligned}
\psi(x,t) = \exp\Bigg[&-\frac{(x-\sigma)^2}{4t} + \left(r+\frac{3}{2}\right)\ln t + \Big(\ln(2c_\infty) - (r+2)\ln(x-\sigma) \\
&\ln\left(\left(e^{\frac{1}{2}c(x-\sigma)} - 1\right)/c\right)\Big) + o(1)\Bigg]
\end{aligned} \tag{5.98}$$

as $t \to 0$ with $x = \sigma + O(1)$. Thus we have via (5.88) and (5.98) that, in region **III(a)**,

$$\begin{aligned}
\alpha(x,t) = 1 - \exp\Bigg[&-\frac{(x-\sigma)^2}{4t} + \left(r+\frac{3}{2}\right)\ln t + \Big(\ln(2c_\infty) - (r+2)\ln(x-\sigma) \\
&\ln\left(\left(e^{\frac{1}{2}c(x-\sigma)} - 1\right)/c\right)\Big) + o(1)\Bigg]
\end{aligned} \tag{5.99}$$

as $t \to 0$ with $x = \sigma + O(1)$. An examination of expansions (5.94) and (5.99) indicates that these expansions remain uniform for $x \gg 1$, and hence the asymptotic structure as $t \to 0$ in $x \geq 0$ is now complete.

The asymptotic structure of the solution to **IVP** as $t \to 0$ when $x \leq 0$ follows after minor modications, that given for $x \geq 0$, and is summarized here for brevity.

**Region II(b)**    $x = -\sigma \pm O(t^{1/2})$,    $t \to 0$

$$\alpha(\hat{\eta}, t) = 1 - t^{1+\frac{r}{2}} \hat{a}_1(\hat{\eta}) + o\left(t^{1+\frac{r}{2}}\right), \tag{5.100}$$

$$\beta(\hat{\eta}, t) = t^{\frac{r}{2}} \hat{\beta}_1(\hat{\eta}) + o\left(t^{\frac{r}{2}}\right), \tag{5.101}$$

with $\hat{\eta} = -\frac{(x+\sigma)}{t^{1/2}} = O(1)$ as $t \to 0$, where $\hat{a}_1(\hat{\eta}), \hat{\beta}_1(\hat{\eta})$ are given by (5.76) and (5.75) (of section **II(a)**) respectively, with $\eta$ replaced by $\hat{\eta}$. Expansions (5.100) and (5.101) become nonuniform when $\hat{\eta} = O(t^{-1/2})$ [that is, $(-x) = \sigma + O(1)$] as $t \to 0$.

**Region III(b)**    $\hat{x} = (-x) = \sigma + O(1)$,    $t \to 0$

The expansions for $\alpha(\hat{x}, t)$ and $\beta(\hat{x}, t)$ in this region are given by (5.99) and (5.94) (with $x$ replaced by $\hat{x}$ and $c$ replaced by $-c$) respectively, with the expansions for $\alpha(\hat{x}, t)$ and $\beta(\hat{x}, t)$ now remaining uniform for $\hat{x} \gg 1$.

The asymptotic structure as $t \to 0$, for $x \leq 0$, is now complete, with the expansions in regions **I**, **II(b)**, and **III(b)** providing a uniform approximation to the solution of **IVP** for $x \leq 0$ as $t \to 0$.

## 5.4.2 Asymptotic Solution as $|x| \to \infty$

We now investigate the asymptotic structure of the solution to **IVP** as $|x| \to \infty$ with $t = O(1)$. We first investigate the structure of the solution to **IVP** as $x \to \infty$ with $t = O(1)$. The form of expansions (5.99) and (5.94) of region **III(a)** for $x \gg 1$ as $t \to 0$ suggests that in this region, which we label as region **IV(a)**, we expand as

$$\alpha(x, t) = 1 - o(1), \tag{5.102}$$

$$\beta(x, t) = e^{-H(x,t)}, \tag{5.103}$$

as $x \to \infty$, with

$$H(x, t) = H_0(t)x^2 + H_1(t)x + H_2(t)\ln x + H_3(t) + O(x^{-1}), \tag{5.104}$$

where $t = O(1)$ as $x \to \infty$. On substituting (5.103) and (5.104) into equation (5.11(b)) and solving at each order in turn, we find (after matching with (5.94) as $t \to 0^+$) that

$$\beta(x,t) = \exp\left(-\left[\frac{x^2}{4t} - \left(\frac{c}{2} + \frac{\sigma}{2t}\right)x + (r+1)\ln x + \left(\frac{c^2-4}{4}\right)t + \frac{\sigma^2}{4t}\right.\right.$$
$$\left.\left. + \frac{c\sigma}{2} - \ln c_\infty - \left(r + \frac{1}{2}\right)\ln t + O(x^{-1})\right]\right) \tag{5.105}$$

as $\xi \to \infty$ with $t = O(1)$. The correction term to expansion (5.102) can, via (5.11(a)), (5.99) and (5.105), be readily determined (the details of which are omitted for brevity), giving

$$\alpha(x,t) = 1 - \exp\left(-\left[\frac{x^2}{4t} - \left(\frac{c}{2} + \frac{\sigma}{2t}\right)x + (r+2)\ln x + \left(\frac{c^2-4}{4}\right)t + \frac{\sigma^2}{4t}\right.\right.$$
$$\left.\left. + \frac{c\sigma}{2} - \ln\left(\frac{2c_\infty}{c}\right) - \left(r + \frac{3}{2}\right)\ln t + O(x^{-1})\right]\right) \tag{5.106}$$

as $x \to \infty$ with $t = O(1)$. Expansions (5.105) and (5.106) will remain uniform for $t \gg 1$ provided that $x \gg t$, but become nonuniform when $x = O(t)$ as $t \to \infty$.

Finally, we investigate the structure of the solution to **IVP** as $x \to -\infty$ with $t = O(1)$. The details in this case follow, after minor modifications, those given above for region **IV(a)** (for $x \to \infty, t = O(1)$) and are summarized here for brevity.

**Region IV(b)** $x \to -\infty$, $\quad t = O(1)$

$$\alpha(\hat{x},t) \sim 1 - \exp\left(-\left[\frac{\hat{x}^2}{4t} - \frac{\sigma\hat{x}}{2t} + (r+2)\ln\hat{x} + \frac{\sigma^2}{4t} - \left(r + \frac{3}{2}\right)\ln t\right.\right.$$
$$\left.\left. - \ln\left(\frac{2c_\infty}{c}\right)\right]\right) + \exp\left(-\left[\frac{\hat{x}^2}{4t} - \left(\frac{\sigma}{2t} - \frac{c}{2}\right)\hat{x} + (r+2)\ln\hat{x} + \frac{\sigma^2}{4t}\right.\right.$$
$$\left.\left. + \left(\frac{c^2-4}{4}\right)t - \left(2r + \frac{7}{2}\right)\ln t - \frac{c\sigma}{2} - \ln\left(\frac{2c_\infty}{c}\right)\right]\right), \tag{5.107}$$

$$\beta(\hat{x},t) = \exp\left(-\left[\frac{\hat{x}^2}{4t} + \left(\frac{c}{2} - \frac{\sigma}{2t}\right)\hat{x} + (r+1)\ln\hat{x} + \left(\frac{c^2-4}{4}\right)t + \frac{\sigma^2}{4t}\right.\right.$$
$$\left.\left. - \frac{c\sigma}{2} - \ln c_\infty - \left(r + \frac{1}{2}\right)\ln t + O(\hat{x}^{-1})\right]\right), \tag{5.108}$$

as $\hat{x} \to \infty$ with $t = O(1)$ where $\hat{x} = -x$. Expansions (5.107) and (5.108) remain uniform for $t \gg 1$ provided that $\hat{x} \gg t$, but become nonuniform when $\hat{x} = O(t)$ as $t \to \infty$.

## 5.4.3 Asymptotic Solution as $t \to \infty$

As $t \to \infty$, the asymptotic expansions (5.105),(5.106) and (5.107),(5.108) of regions **IV(a)** ($x \to \infty, t = O(1)$) and **IV(b)** ($\hat{x} \to \infty, t = O(1)$) respectively, continue to remain uniform provided $x, \hat{x} \gg t$. However, as already noted, a nonuniformity develops when $|x| = O(t)$.

We begin by considering the asymptotic structure as $t \to \infty$ for $x > 0$. To proceed we introduce a new region, region $\mathbf{V(a)}$. To examine region $\mathbf{V(a)}$ we introduce the scaled coordinate $y = \frac{x}{t}$, where $y = O(1)$ as $t \to \infty$, and look for expansions of the form (as suggested by (5.105) and (5.106))

$$\alpha(y, t) = 1 - o(1), \quad \beta(y, t) = e^{-tf(y,t)} \quad \text{as} \quad t \to \infty, \tag{5.109}$$

where

$$f(y, t) = f_0(y) + f_1(y)\frac{\ln t}{t} + f_2(y)\frac{1}{t} + O(t^{-2}), \tag{5.110}$$

as $t \to \infty$ with $y = O(1)$ and $f_0(y) > 0$. It is instructive to consider first the leading order problem in region $\mathbf{V(a)}$. On substituting (5.109(b)) and (5.110) into equation (5.11(b)) (when written in terms of $y$ and $t$) we obtain the leading order problem as

$$f_{0y}^2 - (y - c)f_{0y} + f_0 + 1 = 0, \quad y > 0, \tag{5.111}$$

$$f_0(y) > 0, \quad y > 0, \tag{5.112}$$

$$f_0(y) \sim \frac{(y - c)^2}{4} - 1 \quad \text{as} \quad y \to \infty. \tag{5.113}$$

The final condition, (5.113), arises from matching expansion (5.109(b))($y \gg 1$) with expansion (5.105) ($x = O(t)$). Equation (5.111) has a one-parameter family of linear solutions

$$f_0(y) = a_0(y - a_0) - ca_0 - 1, \quad y > 0, \tag{5.114}$$

for any $a_0 \in \mathbb{R}$, together with the associated envelope solution,

$$f_0(y) = \frac{(y - c)^2}{4} - 1, \quad y > 0. \tag{5.115}$$

Combinations of (5.114) and (5.115) which remain continuous and differentiable also provide solutions to (5.111) (envelope touching solutions). Applying condition (5.113) requires us to select the solution

$$f_0(y) = \frac{(y - c)^2}{4} - 1, \quad y > 0 \tag{5.116}$$

or

$$f_0(y) = \begin{cases} \frac{(y-c)^2}{4} - 1, & y > y_0, \\ \frac{(y_0-c)}{2}\left[y - \frac{(y_0+c)}{2}\right] - 1, & 0 < y \le y_0, \end{cases} \tag{5.117}$$

for any $y_0 > 2 + c$. We next check condition (5.112). We immediately observe that neither (5.116) nor (5.117) can fully satisfy this condition: (5.116) vanishes as $y \to (2 + c)^+$ whilst (5.117) vanishes as $y \to \left(\frac{2}{(y_0-c)} + \frac{(y_0+c)}{2}\right)^+$

($> 2 + c$ for $y_0 > 2 + c$). We conclude that a non-uniformity occurs in expansion (5.109(b)),(5.110) as $y \to y_c (\geq 2 + c)$ where,

$$y_c = \frac{2}{(y_0 - c)} + \frac{y_0 + c}{2} \begin{cases} = 2 + c, \ y_0 = 2 + c, \\ > 2 + c, \ y_0 > 2 + c, \end{cases} \qquad (5.118)$$

for some $y_0 \geq 2 + c$ (when $y_0 = 2 + c$, $f_0(y)$ is given by (5.116), whilst when $y_0 > 2 + c$, $f_0(y)$ is given by (5.117)). A consideration of further terms in (5.109) and (5.110) demonstrates that this nonuniformity occurs when

$$y = y_c + O(1/t) \qquad (5.119)$$

as $t \to \infty$ with $\alpha, \beta = O(1)$ [since at leading order $\beta \sim \exp(-t f_0(y))$ and correspondingly $\alpha \sim 1 - \frac{1}{(cy/2 - (c^2/2 - 1))} \exp(-t f_0(y))$ as $t \to \infty$]. Therefore, we must introduce a further region, which we denote as region **TWR**. In this region we write $y = y_c + \frac{z}{t}$ with $z = O(1)$ as $t \to \infty$, and expand as

$$\alpha(z, t) = \bar{\alpha}(z) + o(1), \quad \beta(z, t) = \bar{\beta}(z) + o(1) \qquad (5.120)$$

as $t \to \infty$ with $z = O(1)$. On substitution of expansions (5.120) into equations (5.11) we obtain the leading order problem as

$$\bar{\alpha}_{zz} + y_c \bar{\alpha}_z - \bar{\alpha}\bar{\beta} = 0, \quad \bar{\beta}_{zz} + (y_c - c)\bar{\beta}_z + \bar{\alpha}\bar{\beta} = 0, \quad -\infty < z < \infty, \ (5.121)$$

$$\bar{\alpha}(z), \bar{\beta}(z) > 0, \quad -\infty < z < \infty, \qquad (5.122)$$

$$\bar{\alpha}(z) \to 1, \quad \bar{\beta}(z) \to 0, \quad \text{as} \quad z \to \infty, \qquad (5.123)$$

$$\bar{\alpha}(z), \bar{\beta}(z) \quad \text{bounded as} \quad z \to -\infty. \qquad (5.124)$$

Conditions (5.123) arise from matching expansions (5.120) (as $z \to \infty$) with expansions (5.109) (as $y \to y_c^+$). We note that (5.121)-(5.124) is the boundary value problem RP$[c, v]$ (with $v$ replaced by $y_c$). Moreover, Propositions (5.4),(5.9),(5.11) and (5.15) allow boundary conditions (5.124) to be replaced by

$$\bar{\alpha}(z) \to 0, \quad \bar{\beta}(z) \to \frac{y_c}{(y_c - c)} \quad \text{as} \quad z \to -\infty. \qquad (5.125)$$

Recalling that we have restricted attention to consideration of $c \geq 0$ (without loss of generality) then the boundary value problem (5.121), (5.122), (5.123) and (5.125) has via Theorem 5.26 a unique solution , $\bar{\alpha}(z) = \bar{\alpha}_T(z, y_c), \bar{\beta}(z) = \bar{\beta}_T(z, y_c)$ for each $y_c \geq 2 + c$ with $c \geq 0$. This solution represents a right permanent form travelling wave structure (RPTW) with speed $v = y_c$. We next match expansions (5.120) (as $z \to \infty$) of region **TWR** to expansions (5.109)(as $y \to y_c^+$) of region **V(a)**. It is convenient to match $A = \ln(1 - \alpha)$ and $B = \ln \beta$ rather than $\alpha$ and $\beta$ themselves. We follow the matching principle of Van Dyke [70], in matching the expansions (5.109), for A and B, to $O(t)$, with expansions (5.120) for A and B, to $O(1)$. Matching then requires that $y_c = 2 + c$ and the RPTW of minimum speed is selected in region **TWR**.

With $y_c (= 2 + c)$ now fixed we have, via (5.118), that $y_0 = 2 + c$ and that $f_0(y) = \frac{(y-c)^2}{4} - 1$, where $(2 + c) + o(1) < y < \infty$. Hence the expansions in both regions **V(a)** and **TWR** are now complete at leading order. We now complete the asymptotic structure of region **V(a)** $((2 + c) + o(1) < y < \infty)$. On substituting (5.109(b)) and (5.110) into equation (5.11(b)) (when written in terms of $y$ and $t$) and solving at each order in turn, we obtain

$$\beta(y,t) = \exp\left(-t\left[\frac{(y-c)^2}{4} - 1 + \frac{1}{2}\frac{\ln t}{t} + \frac{H(y)}{t} + O(t^{-2})\right]\right) \qquad (5.126)$$

as $t \to \infty$ with $(2 + c) + o(1) < y < \infty$. The function $H(y)$ remains undetermined. This is a consequence of using the far field asymptotics $x \gg t \gg 1$ rather than information from the bulk region $x, t = O(1)$ as a basis for the large time asymptotic structure. Matching with expansion (5.105)$(x = O(t))$ as $y \to \infty$ requires that

$$H(y) \sim (r + 1)\ln y - \frac{\sigma}{2}(y - c) - \ln c_\infty \qquad (5.127)$$

as $y \to \infty$. The correction term to expansion (5.109(a)) can via (5.11(a)), (5.106) and (5.126) be readily determined, giving

$$a(y,t) = 1 - \frac{1}{[\frac{cy}{2} - (\frac{c^2}{2} - 1)]}\exp\left(-t\left[\frac{(y-c)^2}{4} - 1 + \frac{1}{2}\frac{\ln t}{t} + \frac{H(y)}{t}\right.\right.$$
$$\left.\left. + O(t^{-2})\right]\right) \qquad (5.128)$$

as $t \to \infty$ with $(2 + c) + o(1) < y < \infty$. Now as $y \to (2 + c)^+$ we move into the wave front region, region **TWR**, where $y = (2 + c) + O(1/t)$ as $t \to \infty$. In region **TWR**, $x \sim s(t)$ and $\alpha$ and $\beta$ (when written in terms of the travelling wave coordinate $z$) have the form (via (5.120))

$$\alpha(z,t) = \alpha_T(z, 2 + c) + o(1), \quad \beta(z,t) = \beta_T(z, 2 + c) + o(1) \qquad (5.129)$$

as $t \to \infty$ with $z = O(1)$, where $z = x - s(t)$ and $s(t) = (2 + c)t + v_1\phi(t) + o(\phi(t))$. Here $\phi(t) = o(t)$ is as yet an undetermined gauge function (to be fixed on matching with region **V(a)** as $z \to \infty$), whilst $\alpha_T(z, 2 + c)$ and $\beta_T(z, 2 + c)$ represent the minimum speed $v = 2 + c$ right permanent form travelling wave solution (RPTW). We recall, from Section 5.2, the following asymptotic properties of $\alpha_T(z, 2 + c)$ and $\beta_T(z, 2 + c)$:

$$\alpha_T(z, 2 + c) \sim 1 - \frac{B^*}{(c+1)}ze^{-z}, \quad \beta_T(z, 2 + c) \sim B^*ze^{-z}, \quad \text{as} \quad z \to \infty \qquad (5.130)$$

and

$$\alpha_T(z, 2 + c) \sim D^*\lambda_3\left(\frac{2(2 + \lambda_3)}{2 + c}\right)e^{\lambda_3 z}, \beta_T(z, 2 + c) \sim \left(1 + \frac{c}{2}\right) - D^*e^{\lambda_3 z} \qquad (5.131)$$

as $z \to -\infty$ where $B^*$ and $D^*$ are positive constants which are, in principle, determined and

$$\lambda_3 = \frac{1}{2}\sqrt{(2+c)(4+c)} - \frac{(2+c)}{2} > 0. \tag{5.132}$$

Matching expansions (5.129)(as $z \to \infty$) with expansions (5.128) and (5.126) of region $\mathbf{V(a)}$ (as $y \to (2+c)^+$), fixes $\phi(t) = \ln t$ and requires that

$$H(y) \sim -\ln B^* - \ln(y - (2+c)) \tag{5.133}$$

as $y \to (2+c)^+$ and $v_1 = -\frac{3}{2}$. The velocity of propagation of the wave front is then given by

$$\dot{s} = (2+c) - \frac{3}{2}\frac{1}{t} + o\left(\frac{1}{t}\right) \tag{5.134}$$

as $t \to \infty$, hence the RPTW which develops asymptotically as $t \to \infty$ propagates with the minimum speed $(2+c)$ and this speed is approached from below as $t \to \infty$.

As $z \to -\infty$, we move into region $\mathbf{VI(a)}$, where $0 < y < (2+c) - o(1)$. The structure of expansions (5.129) as $z \to -\infty$ (obtained via (5.131)), suggests that in region $\mathbf{VI(a)}$ we expand as

$$\alpha(y,t) = D^*\lambda_3\left(\frac{2(2+\lambda_3)}{(2+c)}\right)t^{\frac{3\lambda_3}{2}}\exp([g(y)t + o(1)]), \tag{5.135}$$

$$\beta(y,t) = \left(1 + \frac{c}{2}\right) - D^*t^{\frac{3\lambda_3}{2}}\exp([g(y)t + o(1)]) \tag{5.136}$$

as $t \to \infty$ with $0 < y < (2+c) - o(1)$. On substituting expansions (5.135) and (5.136) into equations (5.11) (when written in terms of $y$ and $t$) and solving at leading order, we obtain (after matching to region $\mathbf{TWR}$ as $y \to (2+c)^-$) that

$$g(y) = \lambda_3(y - (2+c)), \tag{5.137}$$

where $0 < y < (2+c) - o(1)$. A comparison of neglected and retained terms in equation (5.11(b)) (when written in terms of $y$ and $t$) indicates that a nonuniformity develops in expansions (5.135),(5.136) as $y \to c^+$, and a further region will be required when $y = c + O(t^\gamma)(\gamma < 0)$. We label this region, region $\mathbf{CW}$. To investigate region $\mathbf{CW}$, we introduce the scaled variable $\chi$ by

$$\chi = (y - c)t^{-\gamma}, \tag{5.138}$$

with $\gamma < 0$ to be determined, and $\chi = O(1)$ as $t \to \infty$. An examination of (5.135) and (5.136) then determines that $\alpha = o(1)$ and $\beta = O(1)$ in region $\mathbf{CW}$. We now direct our attention to the leading order behaviour of $\beta$ alone. Thus we expand $\beta$ as

$$\beta(\chi,t) = \beta_0(\chi) + o(1) \tag{5.139}$$

as $t \to \infty$ with $\chi = O(1)$. On substituting expansion (5.139) into equation (5.11(b)) (when written in terms of $\chi$ and $t$), gives to obtain the most structured leading order balance that

$$\gamma = -\frac{1}{2}, \qquad (5.140)$$

after which the leading order problem for $\beta_0(\chi)$ in region **CW** is given by

$$\beta_0'' + \frac{\chi}{2}\beta_0' = 0, \quad -\infty < \chi < \infty \qquad (5.141)$$

which is to be solved subject to matching to region **VI(a)** as $\chi \to \infty$, that is,

$$\beta_0(\chi) \to \left(1 + \frac{c}{2}\right) \quad \text{as} \quad \chi \to \infty. \qquad (5.142)$$

The solution to (5.141) and (5.142) is readily obtained as

$$\beta_0(\chi) = \left(1 + \frac{c}{2}\right) + A_0 erfc\left(\frac{\chi}{2}\right), \quad -\infty < \chi < \infty. \qquad (5.143)$$

The constant $A_0$ (which depends on the parameter $c$) remains undetermined at this stage of the analysis and will be determined on matching to the asymptotic structure in $0 < y < c - o(1)$. Hence in region **CW** we have that

$$\alpha(\chi, t) = o(1), \quad \beta(\chi, t) = \left[\left(1 + \frac{c}{2}\right) + A_0 erfc\left(\frac{\chi}{2}\right)\right] + o(1) \qquad (5.144)$$

as $t \to \infty$ with $\chi = O(1)$. As $\chi \to -\infty$ we move out of region **CW** into region **VII(a)** where $0 < y < c - o(1)$ and $\alpha = o(1)$. The order of $\beta$ in this region depends on $A_0$ which is to be determined. We will return to region **VII(a)** when we have developed the large time asymptotic solution to **IVP** in $x < 0$.

We next develop the asymptotic structure as $t \to \infty$ for $x < 0$. As $t \to \infty$, the asymptotic expansions (5.107) and (5.108) of region **IV(b)** continue to remain uniform for $\hat{x} \gg t$. However, as already noted, a nonuniformity develops when $\hat{x} = O(t)$. To proceed we introduce a new region, region **V(b)**. To examine region **V(b)** we introduce the scaled coordinate $\hat{y} = \frac{(-x)}{t}$, where $\hat{y} = O(1)$ as $t \to \infty$, and look for expansions of the form (as suggested by (5.107) and (5.108))

$$\alpha(\hat{y}, t) = 1 - e^{-t\hat{g}(\hat{y},t)}, \quad \beta(\hat{y}, t) = e^{-t\hat{f}(\hat{y},t)} \quad \text{as} \quad t \to \infty, \qquad (5.145)$$

where

$$\hat{f}(\hat{y}, t) = \hat{f}_0(\hat{y}) + \hat{f}_1(\hat{y})\frac{\ln t}{t} + \hat{f}_2(\hat{y})\frac{1}{t} + O(t^{-2}), \qquad (5.146)$$

and

$$\hat{g}(\hat{y}, t) = \hat{g}_0(\hat{y}) + \hat{g}_1(\hat{y})\frac{\ln t}{t} + \hat{g}_2(\hat{y})\frac{1}{t} + O(t^{-2}) \qquad (5.147)$$

as $t \to \infty$ with $\hat{y} = O(1)$ and $\hat{f}_0(\hat{y}), \hat{g}_0(\hat{y}) > 0$. It is instructive to consider first the leading order problem for $\hat{f}_0(\hat{y})$. On substituting (5.145(b)) and (5.146) into equation (5.11(b)) (when written in terms of $\hat{y}$ and $t$) we obtain the leading order problem as

$$\hat{f}_{0\hat{y}}^2 - (\hat{y} + c)\hat{f}_{0\hat{y}} + \hat{f}_0 + 1 = 0, \quad \hat{y} > 0, \tag{5.148}$$

$$\hat{f}_0(\hat{y}) > 0, \quad \hat{y} > 0, \tag{5.149}$$

$$\hat{f}_0(\hat{y}) \sim \frac{(\hat{y} + c)^2}{4} - 1 \quad \text{as} \quad \hat{y} \to \infty. \tag{5.150}$$

The final condition, (5.150), arises from matching expansion (5.145(b)) ($\hat{y} \gg 1$) with expansion (5.108) ($\hat{x} = O(t)$). Equation (5.148) has a one parameter family of linear solutions

$$\hat{f}_0(\hat{y}) = b_0(\hat{y} - b_0) + cb_0 - 1, \quad \hat{y} > 0, \tag{5.151}$$

for any $b_0 \in \mathbb{R}$, together with the associated envelope solution,

$$\hat{f}_0(\hat{y}) = \frac{(\hat{y} + c)^2}{4} - 1, \quad \hat{y} > 0. \tag{5.152}$$

Combinations of (5.151) and (5.152) which remain continuous and differentiable also provide solutions to (5.148)(envelope touching solutions). There are now two cases, which are considered separately.

## (a) $0 \le c < 2$

In this case applying conditions (5.150) requires us to select the solution

$$\hat{f}_0(\hat{y}) = \frac{(\hat{y} + c)^2}{4} - 1, \quad \hat{y} > 0 \tag{5.153}$$

or

$$\hat{f}_0(\hat{y}) = \begin{cases} \frac{(\hat{y}+c)^2}{4} - 1, & \hat{y} > \hat{y}_0, \\ \frac{(\hat{y}_0+c)}{2}\left[\hat{y} - \frac{(\hat{y}_0-c)}{2}\right] - 1, & 0 < \hat{y} \le \hat{y}_0, \end{cases} \tag{5.154}$$

for any $\hat{y}_0 > 2 - c$. We next check condition (5.149). We immediately observe that neither (5.153) nor (5.154) can fully satisfy this condition: (5.153) vanishes as $\hat{y} \to (2 - c)^+$ whilst (5.154) vanishes as $\hat{y} \to \left(\frac{2}{(\hat{y}_0+c)} + \frac{(\hat{y}_0-c)}{2}\right)^+$ ($> 2 - c$ for $\hat{y}_0 > 2 - c$). We conclude that a nonuniformity occurs in expansion (5.145(b)),(5.146) as $\hat{y} \to \hat{y}_c^\beta (\ge 2 - c)$ where

$$\hat{y}_c^\beta = \frac{2}{(\hat{y}_0 + c)} + \frac{(\hat{y}_0 - c)}{2} \begin{cases} = 2 - c, \hat{y}_0 = 2 - c, \\ > 2 - c, \hat{y}_0 > 2 - c, \end{cases} \tag{5.155}$$

for some $\hat{y}_0 \ge 2 - c$ (when $\hat{y}_0 = 2 - c$, $\hat{f}_0(\hat{y})$ is given by (5.153), whilst when $\hat{y}_0 > 2 - c$, $\hat{f}_0(\hat{y})$ is given by (5.154)). On substituting (5.145(a)) and (5.147) into equation (5.11(a)) (when written in terms of $\hat{y}$ and $t$) we obtain the leading order problem for $\hat{g}(\hat{y})$ as

$$\hat{g}_{0\hat{y}}^2 - \hat{y}\hat{g}_{0\hat{y}} + \hat{g}_0 = 0, \quad \hat{y} > 0, \tag{5.156}$$

$$\hat{g}_0(\hat{y}) > 0, \quad \hat{y} > 0, \tag{5.157}$$

$$\hat{g}_0(\hat{y}) \sim \frac{\hat{y}^2}{4} \quad \text{as} \quad \hat{y} \to \infty. \tag{5.158}$$

The final condition, (5.158), arises from matching expansion (5.145(a)) ($\hat{y} \gg 1$) with expansion (5.107) ($\hat{x} = O(t)$). Equation (5.156) has a one parameter family of linear solutions

$$\hat{g}_0(\hat{y}) = A(\hat{y} - A), \quad \hat{y} > 0, \tag{5.159}$$

for any $A \in \mathbb{R}$, together with the associated envelope solution

$$\hat{g}_0(\hat{y}) = \frac{\hat{y}^2}{4}, \quad \hat{y} > 0. \tag{5.160}$$

Combinations of (5.159) and (5.160) which remain continuous and differentiable also provide solutions to (5.156) (envelope touching solutions). Applying condition (5.158) requires us to select the solution,

$$\hat{g}_0(\hat{y}) = \frac{\hat{y}^2}{4}, \quad \hat{y} > 0. \tag{5.161}$$

or

$$\hat{g}_0(\hat{y}) = \begin{cases} \frac{\hat{y}^2}{4}, & \hat{y} > \hat{y}_*, \\ \frac{\hat{y}_*}{2}\left[\hat{y} - \frac{\hat{y}_*}{2}\right], & 0 < \hat{y} \le \hat{y}_*, \end{cases} \tag{5.162}$$

for any $\hat{y}_* > 0$. We next check condition (5.157). We immediately observe that neither (5.161) nor (5.162) can fully satisfy this condition: (5.161) vanishes as $\hat{y} \to 0^+$ whilst (5.162) vanishes as $\hat{y} \to \left(\frac{\hat{y}_*}{2}\right)^+$ ($> 0$ for $\hat{y}_* > 0$). We conclude that a nonuniformity occurs in expansion (5.145(a)),(5.147) as $\hat{y} \to \hat{y}_c^\alpha (\ge 0)$ where

$$\hat{y}_c^\alpha = \frac{\hat{y}_*}{2} \begin{cases} = 0, \hat{y}_* = 0, \\ > 0, \hat{y}_* > 0, \end{cases} \tag{5.163}$$

Hence a nonuniformity first occurs in region **V(b)** as $\hat{y} \to \hat{y}_c^+$ where $\hat{y}_c = \max[\hat{y}_c^\alpha, \hat{y}_c^\beta] > 0$. A consideration of further terms in (5.145) and (5.146) demonstrates that this nonuniformity occurs when

$$\hat{y} = \hat{y}_c + O(1/t) \tag{5.164}$$

as $t \to \infty$ where $\alpha, \beta = O(1)$. Therefore, we must introduce a further region, which we denote as region **TWL**. In this region we write $\hat{y} = \hat{y}_c + \frac{\hat{z}}{t}$ with $\hat{z} = O(1)$ as $t \to \infty$, and expand as

$$\alpha(\hat{z}, t) = \bar{\alpha}(\hat{z}) + o(1), \quad \beta(\hat{z}, t) = \bar{\beta}(\hat{z}) + o(1) \tag{5.165}$$

as $t \to \infty$ with $\hat{z} = O(1)$. On substitution of expansions (5.165) into equations (5.11) we obtain the leading order problem as

$$\bar{\alpha}_{\hat{z}\hat{z}} + \hat{y}_c\bar{\alpha}_{\hat{z}} - \bar{\alpha}\bar{\beta} = 0, \quad \bar{\beta}_{\hat{z}\hat{z}} + (\hat{y}_c + c)\bar{\beta}_{\hat{z}} + \bar{\alpha}\bar{\beta} = 0, \quad -\infty < \hat{z} < \infty, \tag{5.166}$$

$$\bar{\alpha}(\hat{z}), \bar{\beta}(\hat{z}) > 0, \quad -\infty < \hat{z} < \infty, \tag{5.167}$$

$$\bar{\alpha}(\hat{z}) \to 1, \quad \bar{\beta}(\hat{z}) \to 0, \quad \text{as} \quad \hat{z} \to \infty, \tag{5.168}$$

$$\bar{\alpha}(\hat{z}), \bar{\beta}(\hat{z}) \quad \text{bounded as} \quad \hat{z} \to -\infty. \tag{5.169}$$

Conditions (5.168) arise from matching expansions (5.165) (as $\hat{z} \to \infty$) with expansions (5.145) (as $y \to y_c^+$). We note that (5.166)-(5.169) is the boundary value problem LP$[c, v]$, when $v = -\hat{y}_c$ and the independent variable is taken as $(-\hat{z})$. Moreover, the remarks of Section 5.1 allow boundary conditions (5.169) to be replaced by

$$\bar{\alpha}(\hat{z}) \to 0, \quad \bar{\beta}(\hat{z}) \to \frac{\hat{y}_c}{(\hat{y}_c + c)}, \quad \text{as} \quad \hat{z} \to -\infty. \tag{5.170}$$

Following Theorem 5.26 boundary value problem (5.166), (5.167), (5.168) and (5.170) has a unique solution, $\bar{\alpha}(\hat{z}) = \bar{\alpha}_T(\hat{z}, \hat{y}_c), \bar{\beta}(\hat{z}) = \bar{\beta}_T(\hat{z}, \hat{y}_c)$ for each $\hat{y}_c \geq 2 - c$ with $0 \geq c < 2$. This solution represents a left permanent form travelling wave structure (LPTW) with speed $-y_c$.

We match expansions (5.165) (as $\hat{z} \to \infty$) of region **TWL** to expansions (5.145) (as $\hat{y} \to \hat{y}_c^+$) of region **V(b)**. It is convenient to match $A = \ln(1 - \alpha)$ and $B = \ln\beta$ rather than $\alpha$ and $\beta$ themselves. We follow the matching principle of Van Dyke [70], in matching the expansions (5.145), for $A$ and $B$, to $O(t)$, with expansions (5.165) for $A$ and $B$, to $O(1)$. Matching then requires that $\hat{y}_c = 2 - c$ and the LPTW of maximum speed is selected in region **TWL**. With $\hat{y}_c (= 2 - c)$ now fixed we have, via (5.154) that $\hat{y}_0 = 2 - c$ and that $\hat{f}_0(\hat{y}) = \frac{(\hat{y}+c)^2}{4} - 1$, where $\infty > \hat{y} > (2 - c) + o(1)$. Matching to next order will fix $\hat{y}_*$ and hence $\hat{g}(\hat{y})$. Hence the expansions in both regions **V(b)** and **TWL** are now complete at leading order.

We now continue the asymptotic structure for $\beta$ in region **V(b)** ($(2 - c) + o(1) < \hat{y} < \infty$). On substituting (5.145(b)) and (5.146) into equation (5.11(b)) (when written in terms of $\hat{y}$ and $t$) and solving at each order in turn, we obtain

$$\beta(\hat{y}, t) = \exp\left(-t\left[\frac{(\hat{y}+c)^2}{4} - 1 + \frac{1}{2}\frac{\ln t}{t} + \frac{\hat{H}(\hat{y})}{t} + o\left(\frac{1}{t}\right)\right]\right) \tag{5.171}$$

as $t \to \infty$. The function $\hat{H}(\hat{y})$ remains undetermined. This is a consequence of using the far field asymptotics $\hat{x} \gg t \gg 1$ rather than information from the bulk region $\hat{x}, t = O(1)$ as a basis for the large time asymptotic structure. Matching with expansion (5.108) ($\hat{x} = O(t)$) as $\hat{y} \to \infty$ requires that

$$\hat{H}(\hat{y}) \sim (r+1)\ln \hat{y} - \frac{\sigma}{2}(\hat{y}+c) - \ln c_\infty \qquad (5.172)$$

as $\hat{y} \to \infty$. Now as $\hat{y} \to (2-c)^+$ we move into the wave front region, region **TWL**, where $\hat{y} = (2-c) + O(1/t)$ as $t \to \infty$. In region **TWL**, $\hat{x} \sim s(t)$ and $\alpha$ and $\beta$ (when written in terms of the travelling wave coordinate $\hat{z}$) have the form (via (5.165))

$$\alpha(\hat{z},t) = \alpha_T(\hat{z}, 2-c) + o(1), \quad \beta(\hat{z},t) = \beta_T(\hat{z}, 2-c) + o(1) \qquad (5.173)$$

as $t \to \infty$ with $\hat{z} = O(1)$, where $\hat{z} = (-x) - s(t)$ and $s(t) = (2-c)t + \hat{v}_1\hat{\phi}(t) + o(\hat{\phi}(t))$. Here $\hat{\phi}(t) = o(t)$ is as yet an undetermined gauge function (to be fixed on matching with region **V(b)** as $\hat{z} \to \infty$), whilst $\alpha_T(\hat{z}, 2-c)$ and $\beta_T(\hat{z}, 2-c)$ represent the maximum speed LPTW.

We recall, from Section 5.2, the following asymptotic properties of $\alpha_T(\hat{z}, 2-c)$ and $\beta_T(z, 2-c)$:

As $\hat{z} \to \infty$

$$\alpha_T(\hat{z}, 2-c) \sim 1 - \begin{cases} \hat{A}^* e^{-(2-c)\hat{z}}, & 1 < c < 2, \\ \frac{\hat{B}^*}{2} \hat{z}^2 e^{-\hat{z}}, & c = 1, \\ \frac{\hat{B}^*}{(1-c)} \hat{z} e^{-\hat{z}}, & 0 < c < 1, \end{cases} \qquad (5.174)$$

and

$$\beta_T(\hat{z}, 2-c) \sim \hat{B}^* \hat{z} e^{-\hat{z}}. \qquad (5.175)$$

As $\hat{z} \to -\infty$

$$\alpha_T(\hat{z}, 2-c) \sim \hat{D}^* \mu_3 \left(\frac{2(2+\mu_3)}{2-c}\right) e^{\mu_3 \hat{z}}, \beta_T(\hat{z}, 2-c) \sim \left(1 - \frac{c}{2}\right) - \hat{D}^* e^{\mu_3 \hat{z}}, \qquad (5.176)$$

where $\hat{B}^*$, $\hat{A}^*$ and $\hat{D}^*$ are positive constants which are, in principle, determined and

$$\mu_3 = \frac{1}{2}\sqrt{(2-c)(4-c)} - \frac{(2-c)}{2} > 0. \qquad (5.177)$$

Matching expansions (5.176(b)) (as $\hat{z} \to \infty$) with expansion (5.171) of region **Vb)** (as $\hat{y} \to (2-c)^+$), fixes $\hat{\phi}(t) = \ln t$ and requires that

$$\hat{H}(\hat{y}) \sim -\ln \hat{B}^* - \ln(\hat{y} - (2-c)) \qquad (5.178)$$

as $y \to (2-c)^+$ and $\hat{v}_1 = -\frac{3}{2}$. The velocity of propagation of the wave front is given by

$$\dot{s}(t) = (2-c) - \frac{3}{2}\frac{1}{t} + o\left(\frac{1}{t}\right) \qquad (5.179)$$

as $t \to \infty$. Hence in terms of $x$ and $t$, the LPTW propagates with the maximum speed $(c-2)$ which is approached from above as $t \to \infty$. It remains to complete the asymptotic structure for $\alpha$ in region **V(b)**. On substituting (5.145(a)) and (5.147) into equation (5.11(b)) (when written in terms of $\hat{y}$ and $t$) and solving at each order in turn, we obtain

$$\alpha(\hat{y}, t) \sim 1 - \exp\left(-t\left[\hat{g}_0(\hat{y}) + o(1)\right]\right) + \exp\left(-t\left[\frac{(\hat{y} + c)^2}{4} - 1 + o(1)\right]\right).$$
(5.180)

We must now match expansion (5.180) (as $\hat{y} \to (2 - c)$) to expansion (5.173) (as $\hat{z} \to \infty$). On using (5.161), (5.162) and (5.174), matching can only be achieved provided we choose

$$\hat{y}_* = 2(2 - c),$$
(5.181)

and then, via (5.162),

$$\hat{g}_0(\hat{y}) = \begin{cases} \frac{1}{4}\hat{y}^2, & \hat{y} > 2(2 - c), \\ (2 - c)[\hat{y} - (2 - c)], & (2 - c) < \hat{y} \le 2(2 - c), \end{cases}$$
(5.182)

and matching is complete.

In summary, when $0 < c < 2$ we have in terms of the original variables $x, t$ a left permanent form travelling wave [LPTW] solution of **IVP** as $t \to \infty$, where the velocity of propagation of the wave front is given by

$$\dot{s}(t) = (c - 2) + \frac{3}{2}\frac{1}{t} + o\left(\frac{1}{t}\right)$$
(5.183)

as $t \to \infty$. Hence the LPTW propagates with the maximum speed $(c - 2)$ which is approached from above as $t \to \infty$. As $\hat{z} \to -\infty$, we move into region **VII(a)**, where $0 \le \hat{y} < (2 - c) - o(1)$. The structure of expansions (5.173) as $\hat{z} \to -\infty$ (obtained via (5.176)) suggests that in region **VII(a)** we expand as

$$\alpha(\hat{y}, t) = \hat{D}^* \mu_3 \left(\frac{2(2 + \mu_3)}{(2 - c)}\right) t^{\frac{3\mu_3}{2}} \exp([\hat{g}(\hat{y})t + o(1)]),$$
(5.184)

$$\beta(\hat{y}, t) = \left(1 - \frac{c}{2}\right) - \hat{D}^* t^{\frac{3\mu_3}{2}} \exp([\hat{g}(\hat{y})t + o(1)])$$
(5.185)

as $t \to \infty$ with $-c \le \hat{y} < (2-c)-o(1)$. On substituting expansions (5.184) and (5.185) into equations (5.11) (when written in terms of $\hat{y}$ and $t$) and solving at leading order, we obtain (after matching to region **TWL** as $\hat{y} \to (2 - c)^-$) that

$$\hat{g}(\hat{y}) = \mu_3(\hat{y} - (2 - c)),$$
(5.186)

where $-c \le \hat{y} < (2-c)-o(1)$. In fact, a consideration of neglected and retained terms in equation (5.11(b)) (when written in terms of $\hat{y}$ and $t$) demonstrates that expansions (5.184), (5.185) remains uniform until we approach $\hat{y} = -c$ (that is, $y = c$) when we move into region **CW**. Matching between the present region, region **VII(a)** and region **CW** then completes the asymptotic structure as $t \to \infty$. Matching expansion (5.144(b)) (as $\chi \to -\infty$) and expansion (5.185) (with (5.186)) up to $O(1)$ requires

$$A_0 = -\frac{c}{2}.$$
(5.187)

This completes the asymptotic structure in this case.

## (b) $c \geq 2$

In this case, after following arguments similar to those presented for $0 \leq c < 2$, and using the travelling wave theory of Section 5.2 on the existence of LPTWs, we are able to conclude that expansions (5.145)-(5.147) in region **V(b)** require the choice

$$\hat{f}_0(\hat{y}) = \frac{(\hat{y}+c)^2}{4} - 1, \quad \hat{y} > 0, \tag{5.188}$$

and

$$\hat{g}_0(\hat{y}) = \frac{\hat{y}^2}{4}, \quad \hat{y} > 0, \tag{5.189}$$

with expansions (5.145)-(5.147) now remaining uniform in $\hat{y} > 0$, and becoming nonuniform as $\hat{y} \to 0^+$. On continuing to higher order in expansions (5.145)-(5.147) we obtain

$$\alpha(\hat{y}, t) = 1 - \exp\left(-t\left[\frac{\hat{y}^2}{4} + \frac{1}{2}\frac{\ln t}{t} + \frac{\hat{K}(\hat{y})}{t} + o\left(\frac{1}{t}\right)\right]\right), \tag{5.190}$$

$$\beta(\hat{y}, t) = \exp\left(-t\left[\frac{(\hat{y}+c)^2}{4} - 1 + \frac{1}{2}\frac{\ln t}{t} + \frac{\hat{H}(\hat{y})}{t} + O(t^{-2})\right]\right), \tag{5.191}$$

with $\hat{y} > 0$, as $t \to \infty$. The functions $\hat{K}(\hat{y})$ and $\hat{H}(\hat{y})$, remain undetermined. This is a consequence of using the far field asymptotics $(\hat{x} \gg)t \gg 1$ rather than information from the bulk region $x, t = O(1)$ as a basis for the large time asymptotic structure. Matching with expansions (5.107),(5.108) $(\hat{x} = O(t))$, we require

$$\hat{K}(\hat{y}) \sim -\frac{\sigma}{2}\hat{y} + (r+2)\ln\hat{y} - \ln\frac{2c_\infty}{c}, \tag{5.192}$$

and

$$\hat{H}(\hat{y}) \sim -\frac{\sigma}{2}\hat{y} + (r+1)\ln\hat{y} - \frac{\sigma c}{2} - \ln c_\infty, \tag{5.193}$$

as $\hat{y} \to \infty$. An examination of expansions (5.190), (5.191) as $\hat{y} \to 0^+$ reveals that expansions (5.190), (5.191) become non-uniform when $\hat{y} = O(t^{-\frac{1}{2}})$ with $\alpha = O(1)$ and $\beta = o(1)$ as $t \to \infty$ We therefore introduce a further asymptotic region, region **TR**. To examine transition region **TR** we introduce the scaled coordinate $\hat{\chi} = \hat{y}t^{\frac{1}{2}} = O(1)$ as $t \to \infty$. We note that $\beta$ is exponentially small and restrict our attention to the leading order behaviour of $\alpha$ alone, we expand as

$$\alpha(\hat{\chi}, t) = \alpha_0(\hat{\chi}) + o(1) \tag{5.194}$$

as $t \to \infty$ with $\hat{\chi} = O(1)$. On substitution of expansion (5.194) into (5.11(a)) (when written in terms of $\hat{\chi}$ and $t$), and noting that $\beta = o(1)$ as $t \to \infty$ in this region, we obtain at leading order the problem

$$\alpha_0'' + \frac{\hat{\chi}}{2}\alpha_0' = 0, \quad -\infty < \hat{\chi} < \infty, \tag{5.195}$$

$$\alpha_0(\hat{\chi}) \sim 1 \quad \text{as} \quad \hat{\chi} \to \infty, \tag{5.196}$$

$$\alpha_0(\hat{\chi}) = o(1) \quad \text{as} \quad \hat{\chi} \to -\infty, \tag{5.197}$$

Condition (5.196) arises from matching to region **V(b)** as $\hat{\chi} \to \infty$, while condition (5.197) is required to enable matching to region **VII(a)** $(o(1) < y(= \frac{x}{t}) < c - o(1))$ as $\hat{\chi} \to -\infty$. The solution to (5.195)-(5.197) is readily obtained as

$$\alpha_0(\hat{\chi}) = 1 - \frac{1}{2}erfc\left(\frac{\hat{\chi}}{2}\right), \quad -\infty < \hat{\chi} < \infty. \tag{5.198}$$

Hence in region **TR** we have that

$$\alpha(\hat{\chi}, t) = \left[1 - \frac{1}{2}erfc\left(\frac{\hat{\chi}}{2}\right)\right] + o(1), \quad \beta(\hat{\chi}, t) = o(1) \tag{5.199}$$

as $t \to \infty$ with $\hat{\tau} = O(1)$. Matching expansion (5.199(a)) (as $\hat{\chi} \to \infty$) of region **TR** to expansion (5.190) (as $\hat{y} \to 0^+$) of region **V(b)** to next order then requires that

$$\hat{K}(\hat{y}) \sim \ln \hat{y} + \ln \sqrt{\pi}, \quad \hat{y} \to 0^+. \tag{5.200}$$

As $\hat{\chi} \to -\infty$ we move out of region **TR** into region **VII(a)** where $o(1) < y(= \frac{x}{t}) < c - o(1)$. The details of region **VII(a)** in this case are omitted for brevity but we note that $\alpha, \beta = o(1)$ as $t \to \infty$ in this region. Matching to region **CW** (as $y \to c^-$) then requires that

$$A_0 = -\frac{1}{2}\left(1 + \frac{c}{2}\right), \tag{5.201}$$

and completes the asymptotic structure in this case.

### 5.4.4 Summary

In this section we have obtained, via the method of matched asymptotic expansions, the structure of the solution to **IVP** (for $c \geq 0$) as $t \to \infty$. We have established that there are two distinct cases to be considered depending on the value of $c$. The results of this section are summarized as follows:

(i) $0 \leq c < 2$

In this case the solution to **IVP** as $t \to \infty$ admits both left and right permanent form travelling wave solutions. The right permanent form travelling wave (RPTW) propagates with the minimum speed available, that being $(2 + c)$, with this speed being (via (5.134)) approached from below as $t \to \infty$. The left permanent form travelling wave (LPTW) propagates with the maximum speed available, that being $(c - 2)$, with this speed being (via (5.183)) approached from above as $t \to \infty$. We note that the correction to the wave speeds of both the RPTW and LPTW is of $O\left(\frac{1}{t}\right)$ as $t \to \infty$. We further note that in this case a convection wave (which primarily allows for the adjustment in the concentration of $\beta$ which has to adjust from a value of $\left(1 - \frac{c}{2}\right)$ (the value of $\beta$ behind the LPTW) to a value of $\left(1 + \frac{c}{2}\right)$ (the value of $\beta$ behind the RPTW)), which is not of permanent form, propagates with speed $c$ in the positive $x$ direction. A schematic representation of the location and thickness of the asymptotic regions as $t \to \infty$ when $0 \leq c < 2$ is given in Figure 5.3.

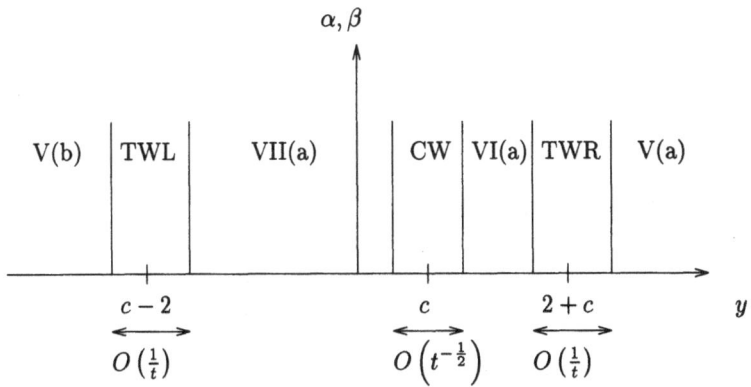

**Fig. 5.3.** Schematic representation of the location and thickness of the asymptotic regions as $t \to \infty$ when $0 \leq c < 2$.

(ii) $c \geq 2$

In this case the solution to **IVP** as $t \to \infty$ admits only the right permanent form travelling wave solution (RPTW) which propagates with the minimum speed available, that being $(2 + c)$, with this speed being (via (5.134)) approached from below as $t \to \infty$. We note that the correction to the wave speed of the RPTW is of $O\left(\frac{1}{t}\right)$ as $t \to \infty$. We also note that in this case a convection wave (which primarily allows for the adjustment in the concentration of $\beta$ which has to adjust from a value of $o(1)$ to a value of $\left(1 + \frac{c}{2}\right)$ (the value of $\beta$ behind the RPTW)), which is not of permanent form, propagates with speed $c$ in the positive $x$ direction. Further, in this case there is an additional important region located about $x = 0$, which primarily allows the adjustment in the concentration of $\alpha$ which has to adjust from $O(1)$ (for $x < 0 - o(1)$) to $o(1)$ (for $x > 0 + o(1)$)) A schematic representation of the location and thickness of the asymptotic regions as $t \to \infty$ when $c \geq 2$ is given in Figure 5.4.

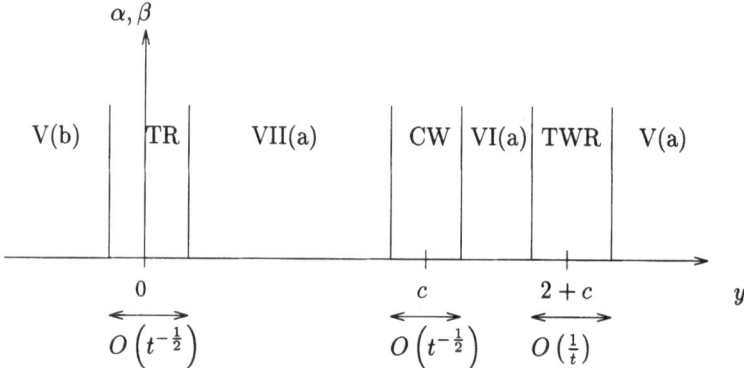

**Fig. 5.4.** Schematic representation of the location and thickness of the asymptotic regions as $t \to \infty$ when $c \geq 2$.

## 5.5 Numerical Solutions

Finally, we present numerical solutions to the initial value problem **IVP** which both support and illustrate the detailed analysis given in the previous sections. We restrict our attention to the case when $\sigma = 1$ and $\beta_0 = \frac{1}{2}$ and consider three values of $c$, namely $c = 0$, $c = 1$ and $c = 3$ which illustrate the distinct types of behaviour. Equations (5.11) were solved using a modified Crank-Nicolson technique with the initial input of the autocatalyst given by

$$g(x) = \cos^2\left(\frac{\pi x}{2}\right), \quad |x| \leq 1. \tag{5.202}$$

In order to accommodate the boundary conditions (5.9) as $|x| \to \infty$, we seek a solution of equations (5.11) for $t > 0$ on the domain $-L \leq x \leq L$, where $L$ is taken sufficiently large, and apply the boundary conditions $\alpha(-L, t) = \alpha(L, t) = 1$, $\beta(-L, t) = \beta(L, t) = 0$ for $t > 0$.

(i) c=0:    Figure 5.5 shows the development of the solution when $c = 0$. As expected, both a RPTW and LPTW of minimum $(v_+ = 2)$ and maximum $(v_- = -2)$ speeds respectively develop with the resulting waves being symmetric about $x = 0$.

(ii) c=1:    Figure 5.6 shows the development of the solution when $c = 1$. Again both a RPTW and a LPTW develop where the RPTW travels with minimum speed $(v_+ = 3)$ while the LPTW travels with maximum speed $(v_- = -2)$.

(iii) c=3:   Figure 5.7 shows the development of the solution when $c = 3$. In this case only a RPTW of minimum speed $(v_+ = 5)$ develops.

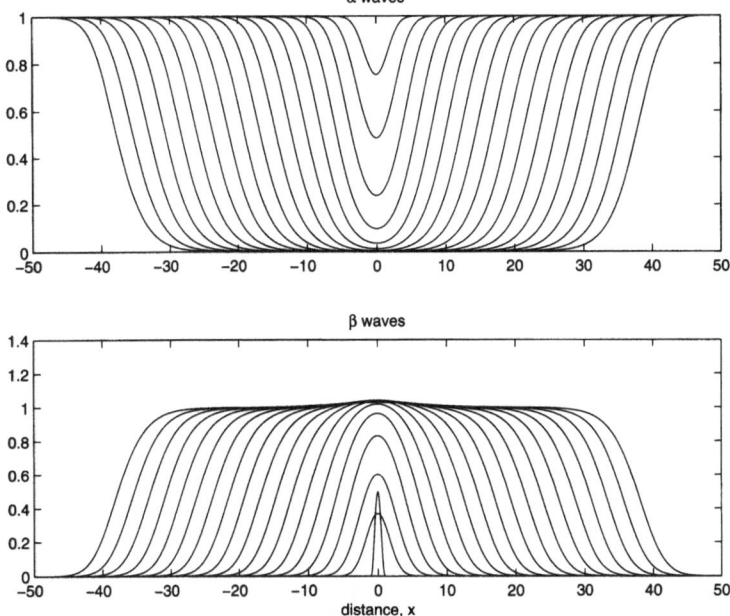

**Fig. 5.5.** $\alpha$ and $\beta$ profiles when $c = 0$. $v_- = -2$, $\beta_{+\infty} = 1$, $v_+ = 2$, $\beta_{-\infty} = 1$.

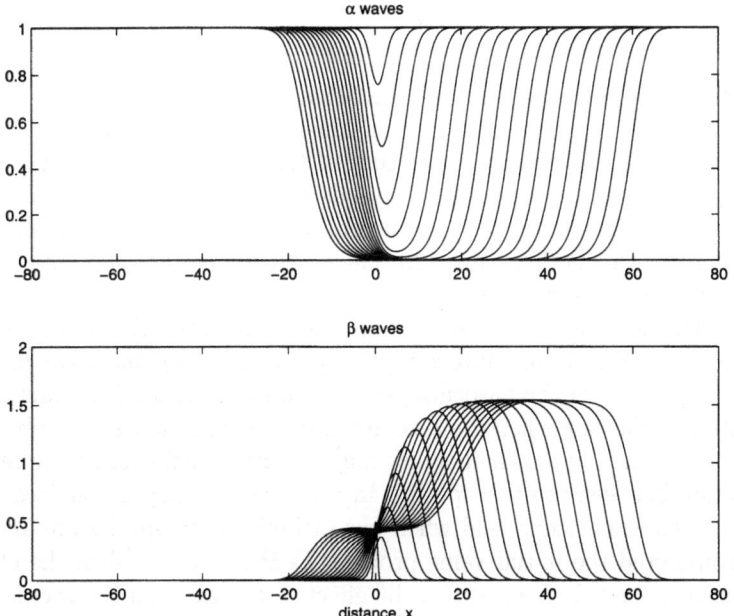

**Fig. 5.6.** $\alpha$ and $\beta$ profiles when $c = 1$. $v_- = -1$, $\beta_{+\infty} = 1.5$, $v_+ = 3$, $\beta_{-\infty} = 0.5$.

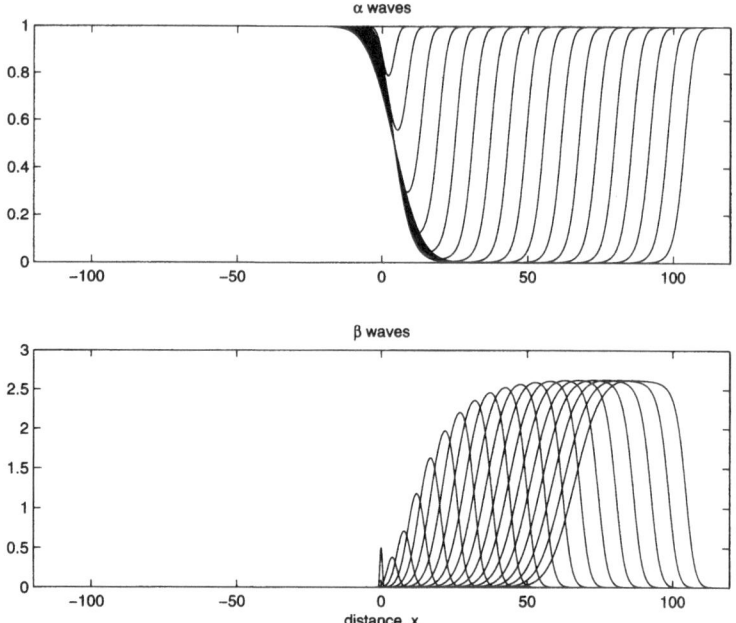

**Fig. 5.7.** $\alpha$ and $\beta$ profiles when $c = 3$. $v_+ = 5$, $\beta_{-\infty} = 2.5$.

## 5.6 Conclusions

In this chapter we have demonstrated that the methodology, based on the method of matched asymptotic expansions, developed in Chapters 2, 3 and 4, to determine the complete structure of the solution of the scalar Fisher-Kolmogorov equation for large-$t$, extends to systems of Fisher-Kolmogorov equations. Specifically, we have obtained the complete large-$t$ structure of the solution to a system of reaction-diffusion equations which form a simple model of an ionic autocatalytic system.

In particular, we have considered the propagating reaction-diffusion waves that develop in a simple autocatalytic system (5.1) by the local introduction of a quantity of the autocatalyst, B, into an expanse of reactant, A. We have studied the case when the reactant and autocatalyst are not only mobile through diffusion, but their mobility may be further enhanced by differential convection between the two species. In practice this may be realized by applying a uniform electric field when one or both of the species are ionic. We have considered the initial value problem **IVP** and established the effect of the parameter $c$, which measures the effect of the external electric field, has on the evolution of LPTWs and RPTWs. Due to the symmetry of the initial value problem, via Theorem 5.29, we need only consider **IVP** with $c \geq 0$. In Section 5.2, via Theorems 5.26 and 5.27, we have determined that for $c \geq 0$ a unique RPTW solution of RP$[c, v]$ exists for each $v \geq 2 + c$. Further, when $0 \leq c < 2$ a unique LPTW solution of LP$[c, v]$ exists for each $v \leq -2 + c$, whilst when $c \geq 2$ no LPTW solution of LP$[c, v]$ exists. The consideration of **IVP** in Section 5.3 indicates that when an RPTW develops it propagates with speed $v_+ = 2 + c$ whilst when an LPTW develops it propagates with speed $v_- = -2 + c$ $(0 \leq c < 2)$. The large-$t$ asymptotic solution to **IVP** is developed, via matched asymptotic expansions, in Section 5.4 for $0 \leq c < 2$ and $c \geq 2$. This analysis both supports the results of the previous sections and establishes that a convective wave, which is not of permanent form, propagates with speed $c$ in the positive $x$ direction. Numerical simulations are presented in Section 5.5 which support and confirm the analysis presented in the previous sections.

# The Analysis of a Class of Singular Scalar Reaction-Diffusion Equations

# Introduction

In Part II of this monograph we consider the analysis of a class of singular reaction-diffusion equations which arise as a model of isothermal, autocatalytic chemical reaction with termination. The scheme is represented formally by the two steps,

$$A \to B, \quad \text{rate } k_1 ab^m \quad \text{(autocatalysis)}, \tag{6.1}$$

$$B \to C, \quad \text{rate } k_2 b^n \quad \text{(decay)}, \tag{6.2}$$

where $a$ and $b$ are the concentrations of the reactant A and the autocatalyst B respectively, $k_1 > 0$ is the rate constant of autocatalysis, $k_2 > 0$ is the rate constant at which the autocatalyst B decays to the inert, stable, product C, and $m$ and $n$ are the (fractional, $0 < m, n < 1$) orders of autocatalysis and decay. We consider the situation arising when a localized input of the autocatalyst B is introduced into an expanse of the reactant A, initially at uniform concentration. The equations that govern the reaction and diffusion of the species A and B under the scheme (6.1) and (6.2) are

$$\frac{\partial a}{\partial \bar{t}} = D \frac{\partial^2 a}{\partial \bar{x}^2} - k_1 ab^m, \tag{6.3}$$

$$\frac{\partial b}{\partial \bar{t}} = D \frac{\partial^2 b}{\partial \bar{x}^2} + k_1 ab^m - k_2 b^n. \tag{6.4}$$

Here $\bar{x}$ and $\bar{t}$ are the coordinates measuring distance and time, respectively, while $D$ is the constant diffusion rate of A and B. The diffusion rates are assumed to be equal, which is a good approximation when the molecular sizes of A and B are comparable. The initial conditions to be considered are

$$a(\bar{x}, 0) \equiv a_0 \quad b(\bar{x}, 0) = \begin{cases} b_0 g(\bar{x}), & |\bar{x}| \le l, \\ 0, & |\bar{x}| > l. \end{cases} \tag{6.5}$$

Here $g(\bar{x})$ is a given, nonnegative function of $\bar{x}$, with a maximum value of unity and is analytic on $|\bar{x}| \le l$ with $g(\pm l) = 0$. The constants $a_0$ and $b_0$ are the positive initial concentration of A and the maximum initial concentration of B, respectively. In addition, we have the boundary conditions

$$a(\bar{x}, \bar{t}) \to a_0, \quad b(\bar{x}, \bar{t}) \to 0 \quad \text{as} \quad |\bar{x}| \to \infty, \quad \bar{t} \geq 0. \tag{6.6}$$

When $k_2 \ll k_1 a_0^{m-n+1}$, the model (6.3)-(6.6) may be simplified. In particular, it may be shown that $a(\bar{x}, \bar{t}) + b(\bar{x}, \bar{t}) \sim a_0$ in (6.3)-(6.6) for $\bar{t} \gg (k_1 a_0^m)^{-1}$, uniformly in $\bar{x}$. This leads us to consider the following scalar reaction–diffusion problem in place of the system (6.3)-(6.6)

$$\bar{u}_{\bar{t}} = D\bar{u}_{\bar{x}\bar{x}} + k_1(a_0 - \bar{u})\bar{u}^m - k_2\bar{u}^n, \quad -\infty < \bar{x} < \infty, \quad \bar{t} > 0, \tag{6.7}$$

$$\bar{u}(\bar{x}, 0) = \begin{cases} b_0 g(\bar{x}), & |\bar{x}| \leq l, \\ 0, & |\bar{x}| > l, \end{cases} \tag{6.8}$$

$$\bar{u}(\bar{x}, \bar{t}) \to 0, \quad \text{as } |\bar{x}| \to \infty, \quad \bar{t} \geq 0, \tag{6.9}$$

with $0 < m, n < 1$. Here, $\bar{u}$ may be regarded as an approximation to $b$ when $k_2 \ll k_1 a_0^{m-n+1}$. It is convenient to introduce dimensionless variables into (6.7)-(6.9) as

$$u = \frac{\bar{u}}{a_0}, \quad t = k_1 a_0^m \bar{t}, \quad x = \left(\frac{k_1 a_0^m}{D}\right)^{\frac{1}{2}} \bar{x}, \tag{6.10}$$

in terms of which (6.7)-(6.9) become

$$u_t = u_{xx} + f(u), \quad -\infty < x < \infty, \quad t > 0, \tag{6.11}$$

$$f(u) = \begin{cases} (1 - u)u^m - ku^n, & u > 0, \\ 0, & u \leq 0, \end{cases} \tag{6.12}$$

$$u(x, 0) = \begin{cases} u_0 g(x), & |x| \leq \sigma, \\ 0, & |x| > \sigma, \end{cases} \tag{6.13}$$

$$u(x, t) \to 0, \quad \text{as } |x| \to \infty, \quad t \geq 0, \tag{6.14}$$

where $0 < m, n < 1$. The problem (6.11)-(6.14) has three positive dimensionless parameters, namely

$$\sigma = \left(\frac{l^2 k_1 a_0^m}{D}\right)^{\frac{1}{2}}, \quad k = \frac{k_2}{k_1 a_0^{m-n+1}}, \quad u_0 = \frac{b_0}{a_0}. \tag{6.15}$$

The parameter $\sigma$ measures the spread of the initial input of autocatalyst, $u_0$ measures the maximum concentration in the initial input of autocatalyst, and $k$ measures the strength of the termination step (6.2) relative to that of the autocatalytic step (6.1). Applications and further details of the scheme (6.1) and (6.2), both when $m, n \geq 1$ and when $0 < m, n < 1$, are extensively discussed and referenced in McCabe, Leach and Needham [38], which should be referred to for a fuller discussion.

For simplicity we choose $g(x)$ to be symmetric about $x = 0$ and replace problem (6.11)-(6.14) by the scalar initial-boundary value problem

$$u_t = u_{xx} + f(u), \quad x, t > 0, \tag{6.16}$$

$$f(u) = \begin{cases} (1-u)\, u^m - ku^n, & u > 0, \\ 0, & u \leq 0, \end{cases} \tag{6.17}$$

$$u(x, 0) = \begin{cases} u_0 g(x), & 0 \leq x \leq \sigma, \\ 0, & x > \sigma, \end{cases} \tag{6.18}$$

$$u(x, t) \to 0 \quad \text{as} \quad x \to \infty, \quad t \geq 0, \tag{6.19}$$

$$u_x(0, t) = 0, \quad t > 0. \tag{6.20}$$

The class of singular scalar reaction-diffusion equations (6.16)-(6.20) will be considered in detail in Chapters 7, 8 and 9 whilst the coupled system (6.3)-(6.6) will be considered in Chapter 10.

It has been established that the fractional order autocatalytic step ((6.1) with $0 < m < 1$) in the absence of termination cannot support permanent form travelling waves (see King and Needham [28]) but, as we demonstrate in Chapter 7, the inclusion of the fractional-order termination step ((6.2) with $0 < n < 1$) restores, in the long time, the capability of the fractional-order autocatalytic step (6.1) to support finite speed travelling waves provided that $m > n$ and $0 < k < k_c$ (where $k_c$ depends on $m$ and $n$). These waves have only semi-infinite support and are of excitable, rather than Fisher–Kolmogorov, type; that is, for fixed $0 < n < m < 1$ and $0 < k < k_c$ there exists exactly one travelling wave with unique wave speed. Further, in Chapter 8, we demonstrate that the initial data (6.18) must exceed a parameter-dependent critical threshold, defined by the steady states of the problem, before propagation can occur, and that the approach, from below, to the constant propagation speed and the contraction of the solution on to the permanent form travelling wave are exponential in $t$.

In the absence of travelling wave structures, solutions to the problem (6.16)-(6.20) display a variety of behaviours. For $(m < n)$ and $(m = n, k < 1)$, we need to relax boundary condition (6.19) and replace it with

$$u(x, t) \to u_\infty(t) \quad \text{as } x \to \infty, \quad t \geq 0, \tag{6.21}$$

where $u_\infty(t)$ is the nontrival solution to the singular initial value problem

$$u'_\infty = f(u_\infty), \quad t > 0,$$

$$u_\infty(0) = 0.$$

In this case we obseve, in (6.16)-(6.20), spatially uniform solutions which grow algebrically in $t$ (time) and approach a non-zero equilibrium state, uniformly in $x \geq 0$, as $t \to \infty$ through terms exponential in $t$. This is the 'lifting at infinity' behaviour associated with $f(u) = u^p, 0 < p < 1$, in (6.16) (see King and Needham [28]). For $(m = n, k > 1)$, the solution exhibits finite $t$ extinction with contracting support, associated with $f(u) = -u^p, 0 < p < 1$, in (6.16) (see Bandle and Stackgold [3] and Grundy and Peletier [23]). Finite $t$ extinction is also observed in (6.16) in the case $m > n$ if $k$ is sufficiently large ($k \geq k_c$) or the initial data is sufficiently small. For ($m = n, k = 1$) the kinetics are regular and the small-$t$ structure is diffusion driven, resulting in a solution with infinite support for all $t > 0$, which decays to zero as $t \to \infty$, uniformly in $x \geq 0$.

All of these phenomena are discussed in Chapter 8 where a detailed study of the qualitative behaviour of the solution to (6.16)-(6.20) is given over all parameter ranges. Further, we develop the full asymptotic structure of the solution to (6.16)-(6.20) as $t \to 0$ and as $t \to \infty$ (for the cases ($m < n$), ($m = n, k = 1$), ($m = n, k < 1$) and (($m > n$), provided that the initial data exceeds a parameter dependent critical threshold)) or as $t \to t_c^-$ (for the cases ($m = n, k > 1$) and (($m > n$), provided that the initial data does not exceed a parameter dependent critical threshold)).

In Chapter 9 we consider initial-boundary value problem (6.16)-(6.20) when the initial data, $u(x, 0)$, has unbounded support with exponential or algebraic decay rates. We demonstrate using the method of matched asymptotic expansions that the solution, $u(x, t)$, develops finite support in infinesimal time. The asymptotic form for the location of the edge of the support as $t \to 0$ is presented in both cases.

In Chapter 10 we shall be concerned with the analysis of the full, coupled system of singular reaction-diffusion equations associated with the steps (6.1) and (6.2). In particular we consider the well stirred system and develop the asymptotic solution as $t \to 0$ over all parameter values. The system closely reproduces the behaviour of the scalar model (its approximation, which we consider in detail in Chapters 7, 8 and 9) and, throughout, we shall draw comparisons with the results of Chapters 7-9.

# Permanent Form Travelling Waves (PTWs)

An important first stage in the study of the initial value problem (6.16)–(6.20) is an investigation of the permanent form travelling wave (PTW) solutions of equation (6.16) which may be generated from the initial value problem (6.16)–(6.20). We study these in this chapter, and to this end we make the following definition.

**Definition 7.1.** *A PTW solution of equation (6.16) is a non-trivial, non-negative classical solution that depends only on the single variable $z = x - \gamma(t)$ (where $\gamma(t)$ is the position of the wavefront) and satisfies the uniform conditions $u \to 0$ as $z \to \infty$ and $u \to u_{-\infty}$ as $z \to -\infty$, where $u_\infty \geq 0$ is constant.*

## 7.1 General Properties of PTW Solutions

The differential equation governing PTW solutions of equation (6.16) is obtained by looking for a solution in the form $u \equiv u(z)$, after which (6.16) becomes

$$u_{zz} + c u_z + f(u) = 0, \quad -\infty < z < \infty, \tag{7.1}$$

where

$$f(u) = \begin{cases} (1-u)\, u^m - k u^n, & u \geq 0, \\ 0, & u < 0, \end{cases} \tag{7.2}$$

and $c(t) = \frac{d\gamma}{dt}$. However, since $u$ is a function of $z$ alone, equation (7.1) requires that the PTW propagation speed, $c$, is constant, after which the symmetry of (7.1) shows that we need only consider $c > 0$. The boundary conditions associated with equation (7.1) are

$$u(z) \to 0 \qquad \text{as} \quad z \to \infty, \tag{7.3a}$$
$$u(z) \to u_{-\infty} \quad \text{as} \quad z \to -\infty, \tag{7.3b}$$
$$u(z) \geq 0 \qquad \text{for all} \quad -\infty < z < \infty. \tag{7.3c}$$

The nonlinear boundary value problem (7.1)–(7.3) may be regarded as an eigenvalue problem for $c$ and we shall denote this problem by $E[m, n, k]$. It is important to note at this stage that the problem $E[m, n, k]$ is singular in the sense that, although $f(u)$ is a continuous function of $u \geq 0$, it is not Lipschitz continuous at $u = 0$ as $m$ and $n$ are both less than unity. This prevents us from applying directly the theory developed for regular problems where $f(u) \in C^1[0, \infty)$ (see, for example Fife [14]).

To proceed, we first examine the nature of $f(u)$ when $u \geq 0$. There are three cases to consider.

(a) $m > n$.  In this case, for $0 < k < k^*$, where

$$k^* = \frac{(m - n)^{m-n}}{(m - n + 1)^{m-n+1}},$$  (7.4)

$f(u)$ has three non-negative zeros, $u = 0$, $u = \eta_1, \eta_2$, with

$$0 < \eta_1 < \eta_2 < 1.$$  (7.5)

In particular,

$$\eta_1, \eta_2 \to \eta_0 = \frac{m - n}{m - n + 1} \quad \text{as } k \to k^*,$$  (7.6)

whilst

$$\eta_2 \to 1, \quad \eta_1 \sim k^{1/(m-n)} \quad \text{as } k \to 0.$$  (7.7)

Both $\eta_1$ and $\eta_2$ behave monotonically with $0 < k < k^*$. We note that $f(u) > 0 [< 0]$ for $\eta_1 < u < \eta_2 [u \in (0, \eta_1) \cup (\eta_2, \infty)]$ respectively.

For $k = k^*$, $f(u)$, has two non-negative zeros at $u = 0$ and $u = \eta_0$, the latter of which is an order two zero.

Finally, for $k > k^*$, $f(u)$ has just a single non-negative zero at $u = 0$ with $f(u) < 0$ for $u > 0$.

A further quantity of interest in this case is

$$H(k) = \int_0^{\eta_2} f(u)\, du, \quad 0 < k < k^*.$$  (7.8)

It is straightforward to show that $H(k) > 0$ as $k \to 0$ whilst $H(k) < 0$ as $k \to k^*$, with $H(k)$ monotone decreasing for $0 < k < k^*$. Thus $H(k)$ has a single zero for $k \in (0, k^*)$ and this occurs at

$$k_c = \left( \frac{n + 1}{m + 1} \right) \left( \frac{m + 2}{m + 1} \right)^{m-n} k^*.$$  (7.9)

We will return to the value $k = k_c$ later in this chapter.

(b) $m = n$.  In this case $f(u)$ has two non-negative zeros for $k < 1$, located at $u = 0$ and $u = 1 - k$. We note that $f(u) > 0 [< 0]$ for $0 < u < 1 - k [u > 1 - k]$ respectively. With $k \geq 1$, $f(u)$ has just a single non-negative zero at $u = 0$ with $f(u) < 0$ for $u > 0$.

**(c) $m < n$.** Here $f(u)$ has two non-negative zeros at $u = 0$ and $u = \nu$ for all $k > 0$. We note that $f(u) > 0$ $[< 0]$ for $0 < u < \nu$ $[u > \nu]$ respectively. In particular,

$$\nu \to 1 \quad \text{as} \quad k \to 0, \qquad \nu \sim k^{-1/(n-m)} \quad \text{as} \quad k \to \infty, \qquad (7.10)$$

with $\nu$ being a monotone function of $k \in (0, \infty)$.

We now consider the possible values of $u_{-\infty}$ in a PTW. We have the following proposition.

**Proposition 7.2.** *A PTW must have $u_{-\infty} > 0$.*

*Proof.* By definition, $u_{-\infty} \geq 0$. Now suppose that $u(z)$ is a PTW with $u_{-\infty} = 0$. We apply the operation $\int_{-A}^{A} \ldots u_z dz$ to equation (7.1) which leads to

$$c \int_{-A}^{A} u_z^2 \, dz \; = \; -\frac{1}{2} \left[ u_z^2 \right]_{-A}^{A} - \left[ G(u(z)) \right]_{-A}^{A}, \qquad (7.11)$$

where $A > 0$ and we define

$$G(u) \; = \; \int_0^u f(s) \, ds. \qquad (7.12)$$

We can take the limit as $A \to \infty$ in both sides of (7.11), which, on using conditions (7.3), results in

$$c \int_{-\infty}^{\infty} u_z^2 \, dz \; = \; 0. \qquad (7.13)$$

Since $c > 0$, this requires $u \equiv 0$, which is the trivial solution and the result follows. $\qquad \square$

We now consider the cases $m > n$, $m < n$, $m = n$, separately.

## 7.2 PTW Solutions when $m > n$

We first have a proposition.

**Proposition 7.3.** *For $m > n$ there are no PTW solutions when $k \geq k^*$.*

*Proof.* (i) $k = k^*$. In this case Proposition 7.2 requires $u_{-\infty} = \eta_0$ in a PTW. Suppose that $u(z)$ is a PTW, then equation (7.1) and conditions (7.3) require that

$$c \int_{-\infty}^{\infty} u_z^2 \, dz \; = \; - \left[ G(u(z)) \right]_{-\infty}^{\infty},$$

$$= \; G(\eta_0) \; < \; 0,$$

via (7.12). This gives a contradiction, and the result follows.

(ii) $k > k^*$. As $u = 0$ is the only non-negative zero of $f(u)$ in this case, the result follows directly from Proposition 7.2. $\qquad \square$

We are left only to consider the situation when $0 < k < k^*$. Firstly we have the following proposition.

**Proposition 7.4.** *For $m > n$, a PTW with $0 < k < k^*$ requires $u_{-\infty} = \eta_2$.*

*Proof.* Via Proposition 7.2, a PTW with $0 < k < k^*$ requires either $u_{-\infty} = \eta_1$ or $u_{\infty} = \eta_2$. Now, suppose that $u(z)$ is a PTW with $u_{-\infty} = \eta_1$, then equation (7.1) and conditions (7.3) require that

$$c \int_{-\infty}^{\infty} u_z^2 \, dz \; = \; G(\eta_1) \; < \; 0,$$

via (7.12), and we obtain a contradiction.    $\square$

Finally, we have the following proposition.

**Proposition 7.5.** *For $m > n$, a necessary condition for the existence of a PTW is $0 < k < k_c$.*

*Proof.* Let $u(z)$ be a PTW with $k \in [k_c, k^*)$. Then, via Proposition 7.4, $u_{-\infty} = \eta_2$, after which equation (7.1) and conditions (7.3) require that

$$G(\eta_2) \; = \; c \int_{-\infty}^{\infty} u_z^2 \, dz \; > \; 0. \tag{7.14}$$

However, via (7.8) and (7.12),

$$G(\eta_2) \; \equiv \; H(k) \; \leq 0, \quad \text{as} \quad k \in [k_c, k^*). \tag{7.15}$$

Thus we arrive at a contradiction through (7.14) and (7.15), and the result follows.    $\square$

We next show that when $m > n$, then $k \in (0, k_c)$ is also a sufficient condition for the existence of a PTW solution. It is convenient to introduce the dependent variable $w \equiv u_z$ and write equation (7.1) as the equivalent planar dynamical system,

$$u_z \; = \; w, \quad w_z \; = \; -cw \; - \; f(u), \quad u \geq 0, \quad -\infty < w < \infty. \tag{7.16}$$

We shall examine the phase portrait of this system (with $c > 0$, $k \in (0, k_c)$) in the right hand half plane of the $(u, w)$ phase plane. Particular attention will be placed on examining the phase portrait close to the equilibrium point at $(0,0)$. The right hand side of the system (7.16) is singular (not Lipschitz continuous) at $(0,0)$ and so neither linearization nor centre manifold theory are able to classify the fully nonlinear and singular behaviour close to this equilibrium point. However, we are able to proceed to classify $(0,0)$ by making suitable transformations which regularize (7.16) in the neighbourhood of $(0,0)$. Thereafter we employ a shooting argument from the regular equilibrium point at $(\eta_2, 0)$.

### 7.2.1 Local Behaviour

The dynamical system (7.16) (for $0 < k < k_c$) has three equilibrium points in $u \geq 0$, at

$$\underline{e}_1 = (0,0), \quad \underline{e}_2 = (\eta_1, 0), \quad \underline{e}_3 = (\eta_2, 0), \qquad (7.17)$$

and a PTW solution to $E[m, n, k]$ requires a directed integral path which remains in $u \geq 0$, and connects $\underline{e}_3$ to $\underline{e}_1$ (via Proposition 7.4). We note that the integral paths of the system (7.16) correspond to solutions of the first order differential equation

$$\frac{dw}{du} = -c - \frac{f(u)}{w}, \quad u \geq 0, \quad -\infty < w < \infty. \qquad (7.18)$$

We begin by examining the local phase portraits in the neighbourhood of the equilibrium points, $\underline{e}_1$, $\underline{e}_2$ and $\underline{e}_3$.

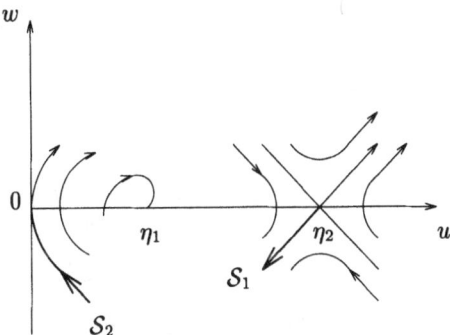

**Fig. 7.1.** The $(u, w)$ phase plane for $n < m, 0 < k < k_c$ and $0 < c < 2[f'(\eta_1)]^{1/2}$.

The equilibrium point $\underline{e}_3$ is a simple hyperbolic equilibrium point, and linearization shows that it is a saddle point, with eigenvalues and eigenvectors given by

$$\underline{\mu}_\pm = [1, \lambda_\pm]^T, \qquad \lambda_\pm = \frac{1}{2}\left[-c \pm \sqrt{c^2 - 4f'(\eta_2)}\right], \qquad (7.19)$$

where $f'(\eta_2) < 0$. Thus, the only path which satisfies (7.3b) is that part of the unstable manifold at $\underline{e}_3$ which is directed into the region $0 \leq u \leq \eta_2$, $w \leq 0$, and we label this as $\mathcal{S}_1$. The remaining part of the unstable manifold at $\underline{e}_3$ is directed into $u \geq \eta_2$, $w \geq 0$, and this region is invariant for the dynamical system (7.16) and so cannot connect to $\underline{e}_1$, and thus is unable to provide a PTW. The local behaviour at $\underline{e}_3$ is illustrated in Figure 7.1.

The equilibrium point $\underline{e}_2$ is also a simple hyperbolic equilibrium point, and linearization establishes that it is a stable spiral for $0 < c < 2\sqrt{f'(\eta_1)}$, a stable degenerate node for $c = 2\sqrt{f'(\eta_1)}$, and a stable node for $c > 2\sqrt{f'(\eta_1)}$

(with $f'(\eta_1) > 0$). The eigenvalues and associated eigenvectors (in the case of a node) are

$$\underline{\nu}_\pm = [1, \gamma_\pm]^T, \qquad \gamma_\pm = \frac{1}{2}\left[-c \pm \sqrt{c^2 - 4f'(\eta_1)}\,\right]. \tag{7.20}$$

The local behaviour at $\underline{e}_2$ is shown in Figure 7.1.

Finally we address the equilibrium point $\underline{e}_1$ at the origin. As $f(u)$ is not Lipschitz continuous at $u = 0$, this equilibrium point is not hyperbolic and cannot be classified via linearization, and a more detailed analysis is required. In order to proceed we consider the local existence of solutions to the initial value problem [P], namely

$$w\,\frac{dw}{du} = -cw - f(u), \quad u \in (0, T], \tag{7.21}$$

$$w(0) = 0 \tag{7.22}$$

for some $T > 0$. Solutions to [P] correspond to invariant manifolds (in $u > 0$) of (7.16) which intersect $\underline{e}_1$. Here $w : [0, T] \to \mathbb{R}$ is continuous on $[0, T]$ and differentiable on $(0, T]$. We wish to investigate local solutions of [P] (i.e. $0 < T \ll 1$), and first observe that when $w(u)$ provides a solution to [P], then

$$w^2(u) = -2c \int_0^u w(s)\,ds - 2\int_0^u f(s)\,ds \tag{7.23}$$

with $u \in [0, T]$, $w(u) \in C[0, T]$. We now have the following lemma.

**Lemma 7.6.** *Let $w(u)$ be a solution to [P], then there exists a $\delta > 0$ such that $w(u)$ is strictly of one sign on $u \in (0, \delta]$.*

*Proof.* Let $u = u^* \in (0, T]$ be a zero of $w(u)$ so that $w(u^*) = 0$. As $w(u)$ is differentiable on $(0, T]$ and $w'(u^*)$ is finite we have, from (7.21), that $f(u^*) = 0$. However, there exists a $\delta > 0$ such that $f(u) < 0$ for all $u \in (0, \delta]$. The result follows. $\qquad\square$

**Lemma 7.7.** *Let $w(u)$ be a solution to [P], then $w(u) \geq O\left(u^{\frac{n+1}{2}}\right)$ as $u \to 0^+$.*

*Proof.* Follows directly as a consequence of Lemma 7.6 and (7.23). $\qquad\square$

**Lemma 7.8.** *The initial value problem [P] has at most one solution which is strictly positive in a sufficiently small neighbourhood of $u = 0$.*

*Proof.* Let $w_1(u)$ and $w_2(u)$ be two solutions of [P], which are strictly positive for $u \in (0, \delta]$, say, for some $\delta > 0$. Let

$$\Theta(u) = (w_2 - w_1)(u), \qquad u \in [0, \delta]. \tag{7.24}$$

From equation (7.21) we have that

$$\frac{1}{2}\frac{d}{du}\left(w_2^2 - w_1^2\right) = -c\left(w_2 - w_1\right), \qquad u \in (0, \delta],$$

which becomes, on using (7.24),

$$\frac{d}{du}\left\{\Theta(u)(w_2 + w_1)\right\} = -2c\Theta(u), \qquad u \in (0, \delta].$$

Now, since $w_1 + w_2 > 0$ on $(0, \delta]$, $\Theta(u)$ satisfies

$$\frac{d\Theta}{du} + \left\{\frac{h'(u)}{h(u)} + \frac{2c}{h(u)}\right\}\Theta = 0, \qquad u \in (0, \delta], \qquad (7.25)$$

with $\Theta(0) = 0$, and $h(u) = (w_1 + w_2)(u)$. We observe that

$$h(u) > 0, \quad \text{for all} \quad u \in (0, \delta], \tag{7.26a}$$
$$h(0) = 0, \tag{7.26b}$$
$$h(u) \geq O\left(u^{\frac{n+1}{2}}\right), \quad \text{as } u \to 0^+. \tag{7.26c}$$

An integration of (7.25) gives

$$\Theta(u) = \frac{A}{h(u)}e^{-2cI(u)}, \qquad u \in (0, \delta], \tag{7.27}$$

where

$$I(u) = \int_0^u \frac{ds}{h(s)}, \qquad u \in (0, \delta],$$

and $A$ is a constant. Now, we require $\Theta(u) \to 0$ as $u \to 0^+$. However, $I(u) \to 0$ as $u \to 0^+$ via (7.26c), and so

$$\Theta(u) \sim \frac{A}{h(u)} \qquad \text{as } u \to 0^+,$$

which then requires $A = 0$. Therefore $\Theta \equiv 0$ for $u \in [0, \delta]$ as required.     □

*Remark 7.9.* We may also show, as above, that [P] has at most one solution which is strictly negative in a sufficiently small neighbourhood of $u = 0$.

We next consider the initial value problem [Q] for $\psi(u)$, namely,

$$\frac{d\psi}{du} = -2cH(\psi) - 2f(u), \qquad u \in (0, T], \tag{7.28}$$

$$\psi(0) = 0, \tag{7.29}$$

where

$$H(\psi) = \begin{cases} \psi^{\frac{1}{2}}, & \psi \geq 0, \\ 0, & \psi < 0. \end{cases} \tag{7.30}$$

We have the following lemma.

**Lemma 7.10.** *The initial value problem [Q] has at least one solution on $u \in [0, \epsilon]$ for some $\epsilon > 0$.*

*Proof.* This follows directly from Picard's local existence theorem (see, for example, [12], Chapter 1) as both $H(\psi)$ and $f(u)$ are continuous functions of $\psi$ and $u$ respectively in a sufficiently small neighbourhood of $\psi = u = 0$.  □

We observe that Picard's theorem does not guarantee uniqueness for [Q] due to the loss of Lipschitz continuity of $H(\psi)$ and $f(u)$ at $\psi = u = 0$. However, we have the following lemma.

**Lemma 7.11.** *Let $\psi(u)$ be a solution of [Q], then $\psi(u) > 0$ or $\psi(u) < 0$ for $u \in (0, \delta]$ for some $\delta > 0$.*

*Proof.* First choose $\delta'$ sufficiently small so that $f(u) < 0$ for all $u \in (0, \delta']$. Now, let $\psi(u)$ be a solution to [Q]. It then follows directly from (7.28) that if $u = u^* \in (0, \delta']$ is a zero of $\psi(u)$, then it is a simple zero. Moreover, it also follows from (7.28) and (7.29) that

$$\psi(u) = -2c \int_0^u H(\psi(s))ds - 2 \int_0^u f(s)ds, \qquad u \in (0, \delta']. \qquad (7.31)$$

Now suppose that $\psi(u)$ is neither strictly positive or strictly negative on $u \in (0, \delta]$ for any $0 < \delta < \delta'$. Then for every $0 < \delta < \delta'$ there exist consecutive zeros of $\psi(u)$, say $u_1, u_2 \in (0, \delta]$ with $u_2 > u_1$, which have

$$\psi(u) < 0, \qquad u \in (u_1, u_2). \qquad (7.32)$$

On using (7.30) and (7.31) we find that

$$0 = \psi(u_2) - \psi(u_1) = -2 \int_{u_1}^{u_2} f(s)ds,$$

which, via (7.30) and (7.32), leads to a contradiction, as $f(u) < 0$ for all $u \in (0, \delta]$ with $0 < \delta < \delta'$. The result follows.  □

Finally we obtain the following lemma.

**Lemma 7.12.** *Every local solution to [Q] is strictly positive sufficiently close to $u = 0$.*

*Proof.* Let $\psi(u)$ be a solution to [Q]. Via Lemma 7.11, we know that $\psi(u) > 0$ or $\psi(u) < 0$, in a sufficiently small neighbourhood of $u = 0$. Suppose that $\psi < 0$, then given any $\epsilon > 0$, there exists $u^* \in (0, \epsilon)$ such that $\psi(u^*) < 0$, with $\psi(u) < 0$ for all $u \in (0, u^*]$. Now set $u = u^*$ in (7.31) to obtain

$$0 > -2 \int_0^{u^*} f(s)ds. \qquad (7.33)$$

We may now select $\epsilon > 0$ sufficiently small so that $f(s) < 0$ for all $s \in (0, \epsilon)$, which leads to a contradiction in (7.33) and completes the proof.  □

We conclude that the initial value problem [Q] has at least one solution $\psi = \psi^*(u)$ on, say, $[0, \delta^*]$ for some $\delta^* > 0$, with $\psi^*(u) > 0$ for all $u \in (0, \delta^*]$. Now set

$$w^*(u) = \sqrt{\psi^*(u)}, \qquad u \in [0, \delta^*]. \tag{7.34}$$

Direct substitution verifies that $w^*(u)$ provides a positive solution to the initial value problem [P] for $u \in [0, \delta^*]$.

We next consider the initial value problem [R] for $\chi(u)$, namely

$$\frac{d\chi}{du} = -2cG(\chi) - 2f(u), \qquad u \in (0, T], \tag{7.35}$$

$$\chi(0) = 0, \tag{7.36}$$

where

$$G(\chi) = \begin{cases} -\chi^{\frac{1}{2}}, \chi \geq 0, \\ 0, \qquad \chi < 0. \end{cases} \tag{7.37}$$

We may readily establish that Lemmas 7.10-7.12 also hold for [R]. Therefore [R] has at least one solution $\chi = \hat{\chi}(u)$ on, say, $u \in [0, \hat{\delta}]$ for some $\hat{\delta} > 0$ with $\hat{\chi}(u) > 0$ for all $u \in (0, \hat{\delta}]$. We set

$$\hat{w}(u) = -\sqrt{\hat{\chi}(u)}, \qquad u \in [0, \hat{\delta}], \tag{7.38}$$

after which direct substitution verifies that $\hat{w}(u)$ provides a negative solution to [P] for $u \in [0, \hat{\delta}]$. We may now establish the following theorem.

**Theorem 7.13.** *The initial value problem [P] has precisely two solutions for* $u \in [0, \bar{\delta}]$, *where* $\bar{\delta} = \min(\delta^*, \hat{\delta})$. *One of these solutions is strictly negative, whilst the other is strictly positive, for all* $u \in (0, \bar{\delta}]$.

*Proof.* Follows via Lemmas 7.6, 7.8 and Remark 7.9, using $w(u) = w^*(u)$ and $w(u) = \hat{w}(u)$ as defined in (7.34) and (7.38). $\qquad\qquad\square$

Via Theorem 7.13 we are now able to conclude that the equilibrium point $\underline{e}_1$ has the structure of a saddle point in $u \geq 0$, with a unique stable manifold (this being $\hat{w}(u)$, $u \in [0, \bar{\delta}]$) and a unique unstable manifold (this being $w^*(u)$, $u \in [0, \bar{\delta}]$). The local phase portrait can now be readily sketched, and is illustrated in Figure 7.1, where we label the unique stable manifold at $\underline{e}_1$ as $S_2$. On denoting this stable manifold by $w = W_s(u)$, we obtain directly from Lemma 7.7 and equation (7.21) that

$$W_s(u) = \begin{cases} -\gamma u^{\frac{n+1}{2}} - cu + o(u), & \frac{n+1}{2} < m < 1, \\ -\gamma u^{\frac{n+1}{2}} + \left(\frac{1}{\gamma} - c\right) u + o(u), & m = \frac{n+1}{2}, \\ -\gamma u^{\frac{n+1}{2}} + \frac{u^p}{\gamma p} - cu + o(u), & 0 < m < \frac{n+1}{2} \end{cases} \tag{7.39}$$

as $u \to 0^+$, with

$$\gamma = \sqrt{\frac{2k}{n+1}}, \quad \text{and} \quad p = m + 1 - \left(\frac{n+1}{2}\right). \tag{7.40}$$

A PTW solution to $E[m, n, k]$ now requires that the integral path leaving $\underline{e}_3$ along the unstable manifold $\mathcal{S}_1$ connects with the origin $\underline{e}_1$ along the unique stable manifold $\mathcal{S}_2$. We will refer to such a (heteroclinic) path as a PTW connection.

### 7.2.2 Existence of a PTW Connection

To establish the existence of a PTW connection, we consider how the global behaviour of $\mathcal{S}_1$ (the unstable manifold leaving $\underline{e}_3$ into $w \leq 0$) changes with the parameter $c \geq 0$ (regarding $m$, $n$, and $k$ as fixed).

Firstly, when $c = 0$, the unstable manifold at $\underline{e}_3$, say $\mathcal{S}_1^0$, has the equation

$$w(u) = -\sqrt{2}\left\{\int_u^{\eta_2} f(s)ds\right\}^{\frac{1}{2}}, \quad 0 \leq u \leq \eta_2. \tag{7.41}$$

Thus, $\mathcal{S}_1^0$ is monotone increasing in $u$ and intersects the $w$ axis as $u \to 0^+$ (for each $0 < k < k_c$) at

$$w^0(k) = -\sqrt{2}[H(k)]^{\frac{1}{2}} \tag{7.42}$$

with $H(k)$ as defined in (7.8). We next define the region $\mathcal{I}$ of the $(u, w)$ phase plane to be that closed, simply connected region bounded by the curves, $\mathcal{S}_1^0$ $(0 \leq u \leq \eta_2)$, $w = 0$ $(0 \leq u \leq \eta_2)$, $u = 0$ $(w^0(k) \leq w \leq 0)$, and observe that the vector field $(u_z, w_z)$ in the dynamical system (7.16) rotates clockwise, at each fixed point in the $(u, w)$ plane, with increasing $c$, tending towards the vertical as $c \to \infty$. We may conclude that for any $c > 0$, $\mathcal{S}_1$ enters $\mathcal{I}$ on leaving $\underline{e}_3$ and must therefore leave the interior of $\mathcal{I}$ (via the Poincaré-Bendixson theorem, see [72]) through the boundary

(i) $w = 0$, $(0 \leq u \leq \eta_1)$    (label as $\mathcal{A}$),    or
(ii) $u = 0$, $(w^0(k) \leq w \leq 0)$    (label as $\mathcal{B}$).

Thus, for each $c \geq 0$, we are able to define a function $d : \mathbb{R}^+ \to \mathbb{R}^+$ such that $d(c)$ measures the distance from the point $(\eta_1, 0)$, *along the coordinate axes*, to that point at which $\mathcal{S}_1$ cuts the boundary $\mathcal{A}$ or the boundary $\mathcal{B}$ of $\mathcal{I}$. The definition of $d(c)$ is shown geometrically in Figure 7.2. We observe, by definition and the construction of $\mathcal{I}$, that

$$0 \leq d(c) \leq \eta_1 - w^0(k) \quad \text{for all} \quad c \in [0, \infty). \tag{7.43}$$

Moreover, we have that

$$d(c) \text{ is continuous for all} \quad c \in [0, \infty), \tag{7.44}$$

which follows from the standard theory of continuity of solutions of dynamical systems with parameters upon which they depend continuously (see for example, [12], Chapter 1). Finally, we have the following lemma.

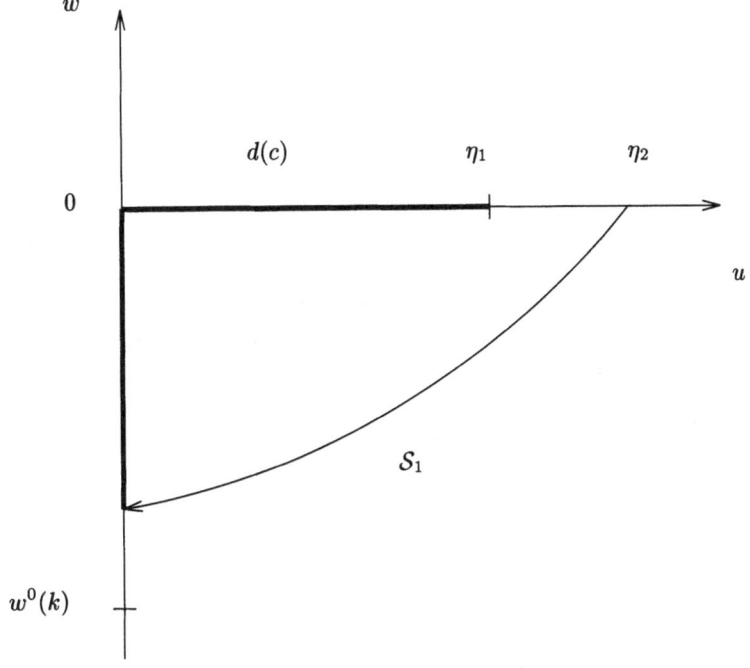

**Fig. 7.2.** The function $d(c)$. Note that $d(c)$ is given by the thicker line.

**Lemma 7.14.** *The function $d(c)$ has the properties:*

*(i) $d(c)$ is strictly monotone decreasing for $c \in [0, \infty)$,*
*(ii) $d(c) \to \eta_1 - w^0(k)$ as $c \to 0$,*
*(iii) $d(c) \to 0$ as $c \to \infty$.*

*Proof.* Property (i) is a direct consequence of the clockwise rotation of the vector field $(u_z, w_z)$ in (7.16) with increasing $c$ at fixed points in the phase plane, together with (7.39). The details follow those given in [5] for a related problem. Property (ii) follows immediately from (7.41), (7.42) and (7.43).

To establish property (iii), we first rescale the dynamical system (7.16) by introducing

$$U = u, \quad W = cw, \quad \xi = c^{-1}z,$$

in terms of which (7.16) becomes

$$\begin{aligned} U_\xi &= W, \\ c^{-2} W_\xi &= -W - f(U), \quad U \geq 0, \quad -\infty < W < \infty. \end{aligned} \tag{7.45}$$

The phase path $\mathcal{S}_1$ is that solution of (7.45) which satisfies the conditions

$$U(\xi) \to \eta_2, \quad W(\xi) \to 0 \text{ as } \xi \to -\infty. \tag{7.46}$$

We also know that $S_1$ remains in $\mathcal{I}$ until it cuts one of the coordinate axes on the boundary of $\mathcal{I}$ ($\mathcal{A}$ or $\mathcal{B}$), so that, on $S_1$,

$$0 \le U(\xi) \le \eta_2. \tag{7.47}$$

An integration of the second of equations (7.45) determines that on $S_1$,

$$W(\xi) = -c^2 e^{-c^2\xi} \int_{-\infty}^{\xi} f(U(s))e^{c^2 s}\,ds \tag{7.48}$$

after using (7.46). However, the bounds on $S_1$, from (7.47), are independent of $c$, and so we may use Watson's lemma to estimate the integral in (7.48) as $c \to \infty$. We have

$$\int_{-\infty}^{\xi} f(U(s))e^{c^2 s}\,ds \sim c^{-2} f(U(\xi))e^{c^2\xi} \tag{7.49}$$

as $c \to \infty$, uniformly in $\xi$. On using (7.49) in (7.48), we obtain, on $S_1$,

$$W(\xi) = -f(U(\xi)) + o(1) \tag{7.50}$$

as $c \to \infty$ uniformly in $\xi$, and then, via (7.45),

$$\begin{aligned} U_\xi(\xi) &= -f(U(\xi)) + o(1), \\ U(\xi) &\to \eta_2 \quad \text{as} \quad \xi \to -\infty. \end{aligned} \tag{7.51}$$

Equations (7.50) and (7.51) are essentially a global centre manifold reduction of equations (7.45) and (7.46) as $c \to \infty$. Allowing $c \to \infty$ in (7.50) and (7.51) determines that $S_1$ approaches the path

$$W = -f(U)$$

as $c \to \infty$, upon which

$$U_\xi = -f(U).$$

It follows, by definition, that $d(c) \to 0$ as $c \to \infty$.    $\square$

We next observe that a PTW connection exists (for given $0 < k < k_c$ and $m > n$) precisely at those values of $c$ for which $d(c) = \eta_1$, and we have the following theorem.

**Theorem 7.15.** *For each $k \in (0, k_c)$, (with $m > n$ fixed), there exists a unique PTW solution. This has propagation speed $c = c^*(k)$, where $c^*(k)$ is the unique positive zero of the function $d(c) - \eta_1$. The PTW is monotone decreasing in $z$. For $k \in [k_c, \infty)$ no PTW solutions exist.*

*Proof.* This follows from (7.43), (7.44) and Lemma 7.14, together with Proposition 7.5. The monotonicity property follows from the observation that $w \le 0$ on $S_1$.    $\square$

### 7.2.3 Properties of the PTW Connection

We have established in Theorem 7.15 that for each $0 < k < k_c$ $(m > n)$ there exists a unique PTW solution of $E[m, n, k]$, say $u = u^*(z, k)$, which has propagation speed $c = c^*(k) > 0$. We now examine the structure of $c^*(k)$, $k \in (0, k_c)$, and the corresponding PTW solution $u^*(z, k)$.

In Section 7.2.2 we defined $c^*$ to be the single, positive solution of $d(c) = \eta_1$, which gives a well-defined value of $c^*(k)$ for each $k \in (0, k_c)$. Moreover, since the right hand sides of the dynamical system (7.16) are continuous functions of both $c$ and $k$ (as well as $u$ and $w$), the integral paths of (7.16) will vary continuously with $c$ and $k$ (see, for example, [12] Chapter 1), and hence $d$ (as defined in Section 7.2.2) may be regarded as a continuous function of both $0 < c < \infty$ and $0 < k < k_c$. In fact, $d(c, k)$ can be regarded as a continuous surface above the rectangle $(0, \infty) \times (0, k_c)$ in the $(c, k)$ plane. If we fix a value of $k \in (0, k_c)$ and slice through this surface, we obtain the graph $d(c)$ for this particular $k$. We observe also that $c^*(k)$ is the locus of $d(c, k) = \eta_1(k)$ in $0 < c < \infty$, $0 < k < k_c$. By the previous properties of $d(c, k)$ and the fact that $d(c, k)$ is continuous, this locus is well defined and continuous for $k \in (0, k_c)$. We have the following corollary.

**Corollary 7.16.** *The function $c^*(k)$ is continuous for all $k \in (0, k_c)$, and is strictly monotone decreasing.*

*Proof.* The continuity of $c^*(k)$ follows from the above. To establish monotonicity, we first consider the change in the vector field $(u_z, w_z)$ in (7.16) when we fix $c \in (0, \infty)$ and increase $k \in (0, k_c)$. An examination of the right hand side of (7.18) reveals $\frac{\partial}{\partial k}\left(\frac{dw}{du}\right) = \frac{u^n}{w} < 0$ in $u > 0$, $w < 0$. Therefore, with $c$ fixed, the vector field rotates clockwise with increasing $k$ in $u > 0$, $w < 0$. We now take $k_1$, $k_2 \in (0, k_c)$ with $k_2 > k_1$. Putting $k = k_1$ and $c = c^*(k_1)$, a PTW connection exists from $\underline{e}_3$ to $\underline{e}_1$ and $d(c^*(k_1), k_1) = \eta_1(k_1)$, by definition. Now, keeping $c$ fixed at $c^*(k_1)$, but increasing $k$ to $k = k_2$, we know that the vector field $(u_z, w_z)$ rotates so that the unstable manifold $S_1$ is no longer connected to $\underline{e}_1$ in the $(u, w)$ phase plane. In fact, as the vector field has rotated clockwise, we have that $d(c^*(k_1), k_2) < \eta_1(k_2)$. Now, fixing $k$ at $k = k_2$, to increase $d$ to the value $\eta_1(k_2)$, it is necessary to reduce $c$ to $c^*(k_2)$. Therefore we conclude that $c^*(k_2) < c^*(k_1)$ for all $k_1, k_2 \in (0, k_c)$ with $k_2 > k_1$. The result follows. $\qquad\square$

We also note that since $d(c) \to 0$ as $c \to \infty$ for each $k \in (0, k_c)$, and that $\eta_1(k) \to 0$ as $k \to 0$, then

$$c^*(k) \to \infty \text{ as } k \to 0^+, \tag{7.52}$$

whilst, from (7.42) and Lemma 7.14, we have

$$c^*(k) \to 0 \text{ as } k \to k_c^-. \tag{7.53}$$

In fact, as will be established later using asymptotic methods, we find that

$$c^*(k) = \begin{cases} O\left[k^{\frac{-(1-m)}{2(m-n)}}\right] & \text{as } k \to 0^+, \\ O\left[k - k_c\right] & \text{as } k \to k_c^-. \end{cases} \tag{7.54}$$

We next observe that in the $(u, w)$ phase plane, the PTW connection has $w = W_s(u)$ for $0 \le u \le \eta_2$. Therefore $u^*(z, k)$ satisfies the first order differential equation

$$\begin{aligned} u_z^* &= W_s(u^*), \quad z > -\infty, \\ u^* &\to \eta_2 \quad \text{as} \quad z \to -\infty. \end{aligned} \tag{7.55}$$

On using the approximation (7.39) in (7.55) as $u \to 0^+$, we observe that the solution to (7.55) has

$$u^*(z, k) \to 0 \quad \text{as} \quad z \to 0^-, \tag{7.56}$$

and

$$u^*(z, k) \equiv 0 \quad \text{for all} \quad z > 0. \tag{7.57}$$

In particular,

$$u^*(z, k) \sim \left[\frac{-\gamma}{2}(1-n)z\right]^{\frac{2}{1-n}} \quad \text{as} \quad z \to 0^-. \tag{7.58}$$

A sketch of $c^*(k)$ is given in Figure 7.3. This completes the analysis in the case $m > n$.


## 7.3 Nonexistence of PTW Solutions when $n > m$

In this case $f(u)$ has two real non-negative zeros at $u = 0$ and $u = \nu\ (> 0)$, for all $k > 0$. Therefore a PTW solution in this case must have

$$u_{-\infty} = \nu, \tag{7.59}$$

via Proposition 7.2. As in Section 7.2, it is convenient to rewrite $E[m, n, k]$ as the dynamical system (7.16), which now has two equilibrium points in the region $u \ge 0$, $-\infty < w < \infty$ of the $(u, w)$ phase plane. These are $\underline{e}_1 = (0, 0)$ and $\underline{e}_2 = (\nu, 0)$. A PTW solution in this case then corresponds to an integral path of (7.16) which leaves $\underline{e}_2$ and enters $\underline{e}_1$, remaining in $u \ge 0$ throughout. Linearization of (7.16) at $\underline{e}_2$ reveals that it is a simple saddle point, and that the only integral path that satisfies condition (7.3b) (with (7.59)) is that part of the unstable manifold at $\underline{e}_2$ which is directed into the region $0 \le u \le \nu$, $w \le 0$, and we label this as S. The remaining part of the unstable manifold at $\underline{e}_2$ is directed into $u \ge \nu$, $w \ge 0$, and this region is invariant for (7.16), and so cannot connect to $\underline{e}_1$, thus being unable to provide a PTW solution. The local behaviour at $\underline{e}_2$ is illustrated in Figure 7.4.

The equilibrium point $\underline{e}_1$ is not hyperbolic due to the lack of Lipschitz continuity of $f(u)$ at $u = 0$, and cannot, therefore, be classified by linearization. A more detailed local analysis is required. In order to proceed we first

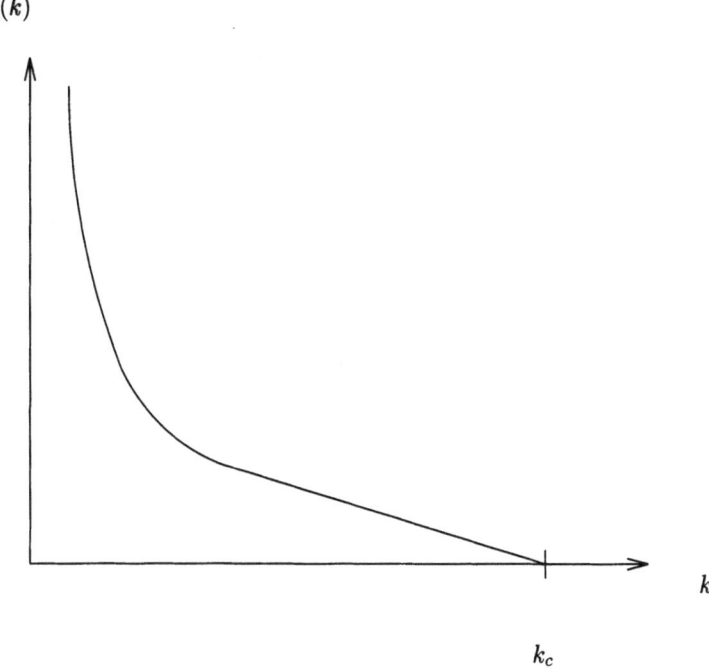

**Fig. 7.3.** A sketch of $c^*(k)$.

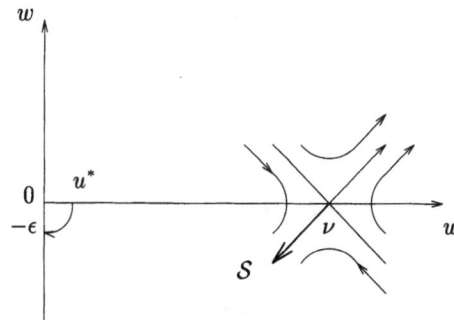

**Fig. 7.4.** The $(u, w)$ phase plane for $n > m$.

consider the nature of local solutions $w(u)$ to the initial value problem, [N], namely

$$\frac{dw}{du} = -c - \frac{f(u)}{w}, \qquad u \in (0, \epsilon], \tag{7.60}$$

$$w(0) = -\epsilon \tag{7.61}$$

for $0 < \epsilon \ll 1$. Solutions to [N] correspond to phase paths of (7.16) starting at $(0, -\epsilon)$ and entering $u > 0$. As $\frac{f(u)}{w}$ is continuous in a neighbourhood of $(0, -\epsilon)$, the existence of a local solution is guaranteed by Picard's local existence theorem (uniqueness is not guaranteed, but this is not required for our present purposes). We first have the following lemma.

**Lemma 7.17.** *Let $w(u)$ be a solution to [N] with $w(u) < 0$ for all $u \in [0, \epsilon]$, then*

$$w(u) > -cu - \epsilon, \quad \text{for all} \quad u \in (0, \epsilon].$$

*Proof.* Suppose

$$w(u) < 0 \quad \text{for all} \quad u \in [0, \epsilon]. \tag{7.62}$$

Now define

$$\phi(u) = w(u) + cu + \epsilon \quad \text{for all} \quad u \in [0, \epsilon]. \tag{7.63}$$

Then

$$\phi(0) = 0 \tag{7.64}$$

and

$$\phi'(u) = w'(u) + c = -\frac{f(u)}{w} \quad \text{for all} \quad u \in (0, \epsilon].$$

Therefore, with $\epsilon$ sufficiently small, we have

$$\phi'(u) > 0 \quad \text{for all} \quad u \in (0, \epsilon]. \tag{7.65}$$

We conclude from (7.65) that $\phi(u)$ is monotone increasing for $u \in [0, \epsilon]$ and since $\phi(0) = 0$, then $\phi(u) > 0$ for all $u \in (0, \epsilon]$, and the proof is complete. □

We next have:

**Proposition 7.18.** *Let $w(u)$ be a solution to [N], then $w(u)$ has at least one zero in $(0, \epsilon]$.*

*Proof.* Let $w(u)$ be a solution to [N] in $[0, \epsilon]$, then

$$w^2(u) = -2c \int_0^u w(s) \, ds - 2 \int_0^u f(s) \, ds + \epsilon^2, \quad u \in [0, \epsilon]. \tag{7.66}$$

Now suppose that $w(u)$ has no zeros in $[0, \epsilon]$, then $w(u) < 0$ for all $u \in [0, \epsilon]$. Consequently, via (7.66) and Lemma 7.17,

$$w^2(u) < 2c \int_0^u (cs + \epsilon) \, ds - 2 \int_0^u f(s) \, ds + \epsilon^2$$

$$= c^2 u^2 + 2\epsilon cu - 2 \int_0^u f(s) \, ds + \epsilon^2$$

$$= c^2 u^2 + 2\epsilon cu - 2 \left( \frac{u^{m+1}}{m+1} - \frac{u^{m+2}}{m+2} - k \frac{u^{n+1}}{n+1} \right) + \epsilon^2$$

for all $u \in [0, \epsilon]$, and so

$$w^2(\epsilon) < c^2\epsilon^2 + 2c\epsilon^2 - 2\left(\frac{\epsilon^{m+1}}{m+1} - \frac{\epsilon^{m+2}}{m+2} - k\frac{\epsilon^{n+1}}{n+1}\right) + \epsilon^2 < 0$$

for $\epsilon > 0$ chosen sufficiently small. Thus we obtain a contradiction, and the result follows.  □

We have now established that the phase path of (7.16) through $(0, -\epsilon)$ for each $\epsilon > 0$ and sufficiently small, enters $u > 0$ with $w < 0$, and intersects the $u$ axis at least once for $u \in (0, \epsilon]$. Denoting the first such intersection as occurring at $u = u^*(> 0)$, a sketch of this phase plane is given in Figure 7.4. We conclude immediately that the phase path S leaving $\underline{e}_2$ cannot enter $\underline{e}_1$ in $u \geq 0$, which establishes the following theorem.

**Theorem 7.19.** *For each $k > 0$ (with $n > m$ fixed) there are no PTW solutions of $E[m, n, k]$.*

This completes the analysis of the case $n > m$.

## 7.4 Nonexistence of PTW Solutions when $n = m$

In this final case, we have the following theorem.

**Theorem 7.20.** *For each $k > 0$ (with $n = m$ fixed) there are no PTW solutions of $E[m, n, k]$.*

*Proof.* We consider the cases $k \geq 1$, $0 < k < 1$ separately.
$k \geq 1$: In this case the only non-negative zero of $f(u)$ is at $u = 0$. The result then follows from Proposition 7.2.
$0 < k < 1$: In this case the result may be established in exactly the same manner as in Section 7.3 for the case $n > m$. We shall not repeat this here.  □

All cases have now been considered. In the next section we examine the structure of $c^*(k)$ (for $m > n$) as $k \to 0$ and as $k \to k_c$, via asymptotic methods.

## 7.5 Asymptotic Forms for $c^*(k)$

Here we use asymptotic methods to determine the structure of $c^*(k)$ $(m > n)$ as $k \to k_c^-$ and as $k \to 0^+$.

### 7.5.1 $k \to 0^+$

We have already established in Section 7.2 that $c^*(k) \to \infty$ as $k \to 0^+$. We now examine the structure of this singular behaviour. The eigenvalue problem which determines $c^*(k)$ is, from (7.1)-(7.3) (on utilizing the results of Section 7.2),

$$u^{*\prime\prime} + c^*(k)\,u^{*\prime} + (1 - u^*)\,u^{*m} - ku^{*n} = 0, \quad -\infty < z < 0, \quad \text{(7.67a)}$$
$$u^*(z), \; u^{*\prime}(z) \to 0 \quad \text{as} \quad z \to 0^-, \qquad\qquad\qquad \text{(7.67b)}$$
$$u^*(z) \to \eta_2(k) \quad \text{as} \quad z \to -\infty, \qquad\qquad\qquad\quad \text{(7.67c)}$$
$$u^*(z) > 0 \quad \text{for all} \quad -\infty < z < 0. \qquad\qquad\qquad \text{(7.67d)}$$

We also have that

$$\eta_2(k) \sim 1 + o(1), \quad [c^*(k)]^{-1} = o(1), \quad \text{as } k \to 0^+. \qquad \text{(7.68)}$$

To obtain a non-trivial balance in equation (7.67a) as $k \to 0^+$, we write

$$u^* = k^\alpha \bar{u}, \quad z = k^\beta \bar{z}, \quad c^* = k^{-\gamma} \bar{c} \qquad\qquad \text{(7.69)}$$

with $\bar{u}, \, \bar{z}, \, \bar{c} = O(1)$ as $k \to 0^+$, and $\alpha, \, \beta, \, \gamma \geq 0$. On substitution from (7.69) into (7.67a) we find that

$$\alpha = \frac{1}{m - n}, \quad \beta = \frac{(1 - m)}{2(m - n)}, \quad \gamma = \frac{(1 - m)}{2(m - n)}. \qquad \text{(7.70)}$$

We now expand

$$\begin{aligned}\bar{u}(\bar{z}, k) &= \bar{u}_0(\bar{z}) + o(1), \\ \bar{c}(k) &= \bar{c}_0 + o(1),\end{aligned} \qquad\qquad \text{(7.71)}$$

as $k \to 0^+$ with $\bar{z} = O(1)$. On substitution from (7.71) into (7.67) we obtain the leading order problem for $\bar{u}_0$ as

$$\begin{aligned}\bar{u}_0^{\prime\prime} + \bar{c}_0 \bar{u}_0^\prime + \bar{u}_0^m - \bar{u}_0^n &= 0, \quad -\infty < \bar{z} < 0, \\ \bar{u}_0(\bar{z}), \; \bar{u}_0^\prime(\bar{z}) &\to 0 \quad \text{as} \quad \bar{z} \to 0^-, \\ \bar{u}_0(\bar{z}) &> 0 \quad \text{for all} \quad -\infty < \bar{z} < 0,\end{aligned} \qquad \text{(7.72)}$$

where now $\prime \equiv \frac{d}{d\bar{z}}$. The boundary condition (7.67c) has been dropped as it is anticipated that expansion (7.71) will develop a non-uniformity when $(-\bar{z}) \gg 1$. The problem (7.72) is readily analyzed in the phase plane, from which the following may be shown:

(i) that there exists a unique value $\bar{c}_0$, say $\bar{c}_0 = c^*(m, n) > 0$, at which (7.72) has a unique solution.
(ii) $\bar{u}_0(\bar{z})$ is monotone decreasing in $\bar{z}$ whilst

$$\bar{u}_0(\bar{z}) \sim \begin{cases} \left[\frac{2(n+1)}{(1-n)^2}\right]^{\frac{1}{n-1}} (-\bar{z})^{\frac{2}{1-n}}, & \bar{z} \to 0^-, \\ \left[\frac{1-m}{(\bar{c}_0)}\right]^{\frac{1}{1-m}} (-\bar{z})^{\frac{1}{1-m}}, & \bar{z} \to -\infty. \end{cases} \qquad \text{(7.73)}$$

It is now clear from (7.73) that expansion (7.71) becomes non-uniform when $(-\bar{z}) \gg 1$, in particular, when

$$(-\bar{z}) = O\left(k^{\frac{-(1-m)}{(m-n)}}\right), \tag{7.74}$$

and then

$$u^* = O(1) \tag{7.75}$$

as $k \to 0^+$. Therefore, to complete the asymptotic structure, we require an outer region, scaled according to (7.74) and (7.75). In the outer region, we introduce the scaled independent variable, $\xi$, via

$$\bar{z} = k^{\frac{-(1-m)}{(m-n)}} \xi \tag{7.76}$$

and expand

$$u^*(\xi, k) = u_0(\xi) + o(1) \tag{7.77}$$

as $k \to 0^+$ with $\xi = O(1)$. After substitution from (7.76), (7.77) into (7.67), the leading order outer problem becomes

$$\begin{aligned}
\bar{c}_0 u_0' &= -(1 - u_0)\, u_0^m, \qquad -\infty < \xi < 0, \\
u_0(\xi) &\sim \left[\frac{(1-m)}{\bar{c}_0}\right]^{\frac{1}{1-m}} (-\xi)^{\frac{1}{1-m}} \quad \text{as} \quad \xi \to 0^-, \\
u_0(\xi) &\to 1 \quad \text{as} \quad \xi \to -\infty, \\
u_0(\xi) &> 0 \quad \text{for all} \quad -\infty < \xi < 0.
\end{aligned} \tag{7.78}$$

Here $\prime \equiv \frac{d}{d\xi}$, and the first condition comes from matching with the inner region. It is readily established that problem (7.78) has a unique solution for each $\bar{c}_0 > 0$. The leading order structure is now complete. However, we observe that the constant $\bar{c}_0$ has not been determined at leading order, and we expect this to be fixed by developing expansions (7.71) and (7.77) to next order as $k \to 0^+$. We do not pursue this further. The present analysis has shown that, as $k \to 0^+$, then $c^*(k) = O\left[k^{-\frac{(1-m)}{2(m-n)}}\right]$, and the PTW solution $u^*(z, k)$ becomes dispersed, with the main variation in concentration occurring over the length scale $z = O\left[k^{-\frac{(1-m)}{2(m-n)}}\right]$.

## 7.5.2 $k \to k_c^-$

We now examine the solution to (7.67a-d) as $k \to k_c^-$. On defining $\epsilon = k_c - k$ ($\ll 1$), we expand in the form

$$\begin{aligned}
u^*(z, \epsilon) &= u_0(z) + \epsilon u_1(z) + O(\epsilon^2), \\
c^*(\epsilon) &= \epsilon c_1 + O(\epsilon^2),
\end{aligned} \tag{7.79}$$

following (7.53) and the form of (7.67a). After substitution from (7.79) into (7.67a-d) we obtain the leading order problem as

$$u_0'' + (1 - u_0) u_0^m - k_c u_0^n = 0, \quad -\infty < z < 0,$$
$$u_0(z), \; u_0'(z) \to 0 \quad \text{as} \quad z \to 0^-,$$
$$u_0(z) \to \eta_2(k_c) \quad \text{as} \quad z \to -\infty, \tag{7.80}$$
$$u_0(z) > 0 \quad \text{for all} \quad -\infty < z < 0.$$

Following an integration, (7.80) can be reduced to the first order equation

$$u_0' = -\sqrt{2} \left\{ \frac{k_c u_0^{n+1}}{n+1} + \frac{u_0^{m+2}}{m+2} - \frac{u_0^{m+1}}{m+1} \right\}^{\frac{1}{2}} \equiv J(u_0), \quad -\infty < z < 0, \tag{7.81}$$

where $J(0) = J(\eta_2(k_c)) = 0$ and $J(u_0) < 0$ for all $u_0 \in (0, \eta_2(k_c))$ (which follows from the definition of $k_c$ in (7.9)). Equation (7.81), subject to the conditions in (7.80), has a unique solution which is monotone decreasing in z, and approaches $\eta_2(k_c)$ as $z \to -\infty$, through terms exponentially small in z.

On proceeding to $O(\epsilon)$, we obtain the linear, inhomogeneous problem

$$u_1'' + \left[ m u_0^{m-1}(1 - u_0) - u_0^m - k_c n u_0^{n-1} \right] u_1 = -c_1 u_0' - u_0^n, \quad -\infty < z < 0, \tag{7.82}$$

$$u_1(z), \; u_1'(z) \to 0 \quad \text{as} \quad z \to 0^-, \tag{7.83}$$

$$u_1(z) \to \eta_2'(k_c) \quad \text{as} \quad z \to -\infty. \tag{7.84}$$

The homogeneous problem associated with (7.82)-(7.84) has the non-trivial solution

$$u_1(z) \equiv A u_0'(z) \tag{7.85}$$

for any constants $A \neq 0$. As the linear operator on the left hand side of (7.82) is self adjoint, it then follows by the Fredholm theory (see, for example, [68]) that (7.82)-(7.84) has a solution provided that the solvability condition

$$\int_{-\infty}^{0} \left[ c_1 u_0'(z) + u_0^n(z) \right] u_0'(z) \, \mathrm{d}z = 0 \tag{7.86}$$

is satisfied. On rearranging (7.86) we determine $c_1$, via (7.81), as

$$c_1 = \frac{\eta_2^{n+1}(k_c)}{n+1} \left\{ \int_0^{\eta_2(k_c)} [-J(x)] \, \mathrm{d}x \right\}^{-1} \tag{7.87}$$

which completes the leading order structure as $k \to k_c^-$.

## 7.6 Conclusions and Further Discussion

In this chapter we have addressed the possibility of permanent form travelling waves (PTWs) developing, in the long time, from the singular reaction-diffusion initial value problem (6.16)-(6.20). We have established that the inclusion of a fractional order termination step ($k \neq 0$) restores the capability of a fractional order autocatalytic step to support finite speed propagating travelling waves provided that

$$m > n \quad \text{and} \quad k \in (0, k_c), \tag{7.88}$$

otherwise finite speed propagation cannot occur. The family of travelling waves which exists when conditions (7.88) are satisfied is not of Fisher-Kolmogorov type, but falls more in to the class of travelling waves which exist in excitable systems; that is with $m, n, k$ fixed to satisfy conditions (7.88), there exists exactly one travelling wave solution, with unique speed $c^*(k)$. This is in contrast to the case when $n, m \geq 1$, (as studied in [52], [44] and [58]) where it is shown that travelling waves are of Fisher-Kolmogorov type. The propagation speed $c^*(k)$ has been shown to be monotone decreasing for $k \in (0, k_c)$, with $c^*(k)$ becoming unbounded as $k \to 0^+$, and vanishing as $k \to k_c^-$, and the asymptotic structure of each limit has been derived. Further, it has been shown that the support of the associated travelling wave solution is only semi-infinite, with the reaction totally dormant ahead of the wavefront, which is again in contrast to the case $m, n \geq 1$, when a weak reaction always persists ahead of the wavefront.

The results presented here will next be extended to consider the evolution of the initial-boundary-value problem (6.16)-(6.20) in Chapter 8.

### 7.6.1 Extension of the Results on Nonexistence of PTW Solutions to a Wider Class of Singular Reaction-Diffusion Problems

We now consider a class of non-dimensional singular initial-boundary value problems, given by

$$u_t = u_{xx} + f^+(u), \quad x > 0, \quad t > 0, \tag{7.89}$$

$$f^+(u) = \begin{cases} f(u), & u \geq 0, \\ 0, & u < 0, \end{cases} \tag{7.90}$$

$$u(x, 0) = \begin{cases} u_0 g(x), & 0 \leq x \leq \sigma, \\ 0, & x > \sigma, \end{cases} \tag{7.91}$$

$$u(x, t) \to 0, \quad \text{as } x \to \infty, \quad t \geq 0, \tag{7.92}$$

$$u_x(0, t) = 0, \quad t > 0, \tag{7.93}$$

where the dimensionless parameters $u_0, \sigma$ and the function $g(x)$ are as defined in Chapter 6 and the non-Lipschitz reaction term, $f(u)$, satisfies the following conditions:

(i) $f(u) \in C[0, \infty)$,
(ii) $f'(u) \in C(0, \infty)$,

(iii) $f(0) = f(\nu) = 0$ for some $\nu > 0$,

(iv) $f(u) > 0$ for all $u \in (0, \nu)$, $f(u) < 0$ for all $u \in (\nu, \infty)$,

(v) $f'(\nu) < 0$,

(vi) $f(u) = O(u^m)$ as $u \to 0$ with $0 < m < 1$.

For the above class of singular reaction-diffusion equations we have the following theorem.

**Theorem 7.21.** *There are no PTW solutions of (7.89)-(7.93) together with conditions (i)-(vi).*

*Proof.* Follows, after minor modifications, the arguments of Section 7.3. See McCabe, Leach and Needham [40] for more details.    □

Notably, the class includes the modified Fisher problem

$$u_t = u_{xx} + u^m(1 - u), \quad x > 0, \quad t > 0, \tag{7.94}$$

with $0 < m < 1$ and (7.91)-(7.93). It is well known that the modified Fisher problem supports permanent form travelling wave solutions if $m \geq 1$ (see Kolmogorov *et al.* [29], Fisher [16] and Needham and Barnes [56]), but, via Theorem 7.21, for $0 < m < 1$ this is no longer the case. Thus we may think of $m = 1$ as a bifurcation point in the behaviour of the solutions to the problem (7.94), (7.91)-(7.93).

# The Initial-Boundary Value Problem

In this chapter, we extend the analysis of Chapter 7 and consider the evolution of the scalar initial-boundary value problem (6.16)-(6.20), namely,

$$u_t = u_{xx} + f(u), \quad x, t > 0, \tag{8.1a}$$

$$f(u) = \begin{cases} (1-u)\, u^m - ku^n, & u > 0, \\ 0, & u \le 0, \end{cases} \tag{8.1b}$$

$$u(x,0) = \begin{cases} u_0 g(x), & 0 \le x \le \sigma, \\ 0, & x > \sigma, \end{cases} \tag{8.1c}$$

$$u(x,t) \to 0 \quad \text{as} \quad x \to \infty, \quad t \ge 0, \tag{8.1d}$$

$$u_x(0,t) = 0, \quad t > 0. \tag{8.1e}$$

where $g(x)$ is a prescribed, positive function for $x < \sigma$ with a maximum value of unity, and which is analytic on $0 \le x \le \sigma$. In particular, $g(x)$ has

$$g(x) \sim g_\sigma \, (\sigma - x)^r \quad \text{as} \quad x \to \sigma^-, \tag{8.2a}$$

$$g(x) \sim g_0 + g_{\tilde{m}} x^{\tilde{m}} + \ldots \quad \text{as} \quad x \to 0^+, \tag{8.2b}$$

for some $r, \tilde{m} \in \mathbb{N}$ and constants $g_{\tilde{m}} \neq 0$, $g_\sigma, g_0, > 0$. The parameters $0 < m, n < 1$ and $u_0, \sigma, k > 0$ are dimensionless. Throughout the chapter, we shall refer to the initial-boundary value problem (8.1) as IBVP, and for convenience of notation we introduce the sets

$$\mathcal{D}_T = (0, \infty) \times (0, T], \qquad \overline{\mathcal{D}}_T = [0, \infty) \times [0, T], \tag{8.3}$$

for any $T > 0$.

In Section 8.1 we discuss existence and uniqueness for IBVP and outline a useful extension of the comparison theorem for scalar parabolic operators. In Sections 8.2, 8.3 and 8.4 we examine the qualitative behaviour of the solution and demonstrate the existence of a parameter dependent critical threshold on the initial data. The small- and large-$t$ behaviour of IBVP, for all values of the parameters, is examined via the method of matched asymptotic expansions in Sections 8.5 and 8.6, respectively. Finally, we present the conclusions in Section 8.7.

## 8.1 Existence, Uniqueness, and the Comparison Theorem

### 8.1.1 Comparison Theorem

We first observe that the comparison theorem for scalar parabolic operators (see, for example, Fife [14], Chapter 4, Theorem 4.1) does not apply in general to IBVP, due to the reaction function $f(u)$ not being Lipschitz continuous at $u = 0$. However, it can be applied in the cases $(m > n)$, $(m = n, k > 1, k = 1)$. In the case $(m = n, k = 1)$ the function $f(u)$ is continuously differentiable and hence the theorem can be applied directly. Whereas, in the cases $(m > n)$, $(m = n, k > 1)$ we note that on any closed bounded interval $I$, with $u_1, u_2 \in I$, then the function

$$h(u_1, u_2) = \begin{cases} \frac{f(u_2)-f(u_1)}{u_2-u_1}, & u_2 \neq u_1, \\ 0, & u_2 = u_1, \end{cases} \tag{8.4}$$

is bounded above. In this case the maximum principle still holds, and the proof of the comparison theorem can be readily completed despite the lack of Lipschitz continuity in $f(u)$ at $u = 0$. However, in the cases $(m < n)$ and $(m = n, k < 1)$ the comparison theorem does not hold. In particular, this can be seen by observing the lack of uniqueness to IBVP in the case of trivial initial data $(g(x) \equiv 0, 0 \leq x \leq \sigma)$ (if the comparison principle did apply, uniqueness for this problem would follow automatically).

### 8.1.2 Existence and Uniqueness for IBVP

We first address the cases $(m < n)$ and $(m = n, k < 1)$. To do this, it is first necessary to relax the boundary condition (8.1d) in IBVP and replace this condition by

$$u(x,t) \text{ remains bounded as } x \to \infty, \quad t \geq 0. \tag{8.5}$$

With this modification to IBVP (referred to as IBVP$'$) it is straightforward to follow the theory of Needham [55] to establish the existence of a non-negative solution to IBVP$'$. Moreover, when $g(x) \not\equiv 0$, then the results of Aguire [2] establish that this solution is unique. In particular the solution to IBVP$'$ has

$$u(x,t) \to u_\infty(t) \text{ as } x \to \infty, \quad t \geq 0, \tag{8.6}$$

where $u_\infty(t)$ is the non-trivial solution to the singular initial value problem

$$u'_\infty = f(u_\infty), \quad t > 0, \tag{8.7}$$

$$u_\infty(0) = 0. \tag{8.8}$$

We may conclude that in these cases (when $g(x) \not\equiv 0$) no solution exists for IBVP (otherwise there would be nonuniqueness in IBVP$'$), whilst a unique solution exists to IBVP'. Thus in these cases we will henceforth replace IBVP by IBVP', allowing temporal growth in $u(x,t)$ as $x \to \infty$, $t \geq 0$. We note, following Needham [55], that in both cases the solution to IBVP' is global.

In the remaining cases, $(m > n)$ and $(m = n, k \geq 1)$, uniqueness of a solution to IBVP follows directly from the comparison theorem, whilst global existence follows from minor modifications to the results of Bandle and Stakgold [3].

## 8.2 Qualitative Behaviour of the Solution to IBVP' for $m < n$

We first recall from Section 8.1 that in this case we have replaced IBVP by IBVP'. Therefore, the details in this case follow (after minor modifications) those given in King and Needham [28], Section 3, with $\gamma$ being replaced by $\nu$ throughout (where, from Chapter 7, $\nu$ is the unique, non-zero, positive real root of $f(u) = 0$, when $m < n$) and the details are omitted for the sake of brevity. Hence, in this case there exists a unique positive global solution to IBVP' which has $u(x,t) \to \nu$ uniformly in $x \geq 0$ as $t \to \infty$.

## 8.3 Qualitative Behaviour of the Solution to IBVP for $m = n$

There are three distinct cases to be considered, these being $k > 1$, $k = 1$ and $k < 1$. We examine these in turn.

### 8.3.1 $k > 1$

We begin by defining the parabolic operator $N[v] \equiv v_t - v_{xx} - \mu v^m + v^{m+1}$, for any suitably differentiable function $v : \overline{\mathcal{D}}_T \to \mathbb{R}$, and where $\mu = 1 - k$. The comparison theorem applies in this case, and it then follows immediately that

$$u(x,t) \geq 0 \quad \text{on } \overline{\mathcal{D}}_T \tag{8.9}$$

for any $T > 0$. We next observe (via Appendix A) that a non-negative solution to the scalar equation

$$w_t = w_{xx} + \mu w^m, \quad x, t > 0 \tag{8.10}$$

exists which has finite support and decays to zero in finite time. This solution is given by

$$w(x,t) \begin{cases} \sim \left[ \frac{-\mu(1-m)^2}{2(1+m)} \right]^{\frac{1}{1-m}} (x_0(t) - x)^{\frac{2}{1-m}}, & 0 \leq x < x_0(t), \\ = 0, & x \geq x_0(t), \end{cases} \tag{8.11}$$

with $0 \leq t \leq t_1^*$ and where $x_0(t) = z_0^*(t_1^* - t)^{\frac{1}{2}}$ is the edge of the support of $w(x,t)$, and $z_0^*$ ($\ll 1$) and $t_1^*$ are fixed constants, with $t_1^*$ chosen sufficiently large so that

$$w(x,0) \geq u_0 g(x) \quad \text{for all} \quad 0 \leq x \leq \sigma. \tag{8.12}$$

The details of this solution may be found in Appendix A. We now have the following proposition.

**Proposition 8.1.** *With* $(x,t) \in [0,\infty) \times [0,T]$ *for any* $0 < T \leq t_1^*$, *we have* $0 \leq u(x,t) \leq w(x,t)$.

*Proof.* We first recall that the comparison theorem holds in this case. The left hand inequality then follows directly. Now observe that

$$N[w] \equiv w_t - w_{xx} - \mu w^m + w^{m+1} = w^{m+1} \geq 0, \ \forall (x,t) \in (0,\infty) \times (0,T] \tag{8.13}$$

for any $0 < T \leq t_1^*$. Moreover, via (8.12), $w(x,0) \geq u(x,0)$, for all $0 \leq x < \infty$, and $w(x,t) \to 0$ as $x \to \infty$ on $t \in [0,T]$ for any $0 < T \leq t_1^*$. Therefore, $w(x,t)$ is an upper solution to IBVP on $\overline{\mathcal{D}}_T$ for any $0 < T \leq t_1^*$. The right hand inequaltity then follows via the comparison theorem, and the proof is complete. □

In particular, we observe immediately from Proposition 8.1 and (8.11) that

$$0 \leq u(x,t) \leq \begin{cases} \left[ \left[ \frac{-\mu(1-m)^2}{2(1+m)} \right]^{\frac{1}{1-m}} z_0^{*\frac{2}{1-m}} (t_1^* - t)^{\frac{1}{1-m}}, & 0 \leq x < x_0(t), \\ 0, & x \geq x_0(t), \end{cases} \tag{8.14}$$

on $\overline{\mathcal{D}}_{t_1^*}$, and so we conclude that

$$u(x,t) \to 0 \text{ as } t \to t_1^{*-} \text{ uniformly for } x \in [0,\infty), \tag{8.15}$$

with $t_1^*$ being the finite extinction time. Moreover, the support of $u(x,t)$ must remain finite. With $x_s(t)$ being the edge of the support of $u(x,t)$ at time $t$, we have, via (8.14), that

$$0 \leq x_s(t) \leq z_0^*(t_1^* - t)^{\frac{1}{2}} \quad \text{for all} \quad 0 \leq t \leq t_1^*, \tag{8.16}$$

and so

$$x_s(t) \to 0 \quad \text{as} \quad t \to t_1^{*-}. \tag{8.17}$$

The support of $u(x,t)$ has contracted to zero as extinction occurs.

In summary, we see in this case that the solution to IBVP undergoes finite time extinction, with contracting support.

### 8.3.2 $k = 1$

We begin in this case by defining the parabolic operator $N[v] \equiv v_t - v_{xx} + v^{m+1}$ for any suitably differentiable function $v : \overline{\mathcal{D}}_T \to \mathbb{R}$, and the function

$$D(x,t) = \bar{u}_0 \int_{-\infty}^{\infty} \bar{g}(s) e^{-s^2 t} e^{isx} \, ds, \qquad 0 \le x < \infty, \ t \ge 0, \tag{8.18}$$

where

$$\bar{g}(s) = \frac{1}{2\pi} \int_{-\sigma}^{\sigma} g(\lambda) e^{-is\lambda} \, d\lambda. \tag{8.19}$$

Here $D(x,t)$ is the unique solution to the pure diffusion problem $D_t = D_{xx}$ in $x, t > 0$, subject to the initial condition

$$D(x,0) = \begin{cases} \bar{u}_0 g(x), & 0 \le x \le \sigma, \\ 0, & x > \sigma, \end{cases} \tag{8.20}$$

where $\bar{u}_0 = \min[1, u_0]$, and the symmetry and boundary conditions (8.1e) and (8.1d). The following properties of $D(x,t)$ are readily established:

(i) $D(x,t) > 0$ for all $x \ge 0$, $t > 0$,
(ii) $D(x,t) \to 0$ as $x \to \infty$, for all $t > 0$,
(iii) $D(x,t) \le 1$ for all $x \ge 0$, $t > 0$,
(iv) $D(x,t) \to 0$ as $t \to \infty$, uniformly in $x \ge 0$.

We can now state the following proposition.

**Proposition 8.2.** *With $(x,t) \in [0, \infty) \times [0, T]$, for any $T > 0$, we have*

$$e^{-t} D(x,t) \le u(x,t) \le \left( mt + \frac{1}{u_0^m} \right)^{-\frac{1}{m}}. \tag{8.21}$$

*Proof.* The comparison theorem again applies in this case.

(a) To establish the left hand inequality, we observe that

$$N\left[e^{-t} D\right] \equiv -e^{-t} D + e^{-t}\left[D_t - D_{xx}\right] + e^{-(m+1)t} D^{m+1}$$
$$= e^{-t} D\left[e^{-mt} D^m - 1\right] \le 0$$

on $\mathcal{D}_T$ for any $T > 0$, via (i)-(iv). Moreover,

$$D(x,0) \le u(x,0) \quad \text{for all} \quad x \ge 0. \tag{8.22}$$

It therefore follows that $e^{-t} D(x,t)$ is a lower solution to IBVP and so via the comparison theorem and (ii) that $e^{-t} D(x,t) \le u(x,t)$ on $\overline{\mathcal{D}}_T$ for any $T > 0$.

(b) To establish the right hand inequality we readily verify that

$$\phi(x,t) \equiv \left( mt + \frac{1}{u_0^m} \right)^{-\frac{1}{m}}, \quad t \geq 0, \tag{8.23}$$

is an upper solution to IBVP on $\overline{\mathcal{D}}_T$, for $T > 0$, and the result follows, via the comparison theorem. □

We observe immediately that, in this case, the support of $u(x,t)$ does not remain finite. In particular, via (i) and Proposition 8.2, the support of $u(x,t)$ extends from $[0,\sigma]$ at $t = 0$ to $[0,\infty)$ in $t > 0$. Also, extinction in finite time cannot occur, via (i), but $u(x,t) \to 0$ as $t \to \infty$, uniformly in $x \geq 0$. In particular, $u(x,t) \leq O\left(t^{-\frac{1}{m}}\right)$ as $t \to \infty$, uniformly in $x \geq 0$.

### 8.3.3 $k < 1$

The comparison theorem does not apply in this case. However, the details in this case follow after minor modification those given in King and Needham [28], Section 3, with $\gamma$ being replaced by $1 - k$ throughout. Hence, referring in this case to IBVP′, there exists a unique positive global solution with $u(x,t) \to 1 - k$, uniformly in $x \geq 0$ as $t \to \infty$.

## 8.4 Qualitative Behaviour of the Solution to IBVP for $m > n$

Here there are four distinct cases to be considered, these being $k > k^*$, $k_c < k \leq k^*$, $k = k_c$ and $0 < k < k_c$, where $k^* = \frac{(m-n)^{m-n}}{(m-n+1)^{m-n+1}}$ and $k_c = \frac{(n+1)}{(m+1)} \left( \frac{m+2}{m+1} \right)^{m-n} k^*$. We begin by defining the parabolic operator $N[v] \equiv v_t - v_{xx} - f(v)$, for any suitably differentiable function $v : \overline{\mathcal{D}}_T \to \mathbb{R}$, and where $f(v) = (1 - v)v^m - kv^n$. We further recall from Chapter 7 that the function $f(v)$ has, when

(i)  $k > k^*$, a single non-negative zero at $v = 0$, with $f(v) < 0$ for $0 < v < \infty$;
(ii)  $k = k^*$, two non-negative zeros, $v = 0$ and $v = \eta_0 = \frac{m-n}{m-n+1}$, with $f(v) \leq 0$ for $0 \leq v < \infty$;
(iii)  $0 < k < k^*$, three non-negative zeros $v = 0$, $\eta_1$ and $\eta_2$ (with $0 < \eta_1 < \eta_2 < 1$), with $f(v) < 0$ for $0 < v < \eta_1$ and $v > \eta_2$ whilst $f(v) > 0$ for $\eta_1 < v < \eta_2$.

Finally, following Section 8.1 the comparison theorem can be applied directly when $m > n$ and it follows immmediately in each case that

$$u(x,t) \geq 0 \quad \text{on } \overline{\mathcal{D}}_T \tag{8.24}$$

for any $T > 0$. We now consider the details of each case in turn.

### 8.4.1 $k > k^*$

We begin by observing (via Appendix A) that a positive solution to the scalar equation

$$w_t = w_{xx} + (k^* - k) w^n, \quad x, t > 0 \tag{8.25}$$

exists which has finite support and decays to zero in finite time. This solution is given by

$$w(x,t) \quad \begin{cases} \sim \left[ \frac{-(k^*-k)(1-n)^2}{2(1+n)} \right]^{\frac{1}{1-n}} (x_0(t) - x)^{\frac{2}{1-n}}, & 0 \le x \le x_0(t), \\ = 0, & x > x_0(t), \end{cases} \tag{8.26}$$

with $0 \le t \le t_2^*$ and where $x_0(t) = z_0^*(t_2^* - t)^{\frac{1}{2}}$ is the edge of the support of $w(x,t)$, and $z_0^* (\ll 1)$ and $t_2^*$ are fixed constants, with $t_2^*$ chosen sufficiently large so that

$$w(x,0) \ge u_0 g(x) \quad \text{for all} \quad 0 \le x \le \sigma. \tag{8.27}$$

The details of this solution may be found in Appendix A. We now have the following proposition.

**Proposition 8.3.** *With* $(x,t) \in [0,\infty) \times [0,T]$ *for any* $0 < T \le t_2^*$ *we have* $0 \le u(x,t) \le w(x,t)$.

*Proof.* We first recall that the comparison theorem holds in this case. The left hand inequality then follows directly. Now observe that

$$\begin{aligned} N[w] &\equiv w_t - w_{xx} - (1 - w)w^m + kw^n \\ &= k^* w^n - (1 - w)w^m \ge 0, \end{aligned} \tag{8.28}$$

on $\overline{\mathcal{D}}_T$, for any $0 < T \le t_2^*$. Moreover, via (8.27), $w(x,0) \ge u(x,0)$, for all $0 \le x < \infty$, and $w_x(0,t) < 0$ with $w(x,t) \to 0$ as $x \to \infty$ on $t \in [0,T]$ for any $0 < T \le t_2^*$. Therefore, $w(x,t)$ is an upper solution to IBVP on $\overline{\mathcal{D}}_T$ for any $0 < T \le t_2^*$. The right hand inequality then follows via the comparison theorem, and the proof is complete. $\qquad \square$

In particular, we observe immediately from Proposition 8.3 and (8.26) that

$$0 \le u(x,t) \le \begin{cases} \left[ \frac{-(k^*-k)(1-n)^2}{2(1+n)} \right]^{\frac{1}{1-n}} z_0^{*\frac{2}{1-n}} (t_2^* - t)^{\frac{1}{1-n}}, & 0 \le x < x_0(t), \\ 0, & x > x_0(t), \end{cases} \tag{8.29}$$

on $\overline{\mathcal{D}}_{t_2^*}$, and so we conclude that $u(x,t) \to 0$ as $t \to t_2^{*-}$ uniformly for $x \in [0,\infty)$, with $t_2^*$ being the extinction time. Moreover, the support of $u(x,t)$ must remain finite on $\overline{\mathcal{D}}_{t_2^*}$. With $x_s(t)$ being the edge of the support of $u(x,t)$ at time $t$, we have, via (8.29), that

$$0 \le x_s(t) \le z_0^*(t_2^* - t)^{\frac{1}{2}} \quad \text{for all } 0 \le t \le t_2^* \tag{8.30}$$

and so

$$x_s(t) \to 0 \quad \text{as} \quad t \to t_2^{*-}. \tag{8.31}$$

The support of $u(x,t)$ has contracted to zero as extinction occurs. In summary, we see in this case that the solution to IBVP undergoes finite time extinction with contracting support.

### 8.4.2 $k_c < k \le k^*$

We first return to the analysis of Chapter 7 and consider possible steady state solutions to IBVP in $x \ge 0$. In particular, we consider steady state solutions to IBVP in $x \ge 0$, say $\tilde{u}(x)$, which satisfy the steady-state problem [SSP], namely

$$\begin{aligned}
&\tilde{u}'' + f(\tilde{u}) = 0, \quad x > 0, \\
&\tilde{u}(0) > 0, \quad \tilde{u}'(0) < 0, \\
&\tilde{u}(x) \to 0 \quad \text{as } x \to \infty, \\
&\tilde{u}(x) \ge 0 \quad \text{for all } x > 0.
\end{aligned}$$

The particular solution to [SSP] which is of use in this section corresponds to the stable manifold in the $(\tilde{u}, \tilde{u}')$ phase plane which intersects the equilibrium point at the origin $(0,0)$. This solution is given by

$$\tilde{u}(x) = \begin{cases} \phi(x), & 0 \le x \le x^*, \\ 0, & x > x^*, \end{cases} \tag{8.32}$$

where $\phi(x)$ is defined implicitly by

$$x = \int_{\phi(x)}^{\tilde{u}_0} \frac{ds}{g(s)}, \quad 0 \le x \le x^*, \tag{8.33}$$

with

$$g(s) = \sqrt{2} \left\{ \frac{s^{m+2}}{(m+2)} + \frac{ks^{n+1}}{(n+1)} - \frac{s^{m+1}}{(m+1)} \right\}^{\frac{1}{2}}, \tag{8.34}$$

for $0 \le s \le \tilde{u}_0$, and

$$x^* = \int_0^{\tilde{u}_0} \frac{ds}{g(s)} \quad (> 0). \tag{8.35}$$

The function (8.32) provides a classical solution to [SSP] for any $\tilde{u}_0 \ge 0$, provided that $k_c < k \le k^*$. We observe the following properties of $\phi(x)$:

(i)  $\phi(0) = \tilde{u}_0, \quad \phi'(0) < 0,$
(ii)  $\phi'(x) < 0$ for all $x \in (0, x^*),$
(iii)  $\phi(x^*) = \phi'(x^*) = \phi''(x^*) = 0.$

We next choose the constant $\tilde{u}_0$ sufficiently large, so that

$$u(x,0) \le \tilde{u}(x) \quad \text{for all} \quad 0 \le x < \infty. \tag{8.36}$$

In addition, we let $\bar{u}(x,t)$ be the solution of the initial-boundary value problem, [MP], given by:

$$N[\bar{u}] = 0, \quad x, t > 0,$$
$$\bar{u}(x,0) = \tilde{u}(x), \quad 0 \leq x < \infty,$$
$$\bar{u}_x(0,t) = 0, \quad t > 0,$$
$$\bar{u}(x,t) \to 0 \quad \text{as} \quad x \to \infty, \quad t \geq 0.$$

Global existence and uniqueness of solutions to [MP] follows directly from the arguments of Section 8.1, whilst the comparison theorem immediately gives

$$\bar{u}(x,t) \geq 0 \quad \text{for all } x, t \geq 0. \tag{8.37}$$

We now have the following proposition.

**Proposition 8.4.** *With $(x,t) \in \overline{\mathcal{D}}_T$, for any $T > 0$, then $\bar{u}(x,t) \geq \bar{u}(x,t+\delta)$ for any $\delta > 0$.*

*Proof.* Fix $\delta > 0$. We first observe, via the comparison theorem, that

$$\bar{u}(x,t) \leq \tilde{u}(x) \text{ on } \overline{\mathcal{D}}_T \tag{8.38}$$

for any $T > 0$. Now define $v^*(x,t) \equiv \bar{u}(x,t+\delta)$ on $\overline{\mathcal{D}}_T$, for any $T > 0$. We observe immediately, via (8.38), that $v^*(x,0) = \bar{u}(x,\delta) \leq \tilde{u}(x) = \bar{u}(x,0)$ for all $0 \leq x < \infty$, and that $v^*(x,t) \to 0$ as $x \to \infty$ for all $t \in [0,T]$. Moreover, $N[v^*] = 0$ on $\overline{\mathcal{D}}_T$, and $v_x^*(0,t) = 0$ for all $t \in [0,T]$. Therefore, $v^*(x,t)$ is a lower solution to [MP] on $\overline{\mathcal{D}}_T$, for any $T > 0$. The result then follows via the comparison theorem. $\square$

We have immediately the following corollary.

**Corollary 8.5.** *On $\overline{\mathcal{D}}_T$ (any $T > 0$), $\bar{u}(x,t)$ is monotone decreasing with $t(\geq 0)$ and bounded below by zero.*

It follows from Corollary 8.5 that

$$\lim_{t \to \infty} \bar{u}(x,t) = l(x), \quad 0 \leq x < \infty \tag{8.39}$$

exists, and must satisfy

$$l'' + f(l) = 0, \quad 0 < x < \infty,$$
$$l'(0) = 0, \tag{8.40}$$
$$0 \leq l(x) \leq \tilde{u}(x), \quad 0 \leq x < \infty.$$

It is straightforward to establish directly that the only solution to (8.40) is $l(x) \equiv 0$, with $0 \leq x < \infty$. Thus, finally, we conclude that

$$\lim_{t \to \infty} \bar{u}(x,t) = 0, \quad 0 \leq x < \infty, \tag{8.41}$$

with the support of $\bar{u}(x,t)$ remaining finite for all $t \geq 0$, and

$$\text{supp}\,\bar{u}(x,t) \subseteq [0,x^*] \times [0,T] \qquad (8.42)$$

on $\overline{\mathcal{D}}_T$ for any $T > 0$. In fact, we may similarly show that $\bar{u}_x \leq 0$ on $\overline{\mathcal{D}}_T$, for any $T > 0$, and so the limit in (8.41) is uniform in $0 \leq x < \infty$. We now have the following proposition.

**Proposition 8.6.** *With* $(x,t) \in [0,\infty) \times [0,T]$, *for any* $T > 0$, *then* $0 \leq u(x,t) \leq \bar{u}(x,t)$, *and,*

*(i)* $u(x,t) \to 0$ *as* $t \to \infty$, *uniformly in* $0 \leq x < \infty$.
*(ii)*$\text{supp}\,u(x,t) \subseteq [0,x^*] \times [0,T]$.

*Proof.* The inequality follows directly from the comparison theorem, (i) follows from (8.41) and (ii) follows from (8.42). □

Now, with $k_c < k \leq k^*$, we may choose $\epsilon > 0$, and sufficiently small, so that

$$f(X) \leq -k_\epsilon X^n \quad \text{for all} \quad X \in [0,\epsilon],$$

with $k_\epsilon > 0$ constant. Also, via Proposition 8.6(i), there exists a $T_\epsilon > 0$ such that

$$0 \leq u(x,t) \leq \epsilon$$

for all $(x,t) \in [0,\infty) \times [T_\epsilon, \infty)$. Following the arguments of Section 8.4.1, we may now establish that

$$w_\epsilon(x,t) \quad \begin{cases} \sim \left[\frac{k_\epsilon(1-n)^2}{2(1+n)}\right]^{\frac{1}{1-n}} (x_\epsilon(t) - x)^{\frac{2}{1-n}}, & 0 \leq x \leq x_\epsilon(t), \\ = 0, & x > x_\epsilon(t), \end{cases} \qquad (8.43)$$

with $x_\epsilon(t) = \hat{z}_0(\hat{t}_2 + T_\epsilon - t)^{\frac{1}{2}}$, $T_\epsilon \leq t \leq T_\epsilon + \hat{t}_2$ and $\hat{z}_0(\ll 1)$ a fixed constant, is an upper solution to IBVP for $(x,t) \in [0,\infty) \times [T_\epsilon, T_\epsilon + \hat{t}_2)$, provided that $\hat{t}_2$ is chosen sufficiently large so that

$$w_\epsilon(x,T_\epsilon) \geq u(x,T_\epsilon) \quad \text{for all} \quad 0 \leq x < \infty. \qquad (8.44)$$

Therefore, we conclude that

$$0 \leq u(x,t) \leq w_\epsilon(x,t) \quad \text{for all} \quad (x,t) \in [0,\infty) \times [T_\epsilon, T_\epsilon + \hat{t}_2). \qquad (8.45)$$

It follows immediately that $u(x,t)$ undergoes finite time extinction, with support contracting to zero.

### 8.4.3 $k = k_c$

We first return to Chapter 7 and consider possible steady state solutions to IBVP in $x \geq 0$. In particular, we consider steady state solutions to IBVP in $x \geq 0$, say $\tilde{u}(x)$, which satisfy the problem [SSP] of Section 8.4.2. The particular solution to [SSP] which is of use in this section corresponds to the stable manifold in the $(\tilde{u}, \tilde{u}')$ phase plane which intersects the equilibrium point at the origin $(0, 0)$. This solution is given by

$$\tilde{u}(x) = \begin{cases} \phi(x), 0 \leq x \leq x^*, \\ 0, \quad x > x^*, \end{cases}$$

where $\phi(x)$ and $x^*$ are given by (8.33) and (8.35) respectively. We observe the following properties of $\phi(x)$:

(i)   $\phi(0) = \tilde{u}_0$,   $\phi'(0) < 0$,
(ii)  $\phi'(x) < 0$ for all $x \in (0, x^*)$,
(iii) $\phi(x^*) = \phi'(x^*) = \phi''(x^*) = 0$.

This solution exists for each $\tilde{u}_0 \in [0, \eta_2)$, with $x^*$ being a monotone increasing function of $\tilde{u}_0$, and $x^* \to 0$ as $\tilde{u}_0 \to 0$, whilst $x^* \to \infty$ as $\tilde{u}_0 \to \eta_2$. When $\tilde{u}_0 = \eta_2$ this solution corresponds to the heteroclinic orbit in the $(\tilde{u}, \tilde{u}')$ phase plane which connects the unstable manifold of the simple saddle point at $(\eta_2, 0)$ to the stable manifold of the singular saddle point at the origin. We first restrict attention to the case when the initial data is such that

$$u_0 g(x) \leq \tilde{u}(x) \quad \text{for some} \quad \tilde{u}_0 \in [0, \eta_2) \tag{8.46}$$

(which certainly requires $u_0 < \eta_2$).

### (a) $u_0 g(x) \leq \tilde{u}(x)$

In this case, the details follow directly those given in Section 8.4.2, and we establish that $u(x, t)$ undergoes finite $t$ extinction, with the support contracting to zero.

### (b) $u_0 g(x) \not\leq \tilde{u}(x)$

We begin by defining $\sigma(t)$ to be the solution of the spatially homogeneous problem, [SHP],

$$\sigma_t = f(\sigma), \quad t > 0,$$

$$\sigma(0) = \max[u_0, \eta_2], \tag{8.47}$$

and note that $f(\sigma) < 0$ for $\eta_2 < \sigma < \infty$, with $f(\eta_2) = 0$. Hence, the solution of [SHP] is monotone decreasing with

$$\sigma(t) \to \eta_2, \quad \text{as} \quad t \to \infty. \tag{8.48}$$

We now have the following proposition.

**Proposition 8.7.** *With $(x,t) \in \overline{\mathcal{D}}_T$, for any $T > 0$, we have $0 \le u(x,t) \le \sigma(t)$.*

*Proof.* The inequalities follow directly from the comparison theorem.    □

Hence, via Proposition 8.7 and (8.48) we have established that $u(x,t)$ is bounded above by $\sigma(t)$ for all $(x,t) \in \overline{\mathcal{D}}_T$, for any $T > 0$, and thus $u(x,t)$ is uniformly bounded on $\overline{\mathcal{D}}_T$ for any $T > 0$. We have been unable to obtain more precise information in this case. However, numerical integrations of IBVP strongly suggest that, again, $u(x,t)$ undergoes finite $t$ extinction, with contracting support.

### 8.4.4 $0 < k < k_c$

We recall from Chapter 7 that for each $k \in (0, k_c)$ there exists a unique permanent form travelling wave solution to IBVP with propagation speed $c(k) > 0$. We denote this by $u_T(z)$, where, $z = x - ct$ is the travelling wave co-ordinate. We note the following properties of $u_T(z)$ (see Chapter 7):

(i)  $u_T(z) \begin{cases} > 0, \ z < 0, \\ \equiv 0, \ z \ge 0, \end{cases}$

(ii)  $u_T'(z) < 0, \quad z < 0,$

(iii)  $u_T(0) = u_T'(0) = u_T''(0) = 0,$

(iv)  $u_T(z) \to \eta_2,$ as $z \to -\infty.$

We now examine the possibility of the solution to IBVP approaching the travelling wave $u_T(z)$ as $t \to \infty$. We first establish that the initial data in IBVP must exceed a certain threshold before the travelling wave structure will develop in large $t$. To do this we return to steady state solutions of IBVP in $x \ge 0$, say $\tilde{u}(x)$, which satisfy [SSPA] (the stationary state problem [SSP] of Section 8.4.2 with the modification that $\tilde{u}'(0) \le 0$). The particular solutions to [SSPA] which are of use here correspond to sections of the homoclinic orbit in the $(\tilde{u}, \tilde{u}')$ phase plane attached to the singular saddle point at the origin. For each $\tilde{u}_0 \in (0, \tilde{a}]$, where $\tilde{a}$ is the smallest positive value such that

$$\int_0^{\tilde{a}} f(s)\, \mathrm{d}s \ = \ 0,$$

($\eta_1 < \tilde{a} < \eta_2$), we have the solution

$$\tilde{u}(x, \tilde{u}_0) \ = \ \begin{cases} H(x, \tilde{u}_0), \ 0 \le x \le x^*, \\ 0, \qquad\quad x > x^*, \end{cases} \tag{8.49}$$

where $H(x, \tilde{u}_0)$ is defined implicitly by

$$x \ = \ \int_{H(x,\tilde{u}_0)}^{\tilde{u}_0} \frac{\mathrm{d}s}{g(s)}, \quad 0 \le x \le x^*,$$

with $g(s)$ as defined by (8.34) for $0 \leq s \leq \tilde{u}_0$, and

$$x^* = \int_0^{\tilde{u}_0} \frac{ds}{g(s)} \quad (> 0).$$

The function (8.49) provides a classical solution to [SSPA], provided $0 < k < k_c$. We observe the following properties of $H(x, \tilde{u}_0)$:

(i)   $H(0, \tilde{u}_0) = \tilde{u}_0, \quad \tilde{u}_0 \in (0, \tilde{a}]$,
(ii)   $H_x(0, \tilde{u}_0) < 0$ for all $\tilde{u}_0 \in (0, \tilde{a})$,
(iii)   $H_x(0, \tilde{a}) = 0$,
(iv)   $H_x(x, \tilde{u}_0) < 0$ for all $x \in (0, x^*)$ and $\tilde{u}_0 \in (0, \tilde{a}]$,
(v)   $H(x^*, \tilde{u}_0) = H_x(x^*, \tilde{u}_0) = H_{xx}(x^*, \tilde{u}_0) = 0$ for all $\tilde{u}_0 \in (0, \tilde{a}]$.

We may now state the following poposition.

**Proposition 8.8.** *Let $u(x, t)$ be the solution to IBVP with initial data $u_0 g(x) \leq \tilde{u}(x, \tilde{a})$, then $0 \leq u(x, t) \leq \tilde{u}(x, \tilde{a})$ on $\overline{\mathcal{D}}_T$ for any $T > 0$.*

*Proof.* This follows immediately from the comparison theorem, on noting that $\tilde{u}(x, \tilde{a})$ provides an upper solution to IBVP on $\overline{\mathcal{D}}_T$ for any $T > 0$.    □

We conclude immediately from Poposition 8.8 that for the travelling wave to develop in IBVP as $t \to \infty$, the initial data $u_0 g(x)$ must exceed $\tilde{u}(x, \tilde{a})$ at least on some subinterval of $x \in [0, \infty)$. Thus we have a threshold which must be exceeded by the initial data before it is possible to generate travelling waves in IBVP as $t \to \infty$. In fact, we can extend this result as in the following proposition.

**Proposition 8.9.** *Let $u(x, t)$ be the solution to IBVP with initial data $u_0 g(x) \leq \tilde{u}(x, \tilde{u}_0)$ for some $\tilde{u}_0 \in (0, \tilde{a})$, then,*

*(i) supp $u(x, t) \subseteq [0, x^*] \times [0, T]$ on $\overline{\mathcal{D}}_T$ for any $T > 0$,*
*(ii) there exists a $t^* > 0$ such that $u(x, t) \to 0$ as $t \to t^*$ uniformly in $x \geq 0$.*

*Proof.* This follows directly the constructions given in Section 8.4.2.    □

The above proposition establishes that with sufficiently small initial data the solution to IBVP will be extinguished in finite $t$, with contracting support.

We next obtain information which indicates that for initial data in which $u_0 g(x) > \tilde{u}(x, \tilde{a})$ for all $x \geq 0$, then the travelling wave will develop in the solution of IBVP as $t \to \infty$. The proofs of these results follow closely those of Section 8.4.2 and, for brevity, we limit ourselves in what follows to sketch proofs. First we consider the following family of functions, for each $\tilde{u}_0 \in (\tilde{a}, \eta_2)$,

$$\hat{u}(x, \tilde{u}_0) = \begin{cases} \psi(x, \tilde{u}_0), & 0 \leq x \leq x^*, \\ 0, & x > x^*, \end{cases}$$

with $\psi(x, \tilde{u}_0)$ defined implicitly by

$$x = \int_{\psi(x,\tilde{u}_0)}^{\tilde{u}_0} \frac{ds}{g(s)}, \quad 0 \le x \le x^*$$

and

$$x^* = \int_0^{\tilde{u}_0} \frac{ds}{g(s)} \quad (> 0).$$

We observe the following:

(i)   $\psi(x,\tilde{u}_0)$ solves the steady state equation in [SSPA] for each $x \in (0,x^*)$,
(ii)  $\psi(0,\tilde{u}_0) = \tilde{u}_0$, $\quad \psi_x(0,\tilde{u}_0) = 0$ for all $u_0 \in (\tilde{a},\eta_2)$,
(iii) $\psi_x(x,\tilde{u}_0) < 0$ for all $x \in (0,x^*]$,
(iv)  $\psi(x^*,\tilde{u}_0) = 0$.

Now consider the solution to IBVP with initial data $u(x,0) = \hat{u}(x,\tilde{u}_0)$ for $x \ge 0$, say $\underline{u}(x,t)$. We can establish the following proposition.

**Proposition 8.10.** *On $\overline{\mathcal{D}}_T$ for any $T > 0$, $\underline{u}(x,t)$ is monotone increasing with $t$, with $\hat{u}(x,\tilde{u}_0) \le \underline{u}(x,t) \le \eta_2$.*

*Proof.* This follows via the comparison theorem, with use of properties (i)-(iv) concerning $\hat{u}(x,\tilde{u}_0)$. □

We conclude from the above proposition that on any compact interval $[0,L]$, then

$$\lim_{t \to \infty} \underline{u}(x,t) = l(x), \quad x \in [0,L]$$

exists (monotone increasing and bounded above), with the function $l(x)$ satisfying

$$l''(x) + f(l) = 0, \quad x \in (0,L],$$

$$l'(0) = 0,$$

$$\hat{u}(x,\tilde{u}_0) \le l(x) \le \eta_2, \quad x \in (0,L],$$

for any $L > 0$. The only solution to this problem has $l(x) \equiv \eta_2$ for $x \in [0,L]$. We conclude that $\underline{u}(x,t) \to \eta_2$ as $t \to \infty$ uniformly over $x \in [0,L]$, for any $L > 0$. However, the convergence is not uniform over all of the real half line $x \ge 0$, as we observe in the following proposition.

**Proposition 8.11.** *On $\overline{\mathcal{D}}_T$, for any $T > 0$, then*

$$\underline{u}(x,t) \le u_T(x - ct - \nu)$$

*for some positive constant $\nu$ chosen sufficiently large.*

*Proof.* We can select $\nu$ sufficiently large so that $\underline{u}(x,0) \le u_T(x - \nu)$ for all $x \ge 0$. We next observe that $u_T(x - ct - \nu)$ is an upper solution to IBVP on $\overline{\mathcal{D}}_T$ for any $T > 0$, and the result follows. □

Therefore, although for fixed $x$, $\underline{u}(x,t) \to \eta_2$ as $t \to \infty$, we have that $\underline{u}(x,t)$ is bounded above by the translated travelling wave $u_T(z - \nu)$. This indicates, in this case, that $\underline{u}(x,t)$ will approach the travelling wave front as $t \to \infty$. We have been unable to achieve a tighter result at present, although numerical simulations support the conjecture given here.

For more general initial data, with

$$\hat{u}(x, \tilde{u}_0^-) \le u_0 g(x) \le \hat{u}(x, \tilde{u}_0^+)$$

for some $\tilde{u}_0^-$, $\tilde{u}_0^+ \in (\tilde{a}, \eta_2)$ then the same conclusions hold for $u(x,t)$, the associated solution to IBVP.

## 8.5 Asymptotic Solution as $t \to 0$ for $0 \le x < \infty$

In this section we develop the formal asymptotic solution to (8.1), for $m < n$, $m = n$ and $m > n$, as $t \to 0$. The behaviour of this solution, in each case, depends critically on the nature of $g(x)$ as $x \to \sigma^-$, (8.2a).

### 8.5.1 $m < n$

In this case the asymptotic solution as $t \to 0$ to IBVP' follows, after minor modifications, that given in King and Needham [28] (with $m$ in the current notation replacing $p$ in the notation of King and Needham). The structure of the asymptotic solution has primarily three asymptotic regions in $x$, labelled as follows:

$$\left. \begin{array}{ll} \text{region } \mathbf{I}: & 0 \le x < \sigma - O(\delta(t)), \ u = O(1) \\ \text{region } \mathbf{II}: & x = \sigma + O(\delta(t)), \quad u = o(1) \\ \text{region } \mathbf{III}: & x > \sigma + O(\delta(t)), \quad u = o(1) \end{array} \right\} \text{ as } t \to 0, \qquad (8.50)$$

where $\delta(t) = o(1)$ (we note that $\delta(t) = t^{\frac{1}{2}}$ $[t^{\frac{1}{r(1-m)}}]$ when $1 \le r \le \frac{2}{1-m}$ $[r > \frac{2}{1-m}]$ respectively ) as $t \to 0$. In particular, the structure establishes formally that IBVP', for $0 < m < n < 1$, has a unique local solution for $0 \le t \le \epsilon$ for some $\epsilon \ll 1$. Moreover, $u_\infty(t) \sim (1 - m)^{1/(1-m)}t^{1/(1-m)}$ and $u(x,t) > (1 - m)^{1/(1-m)}t^{1/(1-m)}$ for all $x \ge 0$, as $t \to 0$.

### 8.5.2 $m = n$

Here there are three distinct cases to be considered, these being $k > 1, k = 1$ and $k < 1$. We examine these in turn.

## (a) $k > 1$

We begin with region **I** where $0 \le x < \sigma - o(1)$ and $u = O(1)$ as $t \to 0$. Since $u(x,0) > 0$ and analytic in region **I** with $u = O(1)$ as $t \to 0$, we expand $u(x,t)$ as a regular power series in $t$. After substitution into equation (8.1a), equating powers of $t$ to zero, and applying initial condition (8.1c), we obtain

$$u(x,t) = u_0 g(x) + t\left\{u_0 g''(x) - (\alpha + u_0 g(x)) u_0^m g^m(x)\right\} + O(t^2) \quad (8.51)$$

as $t \to 0$ with $0 \le x < \sigma - o(1)$, and where $\alpha = k - 1$. Now when $(\sigma - x) \ll 1$, (8.51) becomes

$$u(x,t) \sim u_0 g_\sigma \left[(\sigma - x)^r + \ldots\right] + t\left\{u_0 g_\sigma r(r-1)(\sigma - x)^{r-2}\right.$$
$$\left. - (\alpha + u_0 g_\sigma(\sigma - x)^r) u_0^m g_\sigma^m (\sigma - x)^{rm} + \ldots\right\} + \ldots \quad (8.52)$$

as $t \to 0$, and a nonuniformity develops in expansion (8.51). Clearly, there are three cases to consider.

**Case (i). $1 \le r < \frac{2}{1-m}$.** In this case expansion (8.51) in region **I** becomes non-uniform when $x = \sigma - O(t^{\frac{1}{2}})$, when we observe, via (8.52), that $u = O(t^{\frac{r}{2}})$. We must therefore introduce a further region, which we refer to as region **II**. To examine region **II**, we introduce the scaled coordinate $\eta = (x - \sigma)t^{-\frac{1}{2}}$ and look for an asymptotic expansion of the form

$$u(\eta, t) = t^{\frac{r}{2}} F(\eta) + o(t^{\frac{r}{2}}) \quad \text{as} \quad t \to 0 \quad (8.53)$$

with $\eta = O(1)$. On substitution of (8.53) into equation (8.1a) (when written in terms of $\eta$ and $t$) we obtain at leading order

$$F_{\eta\eta} + \frac{\eta}{2} F_\eta - \frac{r}{2} F = 0, \quad -\infty < \eta < \infty, \quad (8.54)$$

to be solved subject to matching with region **I** as $\eta \to -\infty$, and the initial condition (8.1c) as $t \to 0$, that is,

$$F(\eta) \sim u_0 g_\sigma(-\eta)^r \quad \text{as} \quad \eta \to -\infty, \quad (8.55)$$
$$F(\eta) \to 0 \quad \text{as} \quad \eta \to \infty. \quad (8.56)$$

The boundary value problem (8.54)-(8.56) has been considered in Section 2.2.1. The solution is unique and is readily obtained as (2.15) with (2.16)-(2.18) (with $\breve{u}$ now replaced by $F$). We note from (2.15)(with $\breve{u}$ replaced by $F$) that $F(\eta)$ is positive and monotone decreasing for all $-\infty < \eta < \infty$. From (2.15) we observe that

$$F(\eta) \sim C_\infty \frac{1}{\eta^{r+1}} e^{-\eta^2/4} \quad \text{as} \quad \eta \to \infty, \quad (8.57)$$

with

$$C_\infty = \begin{cases} \dfrac{2u_0 g_\sigma (r!)^2}{\kappa_1 \left[ \left( \frac{1}{2}r \right)! \right]^2}, & r \text{ even,} \\[4mm] \dfrac{2u_0 g_\sigma (r!)^2}{\kappa_2 \left[ \left( \frac{1}{2}(r-1) \right)! \right]^2}, & r \text{ odd.} \end{cases} \tag{8.58}$$

Now, from Section 8.3.1, we have established that, in this case, the support of the solution $u(x,t)$ remains finite in $t \ge 0$. Thus we must conclude, via (8.57), that expansion (8.53) in region **II** must become nonuniform when $\eta \gg 1$. The existence of this additional region can be confirmed in the following way: a typical term retained from equation (8.1a) at leading order in region **II** is $t^{r/2-1}F$, whereas a typical neglected term is $t^{rm/2}F^m$. The ratio of neglected to retained terms is then

$$R(\eta, t) = t^{1 - \frac{r}{2}(1-m)} F^{m-1}(\eta), \tag{8.59}$$

which is of $O\left( t^{1-\frac{r}{2}(1-m)} \right)$ ($R = o(1)$ since $1 - \frac{r}{2}(1-m) > 0$) for $\eta = O(1)$ as $t \to 0$. However, when $\eta \gg 1$, $F(\eta) \sim O\left( \eta^{-(r+1)} e^{-\eta^2/4} \right)$, via (8.57), and so

$$R(\eta, t) \sim t^{1 - \frac{r}{2}(1-m)} \eta^{(r+1)(1-m)} e^{\eta^2(1-m)/4}, \tag{8.60}$$

and $R(\eta, t)$ becomes of $O(1)$ for $\eta$ sufficiently large, confirming the onset of a nonuniformity in expansion (8.53) as $\eta \to \infty$. Further examination of (8.60) reveals that the nonuniformity occurs when

$$\eta^2 = \lambda_0(-\ln t) + \lambda_1 \ln \left[ \lambda_2(-\ln t) \right] + \ldots = C^2(t) \quad \text{as} \quad t \to 0, \tag{8.61}$$

where

$$\lambda_0 = \frac{4}{1-m} \left[ 1 - \frac{r}{2}(1-m) \right], \tag{8.62}$$

and $\lambda_1$, $\lambda_2$ are constants to be fixed via the asymptotic matching procedure. To continue the asymptotic structure, we now introduce region **III**, where, via (8.61), $\eta \sim C(t)$ as $t \to 0$. We define the scaled coordinate $\bar{\eta} = [\eta - C(t)] \chi^{-1}(t)$, where $\bar{\eta} = O(1)$ as $t \to 0$ in region **III**, and look for an expansion of the form

$$u(\bar{\eta}, t) = \psi(t) G(\bar{\eta}) + o(\psi(t)) \quad \text{as} \quad t \to 0, \tag{8.63}$$

where the gauge functions $\chi(t), \psi(t) = o(1)$ as $t \to 0$ are to be determined. On rewriting equation (8.1a) in terms of the scaled variables $\bar{\eta}$ and $G$, we require in this region that reaction and diffusion terms must balance at leading order, to enable the support to remain finite. This requires $\chi(t) = O(C^{-1}(t))$ and $\psi(t) = O\left( (tC^{-2}(t))^{1/(1-m)} \right)$. Thus, without loss of generality, we put,

$$\chi(t) \equiv C^{-1}(t) \quad \text{and} \quad \psi(t) \equiv \left( \frac{t}{C^2(t)} \right)^{\frac{1}{1-m}}. \tag{8.64}$$

At leading order we then obtain the equation

$$G_{\bar{\eta}\bar{\eta}} + \frac{1}{2}G_{\bar{\eta}} - \alpha G^m = 0, \quad -\infty < \bar{\eta} < \bar{\eta}_0, \tag{8.65}$$

where $\bar{\eta}_0$ defines the edge of the support of $u(\bar{\eta}, t)$ and is to be determined. In particular, as the solution is classical, we require $G(\bar{\eta}_0) = G'(\bar{\eta}_0) = 0$. It remains to match the expansion (8.63) in region **III** to the expansion (8.53) in region **II** as $\bar{\eta} \to -\infty$. Matching follows directly giving

$$\lambda_1 = \frac{4}{1-m} - 2(r+1), \quad \lambda_2 = \lambda_0, \tag{8.66}$$

and the matching condition

$$G(\bar{\eta}) \sim C_\infty e^{-\bar{\eta}/2} \quad \text{as} \quad \bar{\eta} \to -\infty. \tag{8.67}$$

Therefore at leading order in region **III** we obtain the boundary value problem

$$G_{\bar{\eta}\bar{\eta}} + \frac{1}{2}G_{\bar{\eta}} - \alpha G^m = 0, \quad -\infty < \bar{\eta} < \bar{\eta}_0, \tag{8.68}$$

$$G(\bar{\eta}) > 0, \quad -\infty < \bar{\eta} < \bar{\eta}_0, \tag{8.69}$$

$$G(\bar{\eta}_0) = 0, \quad G_{\bar{\eta}}(\bar{\eta}_0) = 0, \tag{8.70}$$

$$G(\bar{\eta}) \sim C_\infty e^{-\bar{\eta}/2}, \quad \text{as} \quad \bar{\eta} \to -\infty. \tag{8.71}$$

Equation (8.68) and conditions (8.69)-(8.71) can be considered as an eigenvalue problem for $\bar{\eta}_0$. The rescaling $\tilde{x} = -\frac{1}{2}\bar{\eta}$, $\tilde{g} = (4\alpha)^{\frac{1}{m-1}}G$ transforms this eigenvalue problem into the boundary value problem (problem P) studied in Section 3 of Grundy and Peletier [23], where it is established that a unique solution to this problem exists, with, in principle, $\bar{\eta}_0$ being determined as a function of the single parameter $C_\infty(4\alpha)^{\frac{1}{m-1}}$, that is, $\bar{\eta}_0 = \bar{\eta}_0\left(C_\infty(4\alpha)^{\frac{1}{m-1}}\right)$. The solution is monotone decreasing in $-\infty < \bar{\eta} < \bar{\eta}_0$, with, in particular,

$$G(\bar{\eta}) \sim \left[\frac{\alpha(1-m)^2}{2(1+m)}\right]^{\frac{1}{1-m}} (\bar{\eta}_0 - \bar{\eta})^{\frac{2}{1-m}} \quad \text{as} \quad \bar{\eta} \to \bar{\eta}_0^-. \tag{8.72}$$

A consideration of higher order terms shows that expansion (8.63) remains uniform as $\bar{\eta} \to \bar{\eta}_0^-$ and so no further regions are required. This completes the small time asymptotic structure in this case. In particular, we have shown that the edge of the support of $u(x, t)$, say at $x = s(t)$, has

$$s(t) \sim \sigma + t^{1/2}C(t) + \frac{t^{1/2}}{C(t)}\bar{\eta}_0 + \dots \tag{8.73}$$

as $t \to 0$, which becomes, on using (8.61),

$$s(t) \sim \sigma + t^{\frac{1}{2}}\left[\lambda_0^{\frac{1}{2}}(-\ln t)^{\frac{1}{2}} + \left\{\frac{\lambda_1}{2}\ln[\lambda_2(-\ln t)] + \bar{\eta}_0\right\}\lambda_0^{-\frac{1}{2}}(-\ln t)^{-\frac{1}{2}} + \dots\right] \tag{8.74}$$

as $t \to 0$, with $\lambda_0$, $\lambda_1$ and $\lambda_2$ as given in (8.62) and (8.66). We observe from (8.74) that, in this case, the edge of the support expands initially. More details concerning the eigenvalue problem (8.68)-(8.71) are presented in Appendix B. A schematic representation of the location and thickness of the asymptotic regions as $t \to 0$ is given in Figure 8.1

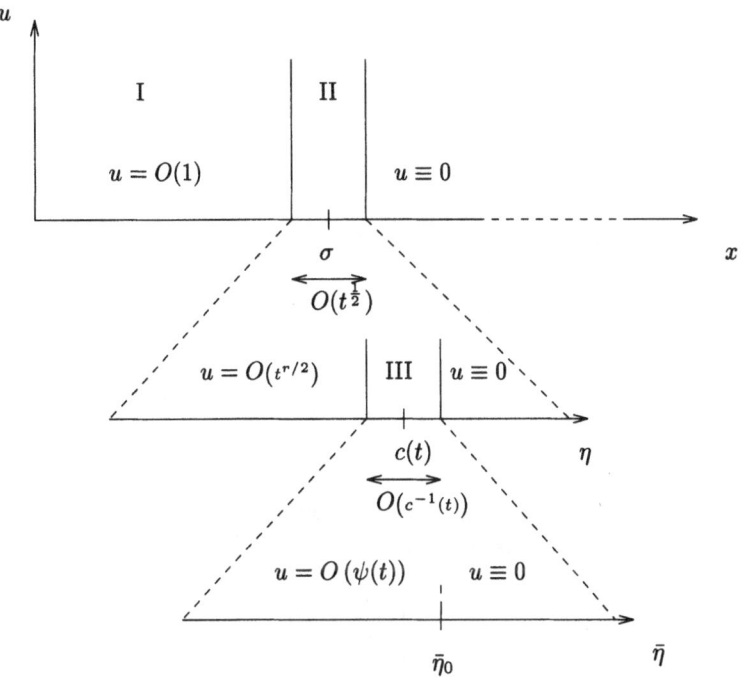

**Fig. 8.1.** Schematic representation of the location and thickness of the asymptotic regions as $t \to 0$ when $m = n, k > 1$ and $1 \le r < \frac{2}{1-m}$.

**Case (ii).** $r = \frac{2}{1-m}$. Again expansion (8.51) in region **I** becomes non-uniform when $x = \sigma - O(t^{\frac{1}{2}})$, but in this case $u = O\left(t^{\frac{1}{1-m}}\right)$ in region **II**. To examine region **II** we again introduce the scaled coordinate $\eta = (x - \sigma)t^{-\frac{1}{2}}$ and look for a solution of the form

$$u(\eta, t) = t^{\frac{1}{1-m}} \bar{F}(\eta) + o\left(t^{\frac{1}{1-m}}\right) \quad \text{as} \quad t \to 0, \qquad (8.75)$$

with $\eta = O(1)$. On substitution of (8.75) into equation (8.1a) (when written in terms of $\eta$ and $t$) we now retain reaction and diffusion at leading order, giving

$$\bar{F}_{\eta\eta} + \frac{\eta}{2}\bar{F}_\eta - \frac{1}{1-m}\bar{F} - \alpha\bar{F}^m = 0, \quad -\infty < \eta < \eta_0. \qquad (8.76)$$

This is to be solved subject to matching with region **I** as $\eta \to -\infty$ and the boundary conditions at the edge of the support when $\eta = \eta_0$ (which is to be determined), namely

$$\bar{F}(\eta) \sim u_0 g_\sigma (-\eta)^{\frac{2}{1-m}} \quad \text{as} \quad \eta \to -\infty, \tag{8.77}$$

$$\bar{F}(\eta_0) = 0, \quad \bar{F}_\eta(\eta_0) = 0. \tag{8.78}$$

We also require that $\bar{F}(\eta) > 0$ for $-\infty < \eta < \eta_0$. The problem (8.76)-(8.78) is an eigenvalue problem for $\eta_0$. As the edge of the support is finite in this region, no further regions are required, and the structure is complete. It remains to analyse (8.76)-(8.78) and this is done in Appendix C. However, we note here, from (8.76)-(8.78) that

$$\bar{F}(\eta) \sim \left[ \frac{\alpha(1-m)^2}{2(1+m)} \right]^{\frac{1}{1-m}} (\eta_0 - \eta)^{\frac{2}{1-m}} \quad \text{as} \quad \eta \to \eta_0^-. \tag{8.79}$$

In particular, we have in this case that the edge of the support of $u(\eta, t)$ behaves as

$$s(t) \sim \sigma + t^{\frac{1}{2}} \eta_0 + o(t^{\frac{1}{2}}) + \dots \quad \text{as } t \to 0. \tag{8.80}$$

In this case, the edge of the support contracts or expands initially, depending upon the sign of $\eta_0$. In particular, it is shown in Appendix C that the sign of $\eta_0$ depends on the single parameter

$$\lambda = u_0 g_\sigma \left[ \frac{2(1+m)}{\alpha(1-m)^2} \right]^{\frac{1}{1-m}},$$

in the sense that $\eta_0 > 0$ when $\lambda > 1$, $\eta_0 = 0$ when $\lambda = 1$, $\eta_0 < 0$ when $0 < \lambda < 1$. A schematic representation of the location and thickness of the asymptotic regions as $t \to 0$ is given in Figure 8.2

**Case (iii).** $r > \frac{2}{1-m}$. Expansion (8.51) in region **I** now becomes nonuniform when $x = \sigma - O\left(t^{\frac{1}{r(1-m)}}\right)$, with $u = O\left(t^{\frac{1}{1-m}}\right)$ in region **II**. To examine region **II**, we introduce the scaled coordinate $\eta = (x - \sigma)t^{-\frac{1}{r(1-m)}}$ and look for an asymptotic expansion of the form

$$u(\eta, t) = t^{\frac{1}{1-m}} \hat{F}(\eta) + o\left(t^{\frac{1}{1-m}}\right) \quad \text{as} \quad t \to 0, \tag{8.81}$$

with $\eta = O(1)$. On substitution of equation (8.81) into equation (8.1a) (when written in terms of $\eta$ and $t$) we obtain at leading order

$$\eta \hat{F}_\eta = r(1-m) \left( \frac{\hat{F}}{1-m} + \alpha \hat{F}^m \right), \quad \eta > -\infty, \tag{8.82}$$

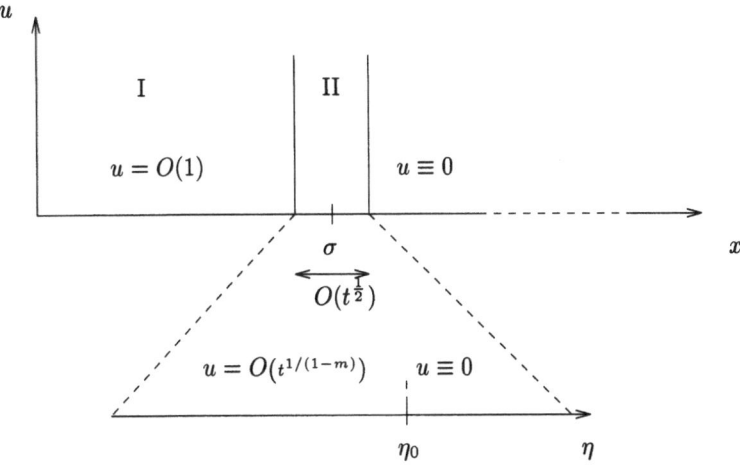

**Fig. 8.2.** Schematic representation of the location and thickness of the asymptotic regions as $t \to 0$ when $m = n, k > 1$ and $r = \frac{2}{1-m}$.

$$\hat{F}(\eta) \sim u_0 g_\sigma (-\eta)^r \quad \text{as} \quad \eta \to -\infty, \tag{8.83}$$

after matching with region **I**. The solution to (8.82), (8.83) may be obtained directly by separation of variables as

$$\hat{F}(\eta) = \left[ (u_0 g_\sigma)^{1-m} (-\eta)^{r(1-m)} - (1-m) \right]^{\frac{1}{(1-m)}}, \quad \eta > -\infty. \tag{8.84}$$

An examination of (8.84) reveals that a weak singularity develops in $\hat{F}(\eta)$ as $\eta \to \eta_c$, where

$$\eta_c = - \left[ \frac{\alpha(1-m)}{(u_0 g_\sigma)^{1-m}} \right]^{\frac{1}{r(1-m)}}, \tag{8.85}$$

with

$$\hat{F}(\eta) \sim \left[ \frac{\alpha r(1-m)^2}{|\eta_c|} \right]^{\frac{1}{1-m}} (\eta_c - \eta)^{\frac{1}{1-m}} \tag{8.86}$$

as $\eta \to \eta_c^-$. Thus the support of $u(\eta, t)$ ends at $\eta = \eta_c$ in this region. However, in (8.81) and (8.86) the degree of $(\eta_c - \eta)$ as $\eta \to \eta_c^-$ (which is $\frac{1}{1-m}$) is too weak, and consideration of further terms in (8.81) reveal a weak nonuniformity as $\eta \to \eta_c^-$. Therefore a further region is required to complete the structure, in which $\eta = \eta_c + o(1)$ as $t \to 0$, and diffusion effects are retained at leading order to enable the appropriate behaviour to be achieved at the edge of the support. We label this region as region **III** and introduce the scaled coordinate $\tilde{\eta}$ by

$$\eta = \eta_c + t^\gamma \tilde{\eta}, \tag{8.87}$$

with $\gamma > 0$ to be determined, and $\tilde{\eta} = O(1)$ as $t \to 0$ in region **III**. An examination of (8.81) and (8.86) then determines that $u = O\left( t^{\frac{1+\gamma}{1-m}} \right)$ in region **III**.

Thus we expand

$$u(\tilde{\eta}, t) = t^{\frac{1+\gamma}{1-m}} H(\tilde{\eta}) + o\left(t^{\frac{1+\gamma}{1-m}}\right) \qquad (8.88)$$

as $t \to 0$, with $\tilde{\eta} = O(1)$. On substituting (8.88) into equation (8.1a), to retain diffusion terms at leading order requires

$$\gamma = \left[1 - \frac{2}{r(1-m)}\right] > 0, \qquad (8.89)$$

after which the leading order problem is

$$H_{\tilde{\eta}\tilde{\eta}} + \frac{\eta_c}{r(1-m)} H_{\tilde{\eta}} - \alpha H^m = 0, \quad -\infty < \tilde{\eta} < \tilde{\eta}_0, \qquad (8.90)$$

$$H(\tilde{\eta}_0) = H_{\tilde{\eta}}(\tilde{\eta}_0) = 0, \qquad (8.91)$$

$$H(\tilde{\eta}) \sim \left\{\frac{\alpha r(1-m)^2}{|\eta_c|}\right\}^{\frac{1}{1-m}} (-\tilde{\eta})^{\frac{1}{1-m}} \quad \text{as} \quad \tilde{\eta} \to -\infty. \qquad (8.92)$$

The problem (8.90)-(8.92) is autonomous and can be studied in the $(H, H_{\tilde{\eta}})$ phase plane. It is established in Appendix D that (8.90)-(8.92) has a unique solution for each $\tilde{\eta}_0$. Thus (8.90)-(8.92) does not fix a unique value of $\tilde{\eta}_0$. In fact, $\tilde{\eta}_0$ will be fixed by matching expansion (8.88) (as $\tilde{\eta} \to -\infty$) to expansion (8.81) (as $\eta \to \eta_c^-$), when expansion (8.81) is taken to next order. We do not pursue this here. We observe that the solution to (8.90)-(8.92) is monotone decreasing in $-\infty < \tilde{\eta} < \tilde{\eta}_0$, and has

$$H(\tilde{\eta}) \sim \left[\frac{\alpha(1-m)^2}{2(1+m)}\right]^{\frac{1}{1-m}} (\tilde{\eta}_0 - \tilde{\eta})^{\frac{2}{1-m}} \quad \text{as} \quad \tilde{\eta} \to \tilde{\eta}_0^-, \qquad (8.93)$$

which has the required decay rate in $(\tilde{\eta}_0 - \tilde{\eta})$ as the edge of the support is approached. Consideration of further terms in this region shows that expansion (8.88) remains uniform as $\tilde{\eta} \to \tilde{\eta}_0^-$, and so no further regions are required and the asymptotic structure is complete. In particular, we have that the edge of the support $x = s(t)$ behaves as

$$s(t) \sim \sigma + \eta_c t^{\frac{1}{r(1-m)}} + \tilde{\eta}_0 t^{\frac{r(1-m)-1}{r(1-m)}} + \dots \qquad (8.94)$$

as $t \to 0$. We observe from (8.94) that, since $\eta_c < 0$, the edge of the support is contracting initially. A schematic representation of the location and thickness of the asymptotic regions as $t \to 0$ is given in Figure 8.3

At this stage the main asymptotic structure is complete in each case. However, we note finally that, with $g(x) \sim g_0 + g_{\tilde{m}} x^{\tilde{m}} + \dots$ as $x \to 0^+$, for some $g_{\tilde{m}} \neq 0$ and $\tilde{m} \in \mathbb{N}$ (as $g(x)$ is analytic in $0 \le x \le \sigma$), the expansion (8.51) in region **I** does not, in general, satisfy boundary condition (8.1e) at $x = 0$ and a further passive region is required in the neighbourhood of $x = 0$ as $t \to 0$. We denote this region as region $\mathbf{I_0}$ and it is readily deduced that in this region $x = O(t^{\frac{1}{2}})$ as $t \to 0$. The details of this region follow after minor modifications, those given for region $\mathbf{I_0}$ in part (a) of Section 2.2.1.

The full asymptotic structure has now been completed in this case.

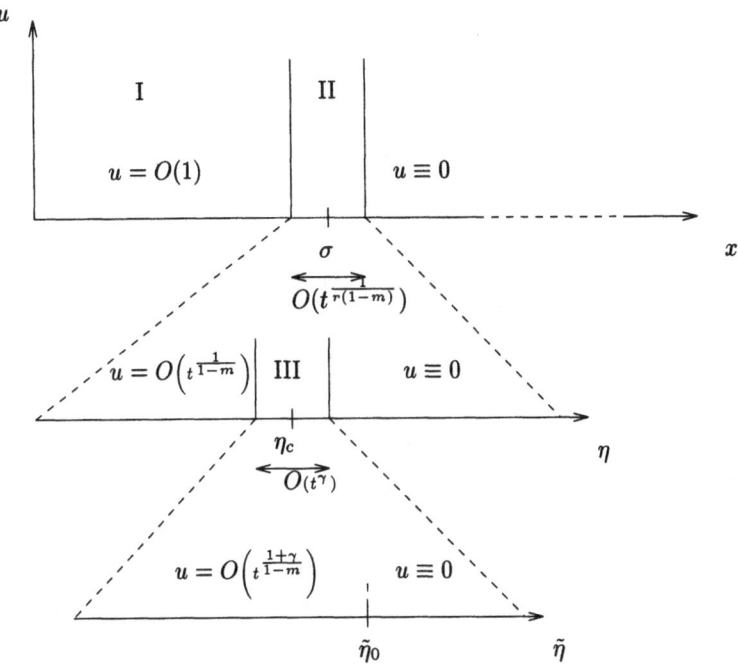

**Fig. 8.3.** Schematic representation of the location and thickness of the asymptotic regions as $t \to 0$ when $m = n, k > 1$ and $r > \frac{2}{1-m}$.

## (b) $k = 1$

In this case we will see that the asymptotic structure as $t \to 0$ is significantly different from the case $k > 1$ just examined. We begin with region **I** where $0 \le x < \sigma - o(1)$ and $u = O(1)$ as $t \to 0$. As in the previous case we expand $u(x,t)$ in region **I** as a regular power series in $t$. After substitution into equation (8.1a), equating powers of $t$ to zero, and applying initial condition (8.1c), we get

$$u(x,t) = u_0 g(x) + t \left\{ u_0 g''(x) - u_0^{m+1} g^{m+1}(x) \right\} + O(t^2) \quad \text{as } t \to 0. \quad (8.95)$$

Now, as $x \to \sigma^-$, (8.95) becomes nonuniform when $x = \sigma - O\left(t^{\frac{1}{2}}\right)$. To continue we introduce region **II** with $x = \sigma \pm O\left(t^{\frac{1}{2}}\right)$ and, from (8.95), $u = O\left(t^{\frac{1}{2}}\right)$. In region **II** we introduce the scaled coordinate $\breve{\eta} = (x - \sigma)t^{-\frac{1}{2}}$ and look for an asymptotic expansion of the form

$$u(\breve{\eta}, t) = t^{\frac{r}{2}} \breve{u}_0(\breve{\eta}) + o(t^{\frac{r}{2}}) \quad \text{as} \quad t \to 0, \quad (8.96)$$

with $\breve{\eta} = O(1)$. On substituting (8.96) into (8.1a) (when written in terms of $\breve{\eta}$ and $t$) we obtain at leading order

$$\check{u}_{0\check{\eta}\check{\eta}} + \frac{\check{\eta}}{2}\check{u}_{0\check{\eta}} - \frac{r}{2}\check{u}_0 = 0, \quad -\infty < \check{\eta} < \infty, \tag{8.97}$$

to be solved subject to initial condition (8.1b) and matching with region **I** as $\check{\eta} \to -\infty$. These matching conditions are readily obtained as

$$\check{u}_0(\check{\eta}) \sim u_0 g_\sigma (-\check{\eta})^r \quad \text{as} \quad \check{\eta} \to -\infty, \tag{8.98}$$

$$\check{u}_0(\check{\eta}) \to 0 \quad \text{as} \quad \check{\eta} \to \infty. \tag{8.99}$$

The solution to the boundary value problem (8.97), (8.98) and (8.99) is unique and is given in Section 2.2.1 by (2.15) with (2.16)-(2.18) (where $\check{u}$ is now replaced by $\check{u}_0$). Now for $\check{\eta} \gg 1$,

$$u(\check{\eta}, t) \sim t^{\frac{r}{2}} \frac{C_\infty e^{-\check{\eta}^2/4}}{\check{\eta}^{r+1}} + \dots \quad \text{as} \quad t \to 0, \tag{8.100}$$

where $C_\infty$ is the constant given by (8.58). Expansion (8.96) remains uniform as $\check{\eta} \to \infty$ and we are left to introduce the final region **III** where $x = \sigma + O(1)$. In region **III**, where $x = \sigma + O(1)$ and $u = o(1)$, motivation from (8.100) leads us to look for a solution of the form

$$u(x, t) = e^{-\check{\psi}(x,t)/t} \quad \text{as} \quad t \to 0, \tag{8.101}$$

where $\check{\psi}(x,t) = O(1)$ and $\check{\psi}(x,t) > 0$ for all $x > \sigma$. Substitution of (8.101) into (8.1a) leads to an expansion for $\check{\psi}$ in the form,

$$\check{\psi}(x,t) = \check{\psi}_0(x) + t\{\check{\psi}_1(x) + \check{\psi}_2(x) \ln t\} + O(t^2) \tag{8.102}$$

as $t \to 0$, with $x = \sigma + O(1)$. On solving at each order in turn we obtain

$$\check{\psi}(x,t) = \frac{1}{4}(x - \sigma)^2 + t\left\{\left(\frac{1}{2} + \check{D}\right)\ln(x - \sigma) + \check{A} - \check{D}\ln t\right\} + O(t^2 \ln^2 t), \tag{8.103}$$

where $\check{A}$ and $\check{D}$ are constants. Finally in region **III** we have, via (8.101) and (8.103),

$$u(x, t) = \exp\left\{-\frac{(x - \sigma)^2}{4t} - \left(\frac{1}{2} + \check{D}\right)\ln(x - \sigma) - \check{A} + \check{D}\ln t + O(t \ln^2 t)\right\} \tag{8.104}$$

as $t \to 0$, with $x = \sigma + O(1)$. Clearly (8.104) satisfies the initial conditions in region **III** (namely $u(x,t) \to 0$ as $t \to 0$), together with the boundary condition as $x \to \infty$ ($u \to 0$ as $x \to \infty$). It remains to match the expansion (8.104) as $x \to \sigma^+$ with expansion (8.96) in region **II** as $\eta \to \infty$. Matching gives

$$\check{A} = -\ln C_\infty, \quad \check{D} = r + \frac{1}{2}. \tag{8.105}$$

This completes the main asymptotic structure in this case. Note again that expansion (8.95) in region **I** does not, in general, satisfy boundary condition

(8.1e) at $x = 0$ and a further passive region is required in the neighbourhood of $x = 0$ as $t \to 0$. The details of this region, which we label $\mathbf{I_0}$, follow, after minor modifications, those given for region $\mathbf{I_0}$ in part (a) of Section 2.2.1. The full asymptotic structure as $t \to 0$ is now complete in this case. We observe in this case that the support of $u(x,t)$ moves to become unbounded in infinitesimal time, in line with the theory of Section 8.3.2.

## (c) $k < 1$

In this case the asymptotic structure as $t \to 0$ follows (after rescaling equation (8.1a) by $u = (1 - k)^{\frac{1}{1-m}} U$) that given in [28] and again has primarily three asymptotic regions given by (8.50). Again, in the current notation, $m$ replaces $p$ in the notation of [28].

### 8.5.3 $m > n$

Here the asymptotic structure as $t \to 0$ follows, after minor modifications, that given in Section 8.5.2 part (a), with $n$ now replacing $m$ throughout, and $\alpha$ being replaced by $k$.

### 8.5.4 Summary

To summarize, we have established the complete asymptotic structure of the solution to the initial-boundary value problem as $t \to 0$, in all cases. For $m < n$, and $m = n$ with $0 < k < 1$, we have seen that the support of $u(x,t)$ extends to infinity in infinitesimal time, with $u(x,t)$ becoming bounded above zero as $x \to \infty$, $0 < t \ll 1$. For $m = n$ and $k = 1$, again, the support of $u(x,t)$ becomes unbounded in infinitesimal time, but in this case $u(x,t) \to 0$ as $x \to \infty$, $0 < t \ll 1$. Finally, for $m > n$, and $m = n$ with $k > 1$, the support of $u(x,t)$ remains finite for $0 < t \ll 1$, and we have determined expressions for the initial motion of the edge of the support of $u(x,t)$.

## 8.6 Asymptotic Solution as $t \to \infty$ or as $t \to t_c^-$

In this section we obtain the asymptotic structure of the solution to IBVP, for $m < n$, $m = n$ and $m > n$ either as $t \to \infty$ or as $t \to t_c^-$ accordingly.

### 8.6.1 $m < n$

It was shown qualitatively in Section 8.2 that for $m < n$, $u(x,t) \to \nu$ as $t \to \infty$ uniformly in $x \geq 0$. This suggests writing

$$u(x,t) = \nu + \tilde{\phi}(x,t) \quad \text{as} \quad t \to \infty, \tag{8.106}$$

where $\tilde{\phi}(x,t) = o(1)$ as $t \to \infty$, uniformly in $x \geq 0$. On substituting (8.106) into equation (8.1a) we obtain at leading order

$$\tilde{\phi}_t = \tilde{\phi}_{xx} + f'(\nu)\tilde{\phi}, \tag{8.107}$$

with $x > 0$ and $t \gg 1$. Equation (8.107) suggests looking for a solution of the form

$$\tilde{\phi}(x,t) = e^{f'(\nu)t} R(x,t). \tag{8.108}$$

Substitution of (8.108) into (8.107) results in the equation

$$R_t = R_{xx}, \quad x > 0, \quad t \gg 1, \tag{8.109}$$

and we require, from conditions at $x = 0$ and as $x \to \infty$, that

$$\begin{aligned} R_x(0,t) &= 0, \quad t \gg 1, \\ R(x,t) &\to -\nu e^{-u_\infty f'(\nu)} \text{ as } x \to \infty, \quad t \gg 1, \end{aligned} \tag{8.110}$$

where

$$u_\infty = \int_0^\nu \left[ \frac{1}{s^m(1-s) - ks^n} - \frac{1}{f'(\nu)(s-\nu)} \right] ds. \tag{8.111}$$

The appropriate solution to (8.109), (8.110) has

$$R(x,t) = -\nu e^{-u_\infty f'(\nu)} + o(1) \quad \text{in } x \geq 0 \tag{8.112}$$

as $t \to \infty$. Thus, we have finally that

$$u(x,t) = \nu - \nu e^{-u_\infty f'(\nu)} e^{f'(\nu)t} + o\left(e^{f'(\nu)t}\right) \tag{8.113}$$

as $t \to \infty$, uniformly in $x \geq 0$.

Hence the solution to IBVP$'$, $u(x,t)$, approaches the equilibrium value $\nu$ through exponentially small terms, uniformly in $x \geq 0$, as $t \to \infty$.

### 8.6.2 $m = n$

Here there are three distinct cases to be considered, these being $k > 1$, $k = 1$ and $k < 1$. We examine these in turn.

### (a) $k > 1$

It was shown qualitatively in Section 8.3.1 that for $m = n$ and $k > 1$ that $u(x,t) \to 0$ uniformly in $x \geq 0$, with contracting finite support, as $t \to t_c^-$, where $0 < t_c < \infty$. Thus we have $u = o(1)$ and the support of $u$ has $x = o(1)$ as $t \to t_c^-$. To examine the asymptotic structure of the solution of IBVP as $t \to t_c^-$ we therefore introduce the scaled variable $\eta$ so that $x = (t_c - t)^\gamma \eta$ and look for an asymptotic expansion in the form

$$u(\eta, t) = (t_c - t)^\delta F(\eta) + o((t_c - t)^\delta) \qquad (8.114)$$

as $t \to t_c^-$, with $F(\eta)$, $\eta = O(1)$. Substitution of (8.114) into equation (8.1a) (when written in terms of $\eta$) requires, for a non-trivial leading order balance, that

$$\gamma = \frac{1}{2}, \qquad \delta = \frac{1}{1-m}, \qquad (8.115)$$

after which equation (8.1a) becomes

$$F''(\eta) - \frac{\eta}{2}F'(\eta) + \frac{1}{1-m}F(\eta) - \alpha F^m(\eta) = 0, \quad 0 < \eta < \eta_0, \qquad (8.116)$$

where $\alpha = k - 1 > 0$ and $\eta_0$ is the edge of the support of $F(\eta)$. Equation (8.116) has to be solved subject to the boundary conditions

$$F(0) = F_0, \qquad F'(0) = 0, \qquad (8.117)$$
$$F(\eta_0) = 0, \qquad F'(\eta_0) = 0, \qquad (8.118)$$

and that

$$F(\eta) > 0, \qquad 0 \le \eta < \eta_0, \qquad (8.119)$$

where $\eta_0$ is the edge of the support of the function $F(\eta)$ and $F(\eta) \equiv 0$ for $\eta > \eta_0$. We now wish to find $F_0 \,(> 0)$ and $\eta_0 \,(> 0)$ such that the boundary value problem (8.116)-(8.119) has a non-trivial solution. It is straightforward to show directly from equation (8.116) and conditions (8.117)-(8.119) that a necessary condition for a solution to (8.116)-(8.119) is

$$[\alpha(1-m)]^{\frac{1}{1-m}} < F_0 < \left[\frac{2\alpha(1-m)}{(1+m)}\right]^{\frac{1}{1-m}}. \qquad (8.120)$$

The boundary value problem (8.116)-(8.119) can readily be investigated numerically. To do this, it is convenient to regard (8.116) and (8.118) as an initial value problem in $0 \le \eta \le \eta_0$, based at $\eta = \eta_0$, and shoot backwards in $\eta$ from $\eta = \eta_0$, whilst adjusting the value of $\eta_0$ so as to make $F(\eta)$ satisfy conditions (8.117) at $\eta = 0$. The numerical integration (Runge–Kutta) is started close to $\eta_0$ (due to the singular structure at $\eta = \eta_0$) using the asymptotic form

$$F(\eta) \sim \left[\frac{\alpha(1-m)^2}{2(1+m)}\right]^{\frac{1}{1-m}} (\eta_0 - \eta)^{\frac{2}{1-m}}, \qquad \text{as } \eta \to \eta_0^-. \qquad (8.121)$$

For a given $0 < m < 1$ and $k > 1$, it was found that there exists a value $\eta_0^* > 0$ such that for each $\eta_0 \ge \eta_0^*$ the solution of the initial value problem (8.116), (8.118) satisfies the boundary conditions (8.117) at $\eta = 0$ and hence provides a non-trivial solution to the boundary value problem (8.116)-(8.119). The corresponding values of $F_0$ were shown to satisfy the inequality (8.120). Numerical calculation suggests that for each $\eta_0 \ge \eta_0^*$ there exists a solution to (8.116)-(8.119), which is monotone in $\eta$, and has a corresponding value $F(0) = F_0(\eta_0)$. Further

(i) $F_0(\eta_0)$ is monotone decreasing in $\eta_0 \geq \eta_0^*$ from a value $F_0^*$ at $\eta = \eta_0^*$.
(ii) $F_0(\eta_0) \to [\alpha(1-m)]^{\frac{1}{1-m}}$ as    $\eta_0 \to \infty$.

Condition (ii) suggests that we should examine the asymptotic structure of the boundary value problem (8.116)-(8.119) for $\eta_0 \gg 1$. To examine this asymptotic structure we introduce the scaled variable

$$y = \frac{\eta}{\eta_0},\tag{8.122}$$

so that $0 \leq y \leq 1$, and look for a solution of the form

$$F(y) = \bar{F}_0(y) + \frac{1}{\eta_0^2}\bar{F}_1(y) + \dots\tag{8.123}$$

as $\eta_0 \to \infty$. On substituting from (8.123) into equation (8.116) (when written in terms of $y$) we obtain, at leading order,

$$-\frac{y}{2}\bar{F}_{0y} + \frac{1}{1-m}\bar{F}_0 - \alpha\bar{F}_0^m = 0.\tag{8.124}$$

After integrating (8.124) we obtain the general solution as

$$\bar{F}_0(y) = (1-m)^{\frac{1}{1-m}}\left(\alpha - C_1 y^2\right)^{\frac{1}{1-m}},\tag{8.125}$$

where $C_1$ is a constant of integration. Condition (8.118) requires $\bar{F}_0(1) = 0$, which gives $C_1 = \alpha$, whilst condition (8.117) at $y = 0$ is automatically satisfied. On continuing the expansion to higher order, we find that (8.123) remains uniform as $\eta_0 \to \infty$ when $0 \leq y \leq 1 - o(1)$, and that

$$F_0(\eta_0) \sim [\alpha(1-m)]^{\frac{1}{1-m}}\left\{1 + \frac{2}{1-m}\eta_0^{-2} + \dots\right\}\tag{8.126}$$

as $\eta_0 \to \infty$. However, we observe, via (8.125) and (8.121), that expression (8.123) develops a nonuniformity as $y \to 1$. We require an inner region close to $y = 1$ to complete the solution. A balancing of terms in equation (8.116) shows that the region has a thickness of $O(\eta_0^{-2})$. Thus we introduce the inner coordinate $\xi$ by

$$y = 1 + \xi\eta_0^{-2},\tag{8.127}$$

where $\xi = O(1)$ as $\eta_0 \to \infty$. An examination of (8.123) and (8.125) (with $C_1 = \alpha$) then determines that $F = O\left(\eta_0^{-\frac{2}{1-m}}\right)$ in the inner region. Thus we expand

$$F(\xi) = \eta_0^{-\frac{2}{1-m}}H(\xi) + o\left(\eta_0^{-\frac{2}{1-m}}\right)\tag{8.128}$$

as $\eta_0 \to \infty$, with $\xi$, $H(\xi) = O(1)$. On substitution of (8.128) into (8.116) (when written in terms of $\xi$) we obtain at leading order the problem

$$H_{\xi\xi} - \frac{1}{2}H_\xi - \alpha H^m = 0, \quad -\infty < \xi < 0, \tag{8.129}$$

$$H(0) = H_\xi(0) = 0, \tag{8.130}$$

$$H(\xi) \sim [2\alpha(1-m)]^{\frac{1}{1-m}} (-\xi)^{\frac{1}{1-m}}, \quad \text{as } \xi \to -\infty, \tag{8.131}$$

with condition (8.131) arising from matching expansion (8.128) (as $\xi \to -\infty$) with expansion (8.123) (as $y \to 1$). It can be shown (after minor modifications to Appendix D) that the problem (8.129)-(8.131) has a unique solution on $-\infty < \xi \le 0$. The solution to (8.129)-(8.131) is monotone decreasing in $-\infty < \xi \le 0$, with

$$H(\xi) = \left[\frac{\alpha(1-m)^2}{2(1+m)^2}\right]^{\frac{1}{1-m}} (-\xi)^{\frac{2}{1-m}} \tag{8.132}$$

as $\xi \to 0$, which is the required decay rate in $(-\xi)$ as the edge of the support is reached, via (8.128). Consideration of further terms in this region shows that expansion (8.128) remains uniform as $\xi \to 0$, and so no further regions are required. This completes the asymptotic structure, at leading order, of the solution to the boundary value problem (8.116)-(8.119), as $\eta_0 \to \infty$.

Therefore, in this case, the asymptotic solution of IBVP as $t \to t_c^-$ is given by

$$u(x,t) = \begin{cases} (t_c - t)^{\frac{1}{1-m}} F\left(\frac{x}{(t_c-t)^{\frac{1}{2}}}\right) + o\left[(t_c - t)^{\frac{1}{1-m}}\right], & 0 \le x \le \eta_0(t_c - t)^{\frac{1}{2}} \\ 0, & x > \eta_0(t_c - t)^{\frac{1}{2}} \end{cases} \tag{8.133}$$

where $\eta_0, t_c > 0$ are not fixed, and will depend on the transient structure of IBVP for $0 \le t < t_c$. Estimates for $\eta_0$ and $t_c$ have been obtained precisely in section 8.3.1. The important observation is that the support of $u(x,t)$ is of $O\left[(t_c - t)^{\frac{1}{2}}\right]$ as $t \to t_c^-$, whilst the magnitude of $u(x,t)$ is of $O\left[(t_c - t)^{\frac{1}{1-m}}\right]$ as $t \to t_c^-$.

## (b) $k = 1$

It was shown qualitatively in Section 8.3.2 that for the case $m = n$ and $k = 1$ then $u(x,t) \to 0$, uniformly in $x \ge 0$, as $t \to \infty$. To examine the asymptotic structure of the solution of IBVP as $t \to \infty$ we introduce the scaled variable $z = \frac{x}{t^{\frac{1}{2}}}$ and look for an asymptotic expansion in the form

$$u(z,t) = t^{-\frac{1}{m}} \tilde{F}(z) + o\left(t^{-\frac{1}{m}}\right) \quad \text{as} \quad t \to \infty, \tag{8.134}$$

where $\tilde{F}(z)$, $z = O(1)$ which is motivated by looking for a non-trivial balance in equation (8.1a). On substituting (8.134) into equation (8.1a) (when written in terms of $z$) we obtain, at leading order,

$$\tilde{F}_{zz} + \frac{z}{2}\tilde{F}_z + \frac{1}{m}\tilde{F} - \tilde{F}^{m+1} = 0, \quad z > 0, \tag{8.135}$$

which is solved subject to the boundary conditions

$$\tilde{F}(0) = \tilde{F}_0 > 0, \tag{8.136}$$

$$\tilde{F}_z(0) = 0, \tag{8.137}$$

$$\tilde{F}(z) \leq O\left(e^{-z^2/4}\right) \text{ as } z \to \infty, \tag{8.138}$$

$$\tilde{F}(z) > 0 \quad \text{for } z > 0. \tag{8.139}$$

The boundary value problem (8.135)-(8.139) is examined in detail in Chapter 2, Section 13 of Samarskii *et al.* [60], where it is established that there exists at least one (and numerical integrations suggest only one) solution, $\tilde{F}(z)$, and this solution is monotone decreasing in $z \geq 0$, with

$$0 < \tilde{F}_0 < \left(\frac{1}{m}\right)^{\frac{1}{m}}. \tag{8.140}$$

Hence the asymptotic structure of the solution of IBVP in this case is given by (8.134) as $t \to \infty$, uniformly in $x \geq 0$. The support of $u(x,t)$ is unbounded and the spread is of $O\left(t^{\frac{1}{2}}\right)$ as $t \to \infty$, whilst the magnitude of $u(x,t)$ is of $O\left(t^{-\frac{1}{m}}\right)$ as $t \to \infty$.

## (c) $k < 1$

In this case the asymptotic structure of the solution of IBVP' as $t \to \infty$ follows that given in Section 8.6.1, with $1 - k$ replacing $\nu$ throughout.

### 8.6.3 $m > n$

It has been demonstrated qualitatively in Section 8.4 that travelling waves with semi-infinite support develop in the solution to IBVP as $t \to \infty$ when $0 < k < k_c$ (where $k_c$ varies with $m$ and $n$) provided that the initial data exceeds a parameter dependent critical threshold, otherwise solutions decay to zero in finite time, $t_c$, with contracting support. The asymptotic structure of the solution in the latter case as $t \to t_c^-$ is given in Section 8.6.2 part (a) (with $n$ and $k$ replacing $m$ and $\alpha$ respectively).

It remains to examine the large time asymptotic development of IBVP when a travelling wave structure evolves as $t \to \infty$. The permanent form travelling wave which evolves from IBVP as $t \to \infty$ has finite support in $x \geq 0$ and travels with speed $c = c^* > 0$ as given in Chapter 7. Thus, we begin the $t \gg 1$ structure in the travelling wave front region, where, $x \sim s(t) \gg 1$, with

$x = s(t)$ being the location of the support edge of the travelling wave front. We refer to this as region **I**, and introduce the co-ordinate $y$, where

$$x = s(t) + y, \qquad (8.141)$$

with $y = O(1)$ in region **I**. The leading order structure in this region is the permanent form travelling wave (PTW) solution discussed in detail in Chapter 7. Thus in region **I** we have

$$u(y,t) = u^*(y) + \hat{\psi}(t)u_1(y) + o\left(\hat{\psi}(t)\right), \qquad (8.142)$$

$$\dot{s}(t) = c^* + c_1\hat{\chi}(t) + o\left(\hat{\chi}(t)\right), \qquad (8.143)$$

as $t \to \infty$ with $y = O(1)$. Here $u^*(y)$ and $c^*$ are as introduced in Chapter 7, and $\hat{\psi}(t)$, $\hat{\chi}(t) = o(1)$ as $t \to \infty$ are gauge functions to be determined. For the purposes of the present analysis, we recall the following properties from Chapter 7, namely

(i) $u^*(y) \equiv 0$    for   $y \geq 0$,

(ii) $u^*(y) > 0$    for    $-\infty < y < 0$,

(iii) $u^*(y) \sim \left[-\frac{\gamma}{2}(1-n)y\right]^{\frac{2}{1-n}}$    as    $y \to 0^-$,

(iv) $u^*(y)$ is monotone decreasing in $-\infty < y \leq 0$,

(v) $u^*(y) \sim \eta_2 - A^*e^{\lambda_+ y}$    as $y \to -\infty$,

where $\gamma = \left[\frac{2k}{n+1}\right]^{\frac{1}{2}}$ and

$$\lambda_+ = \frac{1}{2}\left[-c^* + \{c^{*2} - 4f_u(\eta_2)\}^{\frac{1}{2}}\right] > 0. \qquad (8.144)$$

We further recall from Chapter 7 that $f_u(\eta_2) < 0$ and $A^* > 0$ is fixed in terms of $n, m$ and $k$. Before proceeding to next order in this region, we observe that expansion (8.142) becomes nonuniform for $(-y) \gg 1$, where it fails to satisfy boundary condition (8.1e) on $x = 0$. We must therefore complete the structure as $t \to \infty$, by introducing region **II**, where $x = O(1)$ as $t \to \infty$, with (from (8.142) and (v)) $u \sim \eta_2 + O\left(e^{-\lambda_+ + s(t)}\right)$ as $t \to \infty$. Thus, in region **II** we expand

$$u(x,t) = \eta_2 + \hat{u}(x)e^{-\lambda_+ + s(t)} + o\left(e^{-\lambda_+ + s(t)}\right) \qquad (8.145)$$

as $t \to \infty$ with $x = O(1)$. After substituting (8.145) into equation (8.1a), we obtain the leading order problem as

$$\hat{u}'' - \lambda_+^2 \hat{u} = 0, \quad 0 < x < \infty, \qquad (8.146)$$

$$\hat{u}'(0) = 0, \tag{8.147}$$

$$\hat{u}(x) \sim -A^* e^{\lambda_+ x} \quad \text{as} \quad x \to \infty, \tag{8.148}$$

where the final condition arises from matching expansion (8.142) [to $O(1)$ as $y \to -\infty$] with expansion (8.145) [to $O\left(e^{\lambda_+ s(t)}\right)$ as $x \to \infty$] following the generalized matching principle of Van Dyke [70]. The solution of (8.146)-(8.148) is readily obtained as

$$\hat{u}(x) = -2A^* \cosh(\lambda_+ x), \quad x \ge 0. \tag{8.149}$$

Having determined the leading order term in region **II**, we now return to the correction terms in region **I**. An examination of the form of expansion (8.145), when written in region **I** via (8.141) and (8.149), reveals that

$$\hat{\psi}(t) = e^{-2\lambda_+ s(t)}, \tag{8.150}$$

after which a balancing of terms requires

$$\hat{\chi}(t) = e^{-2\lambda_+ s(t)}. \tag{8.151}$$

On using (8.150), (8.151) in (8.142), (8.143), the correction problem in region **I** becomes

$$u_1'' + c^* u_1' + [f_u[u^*(y)] + 2\lambda_+ c^*] u_1 = -c_1 u^{*'}(y), \quad -\infty < y < 0, \tag{8.152}$$

$$u_1(0) = u_1'(0) = 0, \tag{8.153}$$

$$u_1(y) \le O\left[(-y)^{\frac{2}{1-n}}\right] \quad \text{as} \quad y \to 0^-, \tag{8.154}$$

$$u_1(y) \sim -A^* e^{-\lambda_+ y} \quad \text{as} \quad y \to -\infty. \tag{8.155}$$

Here condition (8.154) is required to maintain the uniformity of expansion (8.142) as $y \to 0$, whilst condition (8.155) is required for matching to region **II**. Now, the general solution to equation (8.152) may be written as

$$u_1(y) = C u_+(y) + D u_-(y) - \frac{c_1}{2\lambda_+ c^*} u^{*'}(y) \tag{8.156}$$

in $-\infty < y \le 0$, where $u_\pm(y)$ are basis functions of the homogeneous part of (8.152), and may be chosen so that

$$\begin{aligned} u_+(y) &\sim (-y)^{\frac{2}{1-n}-1}, \\ u_-(y) &\sim (-y)^{\frac{-2n}{1-n}}, \end{aligned} \quad \text{as} \quad y \to 0^-, \tag{8.157}$$

and

$$\begin{aligned} u_+(y) &\sim -A_\infty e^{-\lambda_+ y}, \\ u_-(y) &\le O\left(e^{-\lambda_+ y}\right), \end{aligned} \quad \text{as} \quad y \to -\infty. \tag{8.158}$$

Here, $C$ and $D$ are arbitrary constants, whilst $A_\infty > 0$ is fixed in terms of $n, m$ and $k$. Now, to satisfy condition (8.153) we require $D = 0$, which gives, via (8.157) and (iii),

$$u_1(y) = C(-y)^{\frac{2}{1-n}-1} + \frac{c_1}{\lambda_+ c^*(1-n)} \left(\frac{\gamma}{2}(1-n)\right)^{\frac{2}{1-n}} (-y)^{\frac{2}{1-n}-1}$$
$$+ O\left[(-y)^{\frac{2}{1-n}}\right] \quad \text{as} \quad y \to 0^-.$$

Thus, we require, via condition (8.154), that

$$C = -\frac{c_1}{\lambda_+ c^*(1-n)} \left(\frac{\gamma}{2}(1-n)\right)^{\frac{2}{1-n}}. \tag{8.159}$$

Finally, we must satisfy condition (8.155). As $y \to -\infty$, we have, from (8.156), (8.158), (8.159) and (v), that

$$u_1(y) \sim \frac{c_1 A_\infty}{\lambda_+ c^*(1-n)} \left(\frac{\gamma}{2}(1-n)\right)^{\frac{2}{1-n}} e^{-\lambda_+ y} \tag{8.160}$$

and so, on comparison with (8.155) we have

$$A^* = -\frac{c_1 A_\infty}{\lambda_+ c^*(1-n)} \left(\frac{\gamma}{2}(1-n)\right)^{\frac{2}{1-n}}, \tag{8.161}$$

and all quantities in (8.161), except $c_1$, are known, in principle, in terms of $n, m$ and $k$. Therefore (8.161) finally fixes $c_1$ as

$$c_1 = -\frac{\lambda_+ A^* c^*(1-n)}{A_\infty} \left(\frac{2}{\gamma(1-n)}\right)^{\frac{2}{1-n}} < 0, \tag{8.162}$$

after which

$$u_1(y) = \frac{A^*}{A_\infty} u_+(y) + \frac{A^*(1-n)}{2A_\infty} \left[\frac{2}{\gamma(1-n)}\right]^{\frac{2}{1-n}} u^{*\prime}(y) \tag{8.163}$$

in $-\infty < y \le 0$.

In summary we see that the contraction of the solution onto the permanent form travelling wave is exponential in $t$, via (8.142) and (8.150), whilst the approach to the constant propagation speed $c^*$ is also exponential in $t$, via (8.143) and (8.151), and the speed $c^*$ is approached from below ($c_1 < 0$) as $t \to \infty$. This is in accord with numerical results.

## 8.7 Conclusions

In this chapter we have extended the analysis of Chapter 7 and examined the evolution of solutions to the singular initial-boundary value problem IBVP (and its modification IBVP$'$) with respect to the reaction parameters, $0 <$ $m, n < 1$, and $k, \sigma, u_0 > 0$. We summarize the results of this chapter as follows.

For $0 < m < n < 1$ the equilibrium state $u \equiv \nu$ is approached uniformly in $x$ as $t \to \infty$, through terms exponential in $t$. This uniform convergence to the 'fully reacted state' arises through the small time behaviour, in which $u$ becomes non-zero immediately at infinite distances. This reproduces the situation described by King and Needham [28] ($f(u) \sim u^m$ for $0 < u \ll 1$) and is due to the vigour of the kinetics at low concentration (though the spreading is due to diffusion).

For equal orders of reaction and decay, $0 < m = n < 1$, the evolution of the problem depends upon the chain branching factor $k$. With $k < 1$, representing a low decay to autocatalysis ratio, the qualitative behaviour of the solution is unchanged from that of the above case, with the equilibrium state $u = 1 - k$ being approached uniformly in $x$ as $t \to \infty$. For $k = 1$, the kinetics are regular, and the small-$t$ evolution is diffusion dominated. In this case the solution decays, uniformly in $x$, to zero as $t \to \infty$, with its support becoming infinite immediately. For $k > 1$ (modelling a powerful termination effect), the initial data undergoes finite $t$ extinction and remains compactly supported throughout, with the edge of the support collapsing to zero in finite $t$.

For $0 < n < m < 1$, we have shown in Chapter 7 that permanent form travelling waves may develop in the problem if $0 < k < k_c$. Here we have extended this analysis and our results indicate that, providing the initial data exceeds a parameter dependent critical threshold, defined by the steady states of the problem, then for $0 < k < k_c$, a permanent form travelling wave evolves as $t \to \infty$. The dependence of these steady state solutions on the parameters $u_0$ and $k$ is given in Section 8.4. Effectively, for a given $k \in (0, k_c)$, the threshold requires $u_0$ to be sufficiently large, in particular, $u_0 > \tilde{a}(k)$, with $\tilde{a}$ as defined in Section 8.4. In combustion terms, this corresponds to supercritical behaviour. The contraction of the solution onto the travelling wave and the approach, from below, to the constant propagation speed are exponential in $t$. However, if $k$ is sufficiently large ($k \geq k_c$), or the initial data sufficiently small, our results indicate that the solution will, again, undergo finite $t$ extinction with contracting support, corresponding to subcritical behaviour in combustion systems.

These results are summarized in Table 8.1. Details of the numerical methods used to solve IBVP and results illustrating all the behaviour described above can be found in McCabe, Leach and Needham [39].

It may be expected that the considerable variation in the behaviour of the solution to the initial-boundary value problem for different parameter sets and the dramatic shift in the long (and short) time behaviour between $m, n \geq 1$ and $0 < m, n < 1$ will be reproduced by the full system of equations and this shall be addressed in Chapter 10 with particular attention to the asymptotic solution as $t \to 0$.

**Table 8.1.** Behaviour of the solution, $u(x,t)$, to IBVP when $(m = n, k \geq 1)$ and $(m > n)$, and IBVP$'$ when $(m < n)$ and $(m = n, k < 1)$ with respect to the parameters $n, m$ and $k$.

| $m > n$ | Travelling wave solutions exist provided that the initial data exceeds a parameter dependent critical threshold; otherwise solutions decay to zero with contracting support. | | $u(x,t)$ undergoes finite $t$ extinction, with support contracting to zero. | |
|---|---|---|---|---|
| $m = n$ | $u(x,t) \to 1 - k$ as $t \to \infty$, uniformly in $0 \leq x < \infty$. | | $u(x,t) \to 0$ as $t \to \infty$, uniformly in $0 \leq x < \infty$ | $u(x,t)$ undergoes finite $t$ extinction, with support contracting to zero. |
| $m < n$ | $u(x,t) \to \nu$ as $t \to \infty$, uniformly in $0 \leq x < \infty$. | | | |
| | $0 < k < k_c$ | $k_c \leq k < 1$ | $k = 1$ | $k > 1$ |

# Asymptotic Solution of IBVP as $t \to 0$ for $0 \le x < \infty$: Initial Data with Exponential or Algebraic Decay Rates

In this short chapter, we develop, via the method of matched asymptotic expansions, the small time asymptotic structure of the solution to IBVP (when $n < m < 1$) when the initial data, $u_0(x)$, is a continuous, analytic, positive and monotone decreasing function in $x \ge 0$, with $u_0(x) \to 0$ as $x \to \infty$. In particular, we consider the following cases:

(a) Initial data that has an algebraic decay rate as $x \to \infty$

$$u_0(x) \sim \begin{cases} u_\infty x^{-\alpha} + EST(x) & \text{as } x \to \infty, \\ \tilde{u}_0 + \sum_{l=1}^{\infty} \tilde{u}_l x^l & \text{as } x \to 0^+, \end{cases} \tag{9.1}$$

where $\alpha, u_\infty, \tilde{u}_0 > 0$, $\tilde{u}_l$ are constants and $EST(x)$ denotes exponentially small terms in $x$ as $x \to \infty$.

(b) Initial data that has an exponential decay rate as $x \to \infty$

$$u_0(x) \sim \begin{cases} u_\infty x^{-\beta} e^{-\sigma x} + O[e^{-f(x)}] & \text{as } x \to \infty, \\ \tilde{u}_0 + \sum_{l=1}^{\infty} \tilde{u}_l x^l & \text{as } x \to 0^+, \end{cases} \tag{9.2}$$

for some $f(x) > O(x)$ as $x \to \infty$, where $u_\infty, \tilde{u}_0, \sigma > 0$ and $\beta, \tilde{u}_l$ are constants.

The existence and uniqueness of the solution to IBVP with $n < m < 1$ (and with $m = n < 1, k \ge 1$) follows directly from Section 8.1. The behaviour of the solution depends critically on the nature of $u_0(x)$ as $x \to \infty$ and we establish that in both cases the support of the solution becomes finite in infinitesimal time. We conclude by presenting the asymptotic form for the location of the edge of the support as $t \to 0$ in both cases. Reference to the cases ($m = n < 1, k = 1$) and ($m = n < 1, k > 1$) is also made.

## 9.1 Initial Data with Algebraic Decay as $x \to \infty$

We first consider region **I**, where $x = O(1)$ as $t \to 0$ and expand the solution to IBVP as

$$u(x,t) = u_0(x) + tu_1(x) + O\left(t^2\right) \tag{9.3}$$

as $t \to 0$. On substitution into equation (8.1a) and applying initial condition (8.1c) we readily obtain

$$u(x,t) = u_0(x) + t\left(u_0''(x) + (1 - u_0(x))u_0^m(x) - ku_0^n(x)\right) + O\left(t^2\right) \tag{9.4}$$

as $t \to 0$. Now, for $x \gg 1$, expansion (9.3), with (9.1), takes the form

$$u(x,t) \sim u_\infty x^{-\alpha} + t\left(u_\infty\alpha(\alpha + 1)x^{-(\alpha+2)} - ku_\infty^n x^{-n\alpha} + \dots\right) + \dots \tag{9.5}$$

as $t \to 0$, and we conclude that expansion (9.4) becomes nonuniform when $x = O\left(t^{-\frac{1}{\alpha(1-n)}}\right)$, when we observe, via (9.4), that $u = O\left(t^{\frac{1}{(1-n)}}\right)$ as $t \to 0$. We must therefore introduce a further region, which we refer to as region **II**. To examine region **II**, we introduce the scaled coordinate $\eta = xt^{\overline{\alpha(1-n)}}$ and look for an asymptotic expansion of the form

$$u(\eta,t) = t^{\frac{1}{(1-n)}} F(\eta) + o\left(t^{\frac{1}{(1-n)}}\right) \tag{9.6}$$

as $t \to 0$, with $\eta = O(1)$. The leading order problem then becomes, on matching with (9.5),

$$\eta F_\eta + \alpha F + k(1 - n)\alpha F^n = 0, \quad 0 < \eta < \infty, \tag{9.7}$$

$$F(\eta) \sim u_\infty \eta^{-\alpha} \quad \text{as} \quad \eta \to 0. \tag{9.8}$$

The solution to (9.7),(9.8) is readily obtained as

$$F(\eta) = \left[u_\infty^{(1-n)}\eta^{-\alpha(1-n)} - k(1 - n)\right]^{\frac{1}{(1-n)}}, \quad \eta > 0. \tag{9.9}$$

An examination of (9.9) reveals that a weak singularity develops in $F(\eta)$ as $\eta \to \eta_c^-$, where

$$\eta_c = \left[\frac{u_\infty^{(1-n)}}{k(1 - n)}\right]^{\frac{1}{\alpha(1-n)}}, \tag{9.10}$$

with

$$F(\eta) \sim \left[\frac{k(1 - n)^2\alpha}{\eta_c}\right]^{\frac{1}{(1-n)}} (\eta_c - \eta)^{\frac{1}{(1-n)}}, \tag{9.11}$$

as $\eta \to \eta_c^-$. Thus the support of $u(\eta,t)$ ends at $\eta = \eta_c$ in this region. However, in (9.6) and (9.11) the degree of $(\eta_c - \eta)$ as $\eta \to \eta_c^-$ (which is $\frac{1}{(1-n)}$) is too weak, and consideration of further terms in (9.6) reveals a weak nonuniformity as $\eta \to \eta_c^-$. Therefore a further region is required to complete the asymptotic structure, in which $\eta = \eta_c + o(1)$ as $t \to 0$, and diffusion effects are retained at leading order to enable the appropriate behaviour to be achieved at the edge of the support. We label this region as region **III** and introduce the scaled coordinate $\bar{\eta}$ by

$$\eta = \eta_c + t^\gamma \bar{\eta} \tag{9.12}$$

with $\gamma > 0$ to be determined, and $\bar{\eta} = O(1)$ as $t \to 0$ in region **III**. An examination of (9.6) and (9.11) then determines that $u = O\left(t^{\frac{\gamma+1}{1-n}}\right)$ in region **III**. Thus we expand

$$u(\bar{\eta}, t) = t^{\frac{\gamma+1}{1-n}} H(\bar{\eta}) + o\left(t^{\frac{\gamma+1}{1-n}}\right) \tag{9.13}$$

as $t \to 0$ with $\bar{\eta} = O(1)$. On substituting (9.13) into equation (8.1a) (when written in terms of $\bar{\eta}$ and $t$), to retain diffusion terms at leading order requires,

$$\gamma = 1 + \frac{2}{\alpha(1-n)} \quad (> 0), \tag{9.14}$$

after which the leading order problem is

$$H_{\bar{\eta}\bar{\eta}} + \frac{\eta_c}{\alpha(n-1)} H_{\bar{\eta}} - kH^n = 0, \quad -\infty < \bar{\eta} < \bar{\eta}_0, \tag{9.15}$$

$$H(\bar{\eta}_0) = H_{\bar{\eta}}(\bar{\eta}_0) = 0, \tag{9.16}$$

$$H(\bar{\eta}) \sim \left[\frac{k(1-n)^2\alpha}{\eta_c}\right]^{\frac{1}{1-n}} (-\bar{\eta})^{\frac{1}{1-n}}, \quad \text{as} \quad \bar{\eta} \to -\infty. \tag{9.17}$$

The problem (9.15)-(9.17) is autonomous and can be studied in the $(H, H_{\bar{\eta}})$ phase plane. It can readily be established (after minor modifications to Appendix D) that (9.15)-(9.17) has a unique solution for each $\bar{\eta}_0$. Thus (9.15)-(9.17) does not fix a unique value of $\bar{\eta}_0$. In fact, $\bar{\eta}_0$ will be fixed by matching expansion (9.13) (as $\bar{\eta} \to -\infty$) to expansion (9.6) (as $\eta \to \eta_c^-$), when expansion (9.6) is taken to next order. We observe that the solution to (9.15)-(9.17) is monotone decreasing in $-\infty < \bar{\eta} < \bar{\eta}_0$, and has

$$H(\bar{\eta}) \sim \left[\frac{k(1-n)^2}{2(1+n)}\right]^{\frac{1}{1-n}} (\bar{\eta}_0 - \bar{\eta})^{\frac{2}{1-n}} \quad \text{as} \quad \bar{\eta} \to \bar{\eta}_0^-, \tag{9.18}$$

which has the required decay rate in $(\bar{\eta}_0 - \bar{\eta})$ as the edge of the support is approached. Consideration of further terms in this region shows that expansion (9.13) remains uniform as $\bar{\eta} \to \bar{\eta}_0^-$, and so no further regions are required and the asymptotic structure is complete. In particular, we have that the edge of the support $x = s(t)$ behaves as

$$s(t) \sim \eta_c t^{-\frac{1}{\alpha(1-n)}} + \bar{\eta}_0 t^{1+\frac{1}{\alpha(1-n)}} + \cdots \tag{9.19}$$

as $t \to 0$. We observe from (9.19) that the edge of the support is contracting initially with speed

$$\dot{s}(t) \sim -\frac{\eta_c}{\alpha(1-n)} t^{-1-\frac{1}{\alpha(1-n)}} < 0 \quad \text{as} \quad t \to 0. \tag{9.20}$$

At this stage the main asymptotic structure is complete. However, for $x \ll 1$, (9.4) with (9.1)(b) takes the form

$$u(x,t) \sim (\tilde{u}_0 + \tilde{u}_1 x + \ldots) + t (2\tilde{u}_2 - k(\tilde{u}_0 + \tilde{u}_1 x + \ldots)^n + \ldots) + \ldots \tag{9.21}$$

as $t \to 0$, and in general, will not satisfy the boundary condition (8.1e) at $x = 0$. Therefore, we will require an inner region, region $\mathbf{I_0}$, when $x = o(1)$ as $t \to 0$, over which this region must have scaling $x = O\left(t^{\frac{1}{2}}\right)$ as $t \to 0$ and we introduce the inner coordinate $\hat{\eta} = O(1)$ as $t \to 0$, where $\hat{\eta} = xt^{-\frac{1}{2}}$. The form of (9.21) then suggests that we expand in the inner region in the form

$$u(\hat{\eta}, t) = \tilde{u}_0 + t^{\frac{1}{2}} \bar{u}_1(\hat{\eta}) + O(t) \tag{9.22}$$

as $t \to 0$, with $\hat{\eta} = O(1)$. The leading order problem then becomes, on matching with (9.21),

$$\bar{u}_1'' + \frac{\hat{\eta}}{2} \bar{u}_1' - \frac{1}{2} \bar{u}_1 = 0, \quad \hat{\eta} > 0, \tag{9.23}$$

$$\bar{u}_1'(0) = 0, \tag{9.24}$$

$$\bar{u}_1(\hat{\eta}) \sim \tilde{u}_1 \hat{\eta} \quad \text{as} \quad \hat{\eta} \to \infty. \tag{9.25}$$

The solution to the boundary value problem (9.23)-(9.25) is readily obtained as

$$\bar{u}_1(\hat{\eta}) = 2\sqrt{\pi}\, \tilde{u}_1\, _1F_1[1, 1/2; \hat{\eta}^2/4] e^{-\hat{\eta}^2/4}, \tag{9.26}$$

where $_1F_1[a, b; z]$ is the confluent hypergeometric function (Abramowitz and Stegun [1]). This completes the asymptotic structure in this case as $t \to 0$, with expansions (9.22) (region $\mathbf{I_0}$), (9.4) (region $\mathbf{I}$), (9.6) (region $\mathbf{II}$) and (9.13) (region $\mathbf{III}$) providing a uniform approximation in $x \geq 0$ to the solution of IBVP with $n < m < 1$ as $t \to 0$. A schematic representation of the location and thickness of the asymptotic regions as $t \to 0$ is given in Figure 9.1.

## 9.2 Initial Data with Exponential Decay as $x \to \infty$

In this case expansion (9.4) in region $\mathbf{I}$ becomes nonuniform when $x = c(t) + O(1)$ where

$$c(t) = -\frac{1}{\sigma(1-n)} \ln t - \frac{\beta}{\sigma} \ln(-\ln t) + o(\ln(-\ln t)) \tag{9.27}$$

as $t \to 0$, with $u = O\left(t^{\frac{1}{(1-n)}}\right)$ in region $\mathbf{II}$. To examine region $\mathbf{II}$, we introduce the co-ordinate $\eta = x - c(t)$ and look for an expansion of the form

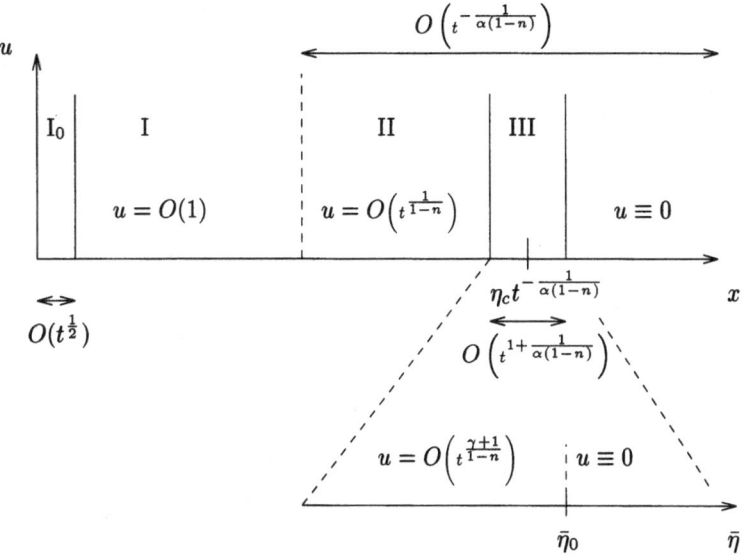

**Fig. 9.1.** Schematic representation of the location and thickness of the asymptotic regions as $t \to 0$ in the case when the initial data has algebraic decay rate as $x \to \infty$. Note that in this case $\gamma = 1 + \frac{2}{\alpha(1-n)}$.

$$u(\eta, t) = t^{\frac{1}{(1-n)}} G(\eta) + o\left(t^{\frac{1}{(1-n)}}\right) \tag{9.28}$$

as $t \to 0$, with $\eta = O(1)$. The leading order problem then becomes, on matching with region **I** (as $\eta \to -\infty$),

$$G_\eta + \sigma G + k\sigma(1-n)G^n = 0, \quad \eta > -\infty, \tag{9.29}$$

$$G(\eta) \sim u_\infty[\sigma(1-n)]^\beta e^{-\sigma\eta} \quad \text{as} \quad \eta \to -\infty. \tag{9.30}$$

The solution to (9.29),(9.30) is readily obtained as

$$G(\eta) = \left[u_\infty^{(1-n)}(\sigma(1-n))^{\beta(1-n)}e^{-\sigma(1-n)\eta} - k(1-n)\right]^{\frac{1}{(1-n)}}, \quad \eta > -\infty. \tag{9.31}$$

An examination of (9.31) reveals that a weak singularity develops in $G(\eta)$ as $\eta \to \eta_c$, where

$$\eta_c = \frac{1}{\sigma(1-n)} \ln\left[\frac{u_\infty^{(1-n)}[\sigma(1-n)]^{\beta(1-n)}}{k(1-n)}\right], \tag{9.32}$$

with

$$G(\eta) \sim \left[k\sigma(1-n)^2\right]^{\frac{1}{(1-n)}} (\eta_c - \eta)^{\frac{1}{(1-n)}} \tag{9.33}$$

as $\eta \to \eta_c^-$. The support of $u(\eta, t)$ ends at $\eta_c$ in this region but as in Section 9.1 the degree of $(\eta_c - \eta)$ (in (9.28)) as $\eta \to \eta_c^-$ is too weak, and consideration of further terms in (9.28) reveal a weak nonuniformity as $\eta \to \eta_c^-$. We therefore introduce a final region, region **III**, to complete the asymptotic structure. The details of this region follow, after minor modifications those given in Section 9.1 with now

$$\eta = \eta_c + \bar{\eta}t \tag{9.34}$$

as $t \to 0$ with $\bar{\eta} = O(1)$, and we look for an expansion of the form

$$u(\bar{\eta}, t) = t^{\frac{2}{(1-n)}} H(\bar{\eta}) + o\left(t^{\frac{2}{(1-n)}}\right) \tag{9.35}$$

as $t \to 0$ with $\bar{\eta} = O(1)$. The leading order problem is then given by

$$H_{\bar{\eta}\bar{\eta}} - \frac{1}{\sigma(1-n)} H_{\bar{\eta}} - kH^n = 0, \quad -\infty < \bar{\eta} < \bar{\eta}_0, \tag{9.36}$$

$$H(\bar{\eta}_0) = H_{\bar{\eta}}(\bar{\eta}_0) = 0, \tag{9.37}$$

$$H(\bar{\eta}) \sim \left[k\sigma(1-n)^2\right]^{\frac{1}{1-n}} (-\bar{\eta})^{\frac{1}{1-n}}, \quad \text{as} \quad \bar{\eta} \to -\infty. \tag{9.38}$$

It can be readily established (after minor modifications to Appendix D) that (9.36)-(9.38) has a unique solution for each $\bar{\eta}_0$ ($\bar{\eta}_0$ will be fixed by matching expansion (9.35) (as $\bar{\eta} \to -\infty$) to expansion (9.28)(as $\eta \to \eta_c^-$), when expansion (9.28) is taken to next order). The solution to (9.36)-(9.38) is monotone decreasing in $-\infty < \bar{\eta} < \bar{\eta}_0$ and has

$$H(\bar{\eta}) \sim \left[\frac{k(1-n)^2}{2(1+n)}\right]^{\frac{1}{1-n}} (\bar{\eta}_0 - \bar{\eta})^{\frac{2}{1-n}} \quad \text{as} \quad \bar{\eta} \to \bar{\eta}_0^-, \tag{9.39}$$

which has the required decay rate in $(\bar{\eta}_0 - \bar{\eta})$ as the edge of the support is approached. Consideration of further terms in this region shows that expansion (9.35) remains uniform as $\bar{\eta} \to \bar{\eta}_0^-$, and the asymptotic structure is complete. In particular, we have that the edge of the support $x = s(t)$, behaves as

$$s(t) \sim -\frac{1}{\sigma(1-n)} \ln t - \frac{\beta}{\sigma} \ln(-\ln t) + \cdots + \eta_c + O(t) \tag{9.40}$$

as $t \to 0$ and we observe from (9.40) that the edge of the support is contracting initially with speed

$$\dot{s}(t) \sim -\frac{1}{\sigma(1-n)t} + \ldots \quad \text{as} \quad t \to 0. \tag{9.41}$$

Finally, we note that expansion (9.4) of region **I**, does not in general, satisfy the boundary condition (8.1e) at $x = 0$ and a further region, region $\mathbf{I_0}$, is required in the neighbourhood of $x = 0$ as $t \to 0$. The details of this region follow those given for region $\mathbf{I_0}$ in Section 9.1. This completes the asymptotic structure in this case as $t \to 0$, with expansions (9.22) (region $\mathbf{I_0}$), (9.4) (region **I**), (9.28) (region **II**) and (9.35) (region **III**) providing a uniform approximation in $x \geq 0$ to the solution of IBVP with $n < m < 1$ as $t \to 0$ in this case. A schematic representation of the location and thickness of the asymptotic regions as $t \to 0$ is given in Figure 9.2.

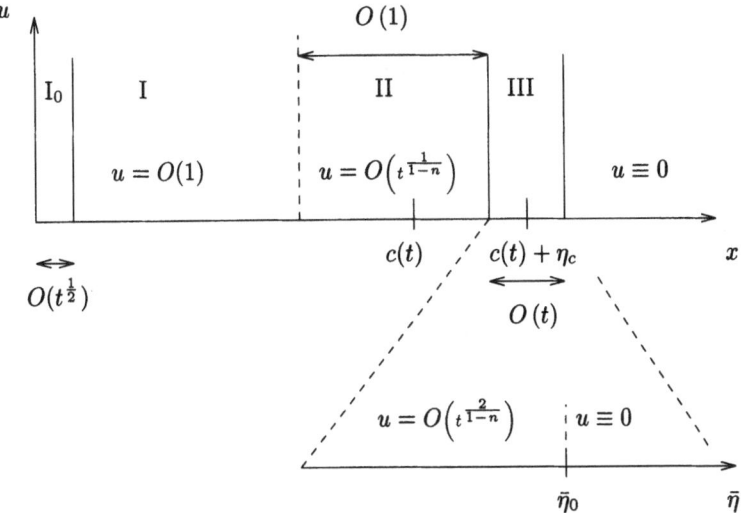

**Fig. 9.2.** Schematic representation of the location and thickness of the asymptotic regions as $t \to 0$ in the case when the initial data has exponential decay rate as $x \to \infty$.

## 9.3 Summary and Further Discussion

In this chapter we have demonstrated directly, via the method of matched asymptotic expansions, that the solution $u(x,t)$ to the initial-boundary value problem IBVP with $n < m < 1$ together with initial data that has infinite support with algebraic or exponential decay as $x \to \infty$ (given by conditions (9.1) and (9.2) respectively) develops finite support in infinitesimal time. In particular, the edge of the support $x = s(t)$, is given, when:

(a) The initial data has algebraic decay rate as $x \to \infty$ by,

$$s(t) \sim \eta_c t^{-\frac{1}{\alpha(1-n)}} + \bar{\eta}_0 t^{1+\frac{1}{\alpha(1-n)}} + \dots \tag{9.42}$$

as $t \to 0$, where the constants $\eta_c$ and $\bar{\eta}_0$ are as described in section 9.1.

(b) The initial data has exponential decay rate as $x \to \infty$ by

$$s(t) \sim -\frac{1}{\sigma(1-n)} \ln t - \frac{\beta}{\sigma} \ln(-\ln t) + \dots + \eta_c + O(t) \tag{9.43}$$

as $t \to 0$, where now $\eta_c$ is given in (9.32).

We note that in both cases the edge of the support is initially contracting. We further note that:

(i) In the case when $m = n$ and $k > 1$ the asymptotic structure of the solution to IBVP as $t \to 0$ follows, after minor modifications, that given in Sections 9.1 and 9.2 with now $-k$ being replaced by $(1 - k)$ throughout.

(ii) In the case when $m = n$ and $k = 1$ expansion (9.4) of region **I** remains uniform for $x \gg 1$.

# Extension to the System of Singular Reaction-Diffusion Equations

In this chapter we shall be concerned with the analysis of the full, coupled system of singular reaction–diffusion equations associated with the steps (6.1) and (6.2). We expect the system to closely reproduce the behaviour of the scalar model (its approximation) and, throughout, we shall draw comparisons with the results of Chapters 7 and 8.

We consider the situation arising when a localized input of the autocatalyst B is introduced into an expanse of the reactant A, initially at uniform concentration. This leads us to examine the initial-boundary value problem

$$\frac{\partial a}{\partial \bar{t}} = D\frac{\partial^2 a}{\partial \bar{x}^2} - k_1 a b^m, \tag{10.1a}$$

$$\frac{\partial b}{\partial \bar{t}} = D\frac{\partial^2 b}{\partial \bar{x}^2} + k_1 a b^m - k_2 b^n, \tag{10.1b}$$

$$a(\bar{x}, 0) \equiv a_0, \quad \bar{x} \geq 0, \tag{10.1c}$$

$$b(\bar{x}, 0) = \begin{cases} b_0 g(\bar{x}), & 0 \leq \bar{x} \leq \ell, \\ 0, & \bar{x} > \ell, \end{cases} \tag{10.1d}$$

$$a(\bar{x}, t) \to \bar{A}(\bar{t}) \quad \text{as} \quad \bar{x} \to \infty, \quad \bar{t} \geq 0, \quad (0 \leq \bar{A}(\bar{t}) \leq a_0), \tag{10.1e}$$

$$b(\bar{x}, t) \to \bar{B}(\bar{t}) \quad \text{as} \quad \bar{x} \to \infty, \quad \bar{t} \geq 0, \quad (0 \leq \bar{B}(\bar{t}) < \infty), \tag{10.1f}$$

$$a_{\bar{x}}(0, \bar{t}) = 0, \quad \bar{t} > 0, \tag{10.1g}$$

$$b_{\bar{x}}(0, \bar{t}) = 0, \quad \bar{t} > 0. \tag{10.1h}$$

Here $\bar{x}$ and $\bar{t}$ are coordinates measuring distance and time respectively, whilst $D$ is the constant diffusion rate of A and B. The prescribed function $g(\bar{x})$ is non-negative and analytic on $0 \leq \bar{x} \leq \ell$ with a maximum value of unity, and the constants $a_0$ and $b_0$ are the positive initial concentration of $A$ and the maximum initial concentration of $B$ respectively. The functions $\bar{A}(\bar{t})$ and $\bar{B}(\bar{t})$ allow for the possible temporal growth in $a(\bar{x}, \bar{t})$ and $b(\bar{x}, \bar{t})$ as $\bar{x} \to \infty$, and are determined in the course of the analysis. For $m, n \geq 1$ equations (10.1a) and (10.1b) together with initial conditions (10.1c) and (10.1d) lead to $\bar{A}(\bar{t}) \equiv a_0$ and $\bar{B}(\bar{t}) \equiv 0$ in $\bar{t} \geq 0$. However, this need not be the case when $0 < m, n < 1$ (as we have observed in the scalar approximation, Chapter 8) due to non-uniqueness in the corresponding well-stirred system when $0 < m, n < 1$ (see Section 10.1).

Following Chapter 6 we introduce the dimensionless variables

$$a = a_0 \alpha, \quad b = a_0 \beta, \quad t = k_1 a_0^m \bar{t}, \quad x = \left( \frac{k_1 a_0^m}{D} \right)^{\frac{1}{2}} \bar{x},$$

in terms of which the model (10.1) becomes

$$\frac{\partial \alpha}{\partial t} = \frac{\partial^2 \alpha}{\partial x^2} - \alpha \beta^m, \tag{10.2a}$$

$$\frac{\partial \beta}{\partial t} = \frac{\partial^2 \beta}{\partial x^2} + \alpha \beta^m - k \beta^n, \quad x, t > 0, \tag{10.2b}$$

$$\alpha(x, 0) = 1, \quad x \geq 0, \tag{10.3a}$$

$$\beta(x, 0) = \begin{cases} \beta_0 g(x), & 0 \leq x \leq \sigma, \\ 0, & x > \sigma, \end{cases} \tag{10.3b}$$

$$\alpha(x, t) \to \alpha_\infty(t) \quad \text{as} \quad x \to \infty, \quad t \geq 0, \quad (0 \leq \alpha_\infty(t) \leq 1), \tag{10.4a}$$

$$\beta(x, t) \to \beta_\infty(t) \quad \text{as} \quad x \to \infty, \quad t \geq 0, \quad (0 \leq \beta_\infty(t) < \infty), \tag{10.4b}$$

$$\alpha_x(0, t) = 0, \quad t > 0, \tag{10.5a}$$

$$\beta_x(0, t) = 0, \quad t > 0. \tag{10.5b}$$

The model has three dimensionless parameters. The parameter $\sigma = \left( \frac{\ell^2 k_1 a_0^m}{D} \right)^{\frac{1}{2}}$ measures the spread of the initial input of autocatalyst, $\beta_0 = b_0 / a_0$ measures the maximum concentration in the input of the autocatalyst, and $k = k_2 / k_1 a_0^{m-n+1}$ measures the strength of the termination step (6.2) relative to that of the autocatalytic step (6.1), and is often referred to as the chain branching factor. The function $g(x)$ is the prescribed, non-negative and analytic on $0 \leq x \leq \sigma$ with a maximum value of unity, In particular, $g(x)$ has

$$g(x) \sim g_\sigma(\sigma - x)^r \quad \text{as} \quad x \to \sigma^-, \tag{10.6a}$$

$$g(x) \sim g_0 + g_p x^p + \dots \quad \text{as} \quad x \to 0^+ \tag{10.6b}$$

for some $r, p \in \mathbb{N}$ and constants $g_p \neq 0, g_\sigma, g_0 > 0$. In what follows we refer to the initial-boundary value problem (10.2)-(10.5) as IBVP.

For $m, n \geq 1$, IBVP has studied extensively in [47],[48], [49] and [58]. However, for $0 < m, n < 1$ the analysis of IBVP is considerably more difficult than that for the regular case $(m, n \geq 1)$ since the kinetic terms, $\pm \alpha \beta^m$ and $-k\beta^n$, are not Lipschitz continuous at $\beta = 0$. In this chapter we examine the well stirred analogue of IBVP and construct, via the method of matched asymptotic expansions, the asymptotic solution as $t \to 0$ of IBVP when $0 < m, n < 1$, over all parameter values. We note that the travelling wave problem for IBVP has been considered in detail by Kay, Needham and Leach [27] while the full initial-boundary value problem is being considered by the authors at present.

## 10.1 The Well-stirred Case

The well stirred (or diffusion free) analogue of IBVP is

$$\alpha_t = -\alpha \beta^m, \qquad \beta_t = \alpha \beta^m - k\beta^n, \qquad \alpha, \beta, t \geq 0, \tag{10.7}$$

subject to the initial conditions

$$\alpha(0) = 1, \qquad \beta(0) = 0 \tag{10.8}$$

or

$$\alpha(0) = 1, \qquad \beta(0) = \beta_0 > 0. \tag{10.9}$$

We shall proceed in this section by examining the initial value problems (10.7), (10.8), which we denote by $\mathrm{IVP}_1$, and (10.7), (10.9), which we denote by $\mathrm{IVP}_2$. The solution to $\mathrm{IVP}_1$ determines the far field functions $\alpha_\infty(t)$ and $\beta_\infty(t)$ $(t \geq 0)$, as defined by (10.4), in IBVP, whilst the solution to $\mathrm{IVP}_2$ will provide information concerning the overall evolution of IBVP.

### 10.1.1 The Phase Portrait

We begin by constructing the phase portrait of equations (10.7) in the quadrant $R = \{(\alpha, \beta) : \alpha, \beta \geq 0\}$, and first note that the $\beta$ axis in $R$ is a phase path of (10.7), whilst the $\alpha$–axis is a line of singular (due to $0 < m, n < 1$) equilibrium points of (10.7) and contains all of the equilibrium points in $R$. There are three distinct cases to consider, depending upon the sign of $m - n$.

### (a) $m < n$

In this case the equation for the phase paths of (10.7) in $R$ is

$$\frac{\mathrm{d}\beta}{\mathrm{d}\alpha} = \frac{k\beta^{n-m}}{\alpha} - 1, \qquad \alpha, \beta \geq 0,$$

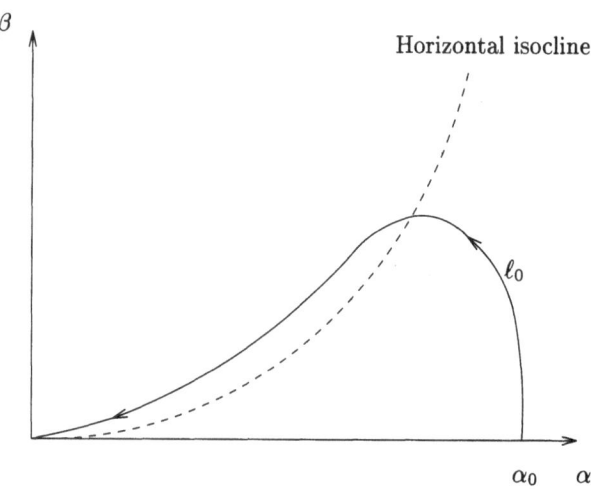

**Fig. 10.1.** A sketch of the $(\alpha, \beta)$ phase portrait for $m < n$.

and the phase portrait is readily obtained on consideration of the isoclines. This is sketched in Figure 10.1. We note that all phase paths in $R$ approach the asymptotically stable equilibrium point at the origin, along the horizontal isocline $\beta = \left(\frac{\alpha}{k}\right)^{\frac{1}{n-m}}$. We label as $\ell_0$ that phase path originating at the unstable equilibrium point $(\alpha_0, 0)$, $\alpha_0 > 0$. On $\ell_0$ we have

$$\beta \sim (\alpha_0 - \alpha) \quad \text{as} \quad \alpha \to \alpha_0^-, \tag{10.10a}$$

$$\beta \sim \left(\frac{\alpha}{k}\right)^{\frac{1}{n-m}} \quad \text{as} \quad \alpha \to 0. \tag{10.10b}$$

**(b) $m = n$**

In this case the equation for the phase paths of (10.7) in $R$ reduces to

$$\frac{d\beta}{d\alpha} = \frac{k}{\alpha} - 1, \quad \alpha, \beta \geq 0,$$

which may be integrated directly to give

$$\beta = k \ln \alpha - \alpha + C_0, \quad \alpha, \beta \geq 0$$

for some constant $C_0$. The phase portrait for equations (10.7) in $R$ is then readily obtained and sketched in Figure 10.2. In $R$, a typical phase path, $\ell_0$, connects an unstable equilibrium point $(\alpha_0, 0)$, with $k < \alpha_0 < \infty$, with an asymptotically stable equilibrium point $(\alpha_s, 0)$, with $0 < \alpha_s < k$. This phase path has a single local maximum on the horizontal isocline $\alpha = k$. In particular, on $\ell_0$ we have

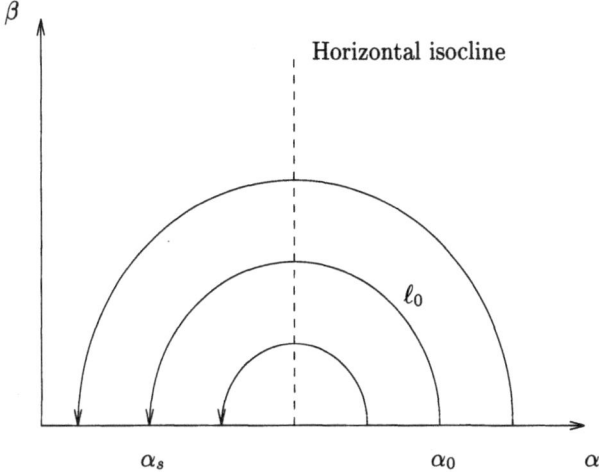

**Fig. 10.2.** A sketch of the $(\alpha, \beta)$ phase portrait for $m = n$.

$$\beta = k \ln\left(\frac{\alpha}{\alpha_0}\right) - \alpha + \alpha_0, \quad \alpha_s \leq \alpha \leq \alpha_0, \tag{10.11a}$$

$$\beta \sim \frac{(\alpha_0 - k)}{\alpha_0}(\alpha_0 - \alpha) \quad \text{as} \quad \alpha \to \alpha_0^-, \tag{10.11b}$$

$$\beta \sim \frac{(k - \alpha_s)}{\alpha_s}(\alpha - \alpha_s) \quad \text{as} \quad \alpha \to \alpha_s^+, \tag{10.11c}$$

with $\alpha_s$ being the smallest positive zero of (10.11a).

**(c) $m > n$**

Here the equation for the phase paths of (10.7) in $R$ is

$$\frac{\mathrm{d}\beta}{\mathrm{d}\alpha} = \frac{k}{\alpha\beta^{m-n}} - 1, \quad \alpha, \beta \geq 0,$$

and now the horizontal isocline is $\beta = \left(\frac{\alpha}{k}\right)^{\frac{-1}{m-n}}$. The phase portrait is readily obtained on consideration of the isoclines and is sketched in Figure 10.3. A typical phase path, $\ell_0$, begins, for large $\alpha$, asymptotic from above to the horizontal isocline and terminates at the asymptotically stable equilibrium point $(\alpha_s, 0)$, $\alpha_s > 0$. On $\ell_0$ we have

$$\beta \sim \left[\frac{k(m - n + 1)}{\alpha_s}(\alpha - \alpha_s)\right]^{\frac{1}{m-n+1}} \quad \text{as} \quad \alpha \to \alpha_s^+, \tag{10.12a}$$

$$\beta \sim \left(\frac{\alpha}{k}\right)^{\frac{-1}{m-n}} \quad \text{as} \quad \alpha \to \infty. \tag{10.12b}$$

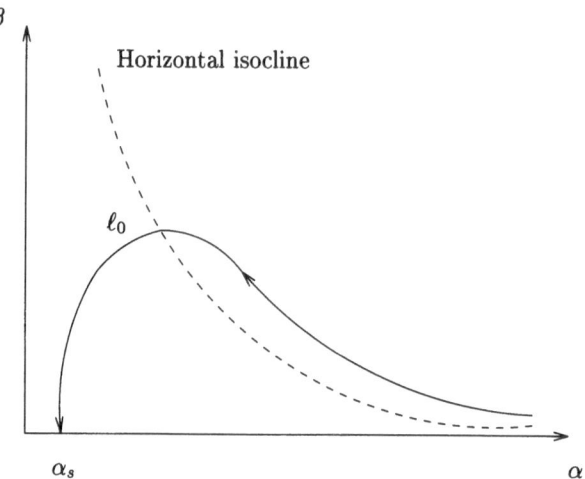

**Fig. 10.3.** A sketch of the $(\alpha, \beta)$ phase portrait for $m > n$.

### 10.1.2 IVP₁

Here, we consider solutions to $IVP_1$ and, again, there are three distinct cases to consider, depending on the sign of $m - n$.

### (a) $m < n$

In this case (due to the loss of Lipschitz continuity in $IVP_1$ at $\beta = 0$) $IVP_1$ does not have a unique solution. Initially, for $0 < t \ll 1$, we have via (10.10a), $\beta \sim (1 - \alpha)$. Therefore

$$\alpha_t \sim -(1 - \alpha)^m, \qquad \alpha(0) = 1$$

as $t \to 0$, and so

$$\alpha(t) \equiv 1 \quad \text{or} \quad \alpha(t) \sim 1 - (1 - m)^{\frac{1}{1-m}} t^{\frac{1}{1-m}} \quad \text{as} \quad t \to 0.$$

Hence $IVP_1$ has the two possible solutions (up to shifts in the $t$ origin)

$$(\alpha(t), \beta(t)) \equiv (1, 0)$$

or

$$(\alpha(t), \beta(t)) \equiv (A(t), B(t)),$$

where $A(t)$ is monotone decreasing in $t$ and has

$$A(t) \sim 1 - (1 - m)^{\frac{1}{1-m}} t^{\frac{1}{1-m}} \quad \text{as} \quad t \to 0, \qquad (10.13a)$$

$$A(t) \to 0 \quad \text{as} \quad t \to \infty. \qquad (10.13b)$$

Correspondingly, via Section 10.1.1, $B(t)$ is initially increasing and then decreasing with $t$ and has

$$B(t) \sim (1-m)^{\frac{1}{1-m}} t^{\frac{1}{1-m}} \quad \text{as} \quad t \to 0, \tag{10.14a}$$
$$B(t) \to 0 \quad \text{as} \quad t \to \infty. \tag{10.14b}$$

## (b) $m = n$

The solution to $IVP_1$ now depends critically upon the chain branching factor $k$.

For $k < 1$ there is, again, nonuniqueness in $IVP_1$ and there are two possible solutions:

$$(\alpha(t), \beta(t)) \equiv (1, 0)$$

or

$$(\alpha(t), \beta(t)) \equiv \left( \hat{A}(t), \hat{B}(t) \right),$$

where $\hat{A}(t)$ is monotone decreasing in $t$ and has

$$\hat{A}(t) \sim 1 - (1-m)^{\frac{1}{1-m}} (1-k)^{\frac{m}{1-m}} t^{\frac{1}{1-m}} \quad \text{as} \quad t \to 0, \tag{10.15a}$$
$$\hat{A}(t) \to \alpha_s \quad \text{as} \quad t \to t_c, \tag{10.15b}$$
$$\hat{A}(t) \equiv \alpha_s \quad \text{in} \quad t \ge t_c. \tag{10.15c}$$

Here $\alpha = \alpha_s$ is the smallest positive root of $k \ln \alpha - \alpha + 1 = 0$ and the finite extinction time, $t_c(> 0)$, is given by

$$t_c = \int_{\alpha_s}^1 \frac{d\lambda}{\lambda [k \ln \lambda - \lambda + 1]^m}.$$

Correspondingly $\hat{B}(t)$ is initially increasing and then decreasing with $t$ and has

$$\hat{B}(t) \sim (1-k)^{\frac{1}{1-m}} (1-m)^{\frac{1}{1-m}} t^{\frac{1}{1-m}} \quad \text{as} \quad t \to 0, \tag{10.16a}$$
$$\hat{B}(t) \to 0 \quad \text{as} \quad t \to t_c, \tag{10.16b}$$
$$\hat{B}(t) \equiv 0 \quad \text{in} \quad t \ge t_c. \tag{10.16c}$$

For $k \ge 1$ uniqueness is restored in $IVP_1$ and so

$$(\alpha(t), \beta(t)) \equiv (1, 0) \quad \text{for all} \quad t \ge 0.$$

## (c) $m > n$

In this case $IVP_1$ has the unique solution

$$(\alpha(t), \beta(t)) \equiv (1, 0) \quad \text{for all} \quad t \ge 0.$$

**Table 10.1.** Possible forms for $\alpha_\infty(t)$ and $\beta_\infty(t)$ in IBVP.

| | $m < n$ | $m = n$ | | | $m > n$ |
|---|---|---|---|---|---|
| | | $k < 1$ | $k = 1$ | $k > 1$ | |
| $\alpha_\infty(t)$ | $A(t)$ | $\hat{A}(t)$ | 1 | 1 | 1 |
| $\beta_\infty(t)$ | $B(t)$ | $\hat{B}(t)$ | 0 | 0 | 0 |

The above results, together with those associated with the scalar approximation (see Chapter 8), suggest possible forms for $\alpha_\infty(t)$ and $\beta_\infty(t)$ in IBVP, as summarized in Table 10.1.

By comparison with the scalar approximation we expect that a necessary condition for the development of permanent form travelling waves (PTW) in IBVP is that $\alpha_\infty(t) \equiv 1$, $\beta_\infty(t) \equiv 0$ in (10.4) and so clearly we do not anticipate PTW in the cases $(m < n)$ and $(m = n, k < 1)$.

### 10.1.3 IVP₂

Here, we consider solutions to IVP$_2$ for the three cases $m < n$, $m = n$ and $m > n$. We note that in each case IVP$_2$ has a unique solution since $(\alpha(0), \beta(0)) = (1, \beta_0)$ is a regular point of the system (10.7).

### (a) $m < n$

We recall from Section 10.1.1 that $\alpha$ is monotone decreasing in $t$ whilst $\beta$ increases initially (if $\beta_0$ is sufficiently small) then decreases with $t$. Thus $(\alpha(t), \beta(t)) \to (0,0)$ either as $t \to \infty$ or in some finite $t$. We observe from Section 10.1.1 that phase paths in $R$ approach $(0,0)$ along the horizontal isocline. That is, $\beta \sim \left(\frac{\alpha}{k}\right)^{\frac{1}{n-m}}$ as $\alpha \to 0$, and so

$$\alpha_t \sim -k^{\frac{-m}{n-m}} \alpha^{\frac{n}{n-m}} \quad \text{as} \quad \alpha \to 0. \tag{10.17}$$

Thus we conclude, via an integration of (10.17), that $(\alpha(t), \beta(t)) \to (0,0)$ as $t \to \infty$.

### (b) $m = n$

Again, in this case, $\alpha$ is monotone decreasing in $t$ whereas the behaviour of $\beta$ depends upon $k$. Specifically, if $k < 1$, $\beta$ increases initially then decreases with $t$, but if $k \geq 1$, $\beta$ is monotone decreasing in $t$. In both cases $(\alpha(t), \beta(t)) \to (\alpha_s, 0)$ either as $t \to \infty$ or in finite $t$ (recall that $\alpha_s < k$). Now,

$$\beta \sim \frac{(k - \alpha_s)}{\alpha_s} (\alpha - \alpha_s) \qquad \text{as} \qquad \alpha \to \alpha_s^+,$$

and so

$$\alpha_t \sim -\alpha_s^{1-m} (k - \alpha_s)^m (\alpha - \alpha_s)^m \qquad \text{as} \qquad \alpha \to \alpha_s^+. \tag{10.18}$$

A direct integration of (10.18) shows immediately that $(\alpha(t), \beta(t)) \to (\alpha_s, 0)$ in finite $t$.

### (c) $m > n$

Here, again, $\alpha$ is monotone decreasing in $t$ whilst $\beta$ first increases then decreases with $t$. Hence $(\alpha(t), \beta(t)) \to (\alpha_s, 0)$ either as $t \to \infty$ or in finite $t$. From (10.12a) we have

$$\beta \sim \left[ \frac{k(m-n+1)}{\alpha_s} (\alpha - \alpha_s) \right]^{\frac{1}{m-n+1}} \qquad \text{as} \qquad \alpha \to \alpha_s^+,$$

and so

$$\alpha_t \sim -\alpha_s \left[ \frac{k(m-n+1)}{\alpha_s} (\alpha - \alpha_s) \right]^{\frac{m}{m-n+1}} \qquad \text{as} \qquad \alpha \to \alpha_s^+,$$

an integration of which demonstrates that $(\alpha(t), \beta(t)) \to (\alpha_s, 0)$ in finite $t$.

### 10.1.4 Summary

The results of this section together with those presented in Chapter 8 for the scalar approximation suggest the following behaviour in IBVP.

(i) $m < n$. "Lifting at infinity" after which $\alpha \to 0$, $\beta \to 0$ as $t \to \infty$ uniformly in $x$.

(ii) $m = n$. $k < 1$: "Lifting at infinity" after which $\alpha \to \alpha_s$, $\beta \to 0$ in finite $t$ uniformly in $x$.

$k \geq 1$: No "lifting at infinity" and no trigger for the growth of $\beta$. Here we expect $\beta \to 0$ in finite $t$ (with a possibility of infinite $t$ extinction if $k = 1$) uniformly in $x$, whereas $\alpha(x, t)$ (with $\alpha_s \leq \alpha(x, t) \leq 1$ for all $(x, t) \in \mathcal{D}_T$ and any $T > 0$) will diffuse back to unity as $t \to \infty$ uniformly in $x$.

(iii) $m > n$. There is a trigger for the growth of $\beta$ if $\beta_0$ is sufficiently large but no "lifting at infinity" . The possibilities are as follows:

(a) $\beta \to 0$ in finite $t$ uniformly in $x$, whereas $\alpha \to 1$ as $t \to \infty$ uniformly in $x$, via diffusion.

(b) Travelling waves develop in $\alpha$ and $\beta$. The wave in $\alpha$ will be monotone increasing in $x$ with $\alpha \to 1$ ahead of the wave and $\alpha \equiv \alpha_s$ behind the wave with $\alpha_s > 0$ a constant. The wave in $\beta$ will have a single crest with

$\beta \to 0$ ahead of the wave and $\beta \equiv 0$ behind the wave.

The results of Chapters 7 and 8 indicate that both (a) and (b) are possible, with (b) occuring for sufficiently small $k$ and sufficiently large $\beta_0$ ($\beta_0 > k^{\frac{1}{m-n}}$ is necessary).

## 10.2 Asymptotic Solution as $t \to 0$

In this section we develop the formal asymptotic solution to IBVP for $m < n$, $m = n$ and $m > n$, as $t \to 0$ via the method of matched asymptotic expansions. The behaviour of this solution, in each case, depends critically on the nature of $g(x)$ as $x \to \sigma^-$, (10.6a). Moreover, the structure of the solution as $t \to 0$ confirms the conjectures made from the well-stirred analogue of Section 10.1, given in Table 10.1.

### 10.2.1 $m < n$

Following Section 8.5, we anticipate at least three primary asymptotic regions in $x \geq 0$, labelled as follows:

$$\left.\begin{array}{ll} \text{region I}: & 0 \leq x < \sigma - O(\delta(t)), \, \alpha = O(1), \beta = O(1) \\ \text{region II}: & x = \sigma + O(\delta(t)), \quad \alpha = O(1), \beta = o(1) \\ \text{region III}: & x > \sigma + O(\delta(t)), \quad \alpha = O(1), \beta = o(1) \end{array}\right\} \text{ as } t \to 0, \quad (10.19)$$

where $\delta(t) = o(1)$ as $t \to 0$. We begin in region **I** where $0 \leq x < \sigma - o(1)$. Since $\alpha(x,0), \beta(x,0) > 0$ are analytic in region **I** with $\alpha, \beta = O(1)$ as $t \to 0$, we expand $\alpha(x,t)$ and $\beta(x,t)$ as regular power series in $t$. After substitution into equations (10.2), equating powers of $t$ to zero, and applying initial conditions (10.3), we obtain

$$\alpha(x,t) = 1 - t\left\{\beta_0^m g^m(x)\right\} + O\left(t^2\right), \quad (10.20a)$$

$$\beta(x,t) = \beta_0 g(x) + t\left\{\beta_0 g_{xx}(x) + \beta_0^m g^m(x) - k\beta_0^n g^n(x)\right\} + O\left(t^2\right), \quad (10.20b)$$

as $t \to 0$ with $0 \leq x < \sigma - o(1)$. Now when $(\sigma - x) \ll 1$, expansions (10.20) become

$$\alpha(x,t) \sim 1 - t\left\{\beta_0^m g_\sigma^m (\sigma - x)^{rm} + \ldots\right\} + \ldots, \quad (10.21a)$$

$$\beta(x,t) \sim \beta_0 g_\sigma\left\{(\sigma - x)^r + \ldots\right\} + t\left\{\beta_0 g_\sigma r(r-1)(\sigma - x)^{r-2} + \beta_0^m g_\sigma^m(\sigma - x)^{rm} - k\beta_0^n g_\sigma^n(\sigma - x)^{rn} + \ldots\right\} + \ldots \quad (10.21b)$$

as $t \to 0$, and a nonuniformity develops first in expansion (10.20b). Clearly, there are three distinct cases to consider depending upon the value of $r$.

**(a) $1 \le r < \frac{2}{1-m}$**

In this case, expansion (10.20b) in region **I** becomes nonuniform when $x = \sigma - O\left(t^{\frac{1}{2}}\right)$, when we observe, via (10.21), that $\alpha = 1 - O\left(t^{1+\frac{rm}{2}}\right)$ and $\beta = O\left(t^{\frac{r}{2}}\right)$. To examine region **II**, we introduce the scaled coordinate $\eta = (x - \sigma)t^{-\frac{1}{2}}$ and look for asymptotic expansions of the form

$$\alpha(\eta, t) = 1 - t^{1+\frac{rm}{2}}\,\bar{\alpha}_0(\eta) + o\left(t^{1+\frac{rm}{2}}\right), \tag{10.22a}$$

$$\beta(\eta, t) = t^{\frac{r}{2}}\,\bar{\beta}_0(\eta) + o\left(t^{\frac{r}{2}}\right) \tag{10.22b}$$

as $t \to 0$ with $\eta = O(1)$. On substitution of expansions (10.22) into equations (10.2) (when written in terms of $\eta$ and $t$) we obtain at leading order

$$\bar{\alpha}_{0\eta\eta} + \frac{\eta}{2}\bar{\alpha}_{0\eta} - R\bar{\alpha}_0 = -\bar{\beta}_0^m, \tag{10.23a}$$

$$\bar{\beta}_{0\eta\eta} + \frac{\eta}{2}\bar{\beta}_{0\eta} - \frac{r}{2}\bar{\beta}_0 = 0, \tag{10.23b}$$

where $-\infty < \eta < \infty$ and $R = 1 + \frac{rm}{2}$. Equations (10.23) are to be solved subject to matching with region **I** as $\eta \to -\infty$, and region **III** as $\eta \to \infty$, that is,

$$\bar{\alpha}_0(\eta) \sim \beta_0^m g_\sigma^m (-\eta)^{rm} \quad \text{as} \quad \eta \to -\infty, \tag{10.24a}$$

$$\bar{\beta}_0(\eta) \sim \beta_0 g_\sigma (-\eta)^r \quad \text{as} \quad \eta \to -\infty \tag{10.24b}$$

and

$$\bar{\alpha}_0(\eta), \; \bar{\beta}_0(\eta) \quad \text{remain bounded as} \quad \eta \to \infty. \tag{10.25}$$

The solution to the boundary value problem (10.23b), (10.24b) and (10.25) is unique and is readily obtained as

$$\bar{\beta}_0(\eta) = \begin{cases} \dfrac{\beta_0 g_\sigma r!}{\left(\frac{1}{2}r\right)! \, \kappa_1} A(\eta) \displaystyle\int_\eta^\infty \dfrac{e^{-s^2/4}}{A^2(s)}\, ds, & r \text{ even}, \\[2em] \dfrac{\beta_0 g_\sigma r!}{\left(\frac{1}{2}(r-1)\right)! \, \kappa_2} A(\eta) \left[\dfrac{1}{\eta} - \displaystyle\int_\eta^\infty \left\{\dfrac{1}{s^2} - \dfrac{e^{-s^2/4}}{A^2(s)}\right\} ds\right], & r \text{ odd}, \end{cases}$$
$$\tag{10.26}$$

where $A(\eta)$, $k_1$ and $k_2$ are as given by (2.16), (2.17) and (2.18) of Section 2.2.1 respectively. The unique solution to the boundary value problem (10.23a), (10.24a) and (10.25), can then be written in terms of $\bar{\beta}_0(\eta)$ as

$$\bar{\alpha}_0(\eta) = e^{-\eta^2/4} \left\{ \kappa_3 u_1(\eta) + \kappa_4 u_2(\eta) + u_1(\eta) \int_0^\eta u_2(s) \bar{\beta}_0^m(s)\, ds \right.$$

$$\left. - u_2(\eta) \int_0^\eta u_1(s) \bar{\beta}_0^m(s)\, ds \right\}, \tag{10.27}$$

where

$$u_1(\eta) = 1 + \sum_{p=1}^{\infty} \frac{(R+\frac{1}{2})\dots(R+\frac{(2p-1)}{2})}{(2p)!}\,\eta^{2p},$$

$$u_2(\eta) = u_1(\eta) \int_{-\infty}^{\eta} \frac{e^{s^2/4}}{u_1^2(s)}\,ds$$

and

$$\kappa_3 = \int_{-\infty}^{0} u_2(s)\bar{\beta}_0^m(s)\,ds,$$

$$\kappa_4 = \frac{1}{\kappa_5}\left[\int_0^{\infty}\left(\kappa_5 u_1(s) - u_2(s)\right)\bar{\beta}_0^m(s)\,ds - \kappa_3\right],$$

$$\kappa_5 = \int_{-\infty}^{\infty} \frac{e^{s^2/4}}{u_1^2(s)}\,ds.$$

We note from (10.27) and (10.26) that $\bar{\alpha}_0(\eta)$ and $\bar{\beta}_0(\eta)$ are positive and monotone decreasing for all $-\infty < \eta < \infty$ and, in particular, we observe (after noting that $u_1(\eta) = O\left(|\eta|^{2+rm}e^{\frac{\eta^2}{4}}\right)$ as $|\eta| \to \infty$) that

$$\bar{\alpha}_0(\eta) \sim \frac{4C_\infty^m}{m(1-m)}\,\eta^{-(2+m(r+1))}\,e^{-m\eta^2/4} \quad \text{as} \quad \eta \to \infty, \quad (10.28a)$$

$$\bar{\beta}_0(\eta) \sim C_\infty\,\eta^{-(r+1)}\,e^{-\eta^2/4} \quad \text{as} \quad \eta \to \infty, \quad (10.28b)$$

where

$$C_\infty = \begin{cases} \dfrac{2\beta_0 g_\sigma (r!)^2}{\kappa_1\left[(\frac{1}{2}r)!\right]^2}, & r \text{ even}, \\[4mm] \dfrac{2\beta_0 g_\sigma (r!)^2}{\kappa_2\left[(\frac{1}{2}(r-1))!\right]^2}, & r \text{ odd}. \end{cases} \quad (10.29)$$

We then have, from (10.22) and (10.28), that for $\eta \gg 1$, as we move into region **III**,

$$\alpha(\eta, t) \sim 1 - t^{1+\frac{rm}{2}}\frac{4C_\infty^m}{m(1-m)}\,\eta^{-(2+m(r+1))}\,e^{-m\eta^2/4} + \dots, \quad (10.30a)$$

$$\beta(\eta, t) \sim t^{\frac{r}{2}}C_\infty\,\eta^{-(r+1)}\,e^{-\eta^2/4} + \dots \quad (10.30b)$$

as $t \to 0$. This is consistent with the requirements of region **III** in (10.19). Moreover, it suggests that $\alpha(x, t) - 1$ and $\beta(x, t)$ are exponentially small $\left(\text{of } O\left(e^{-\frac{1}{t}}\right)\right)$ as $t \to 0$ with $x = \sigma + O(1)$ in region **III**. However, with $x = \sigma + O(1)$ as $t \to 0$, there is no exponential balance in equations (10.2) which has $\alpha(x, t) - 1$ and $\beta(x, t)$ of $O\left(e^{-\frac{1}{t}}\right)$ as $t \to 0$. The only balance in equations (10.2) with $\alpha(x, t) = 1 - o(1)$, $\beta(x, t) = o(1)$ as $t \to 0$ is $\alpha_t$ with the reaction term $-\alpha \beta^m$ in equation (10.2a) and $\beta_t$ with the reaction terms $\alpha \beta^m - k \beta^n$ in equation (10.2b) which leads to $\alpha(x, t) - 1$ and $\beta(x, t)$ being algebraically small in $t$ as $t \to 0$. Thus in region **III** we look for expansions of the form

$$\alpha(x, t) = \alpha_h(t) - \psi(x, t), \tag{10.31a}$$
$$\beta(x, t) = \beta_h(t) + \phi(x, t), \tag{10.31b}$$

as $t \to 0$ with $x = \sigma + O(1)$, where $(\alpha_h(t), \beta_h(t))$ is the solution to the spatially homogeneous system (10.7), with in this case,

$$\alpha_h(t) \sim 1 - (1 - m)^{\frac{1}{1-m}} t^{\frac{1}{1-m}} + O\left(t^{\frac{(n-m+1)}{1-m}}\right), \tag{10.32a}$$

$$\beta_h(t) \sim (1 - m)^{\frac{1}{1-m}} t^{\frac{1}{1-m}} + O\left(t^{\frac{(n-m+1)}{1-m}}\right), \tag{10.32b}$$

as $t \to 0$ with $x = \sigma + O(1)$. The gauge functions $\psi(x, t), \phi(x, t) = o(1)$ are exponentially small in $t$ as $t \to 0$ and are to be determined. On substitution of expansions (10.31) into equations (10.2) we obtain, at leading order, the equations

$$\psi_t = \psi_{xx} + \frac{m}{(1-m)t} \phi, \tag{10.33a}$$

$$\phi_t = \phi_{xx} + \frac{m}{(1-m)t} \phi, \tag{10.33b}$$

as $t \to 0$ with $x = \sigma + O(1)$. On noting that equation (10.33b) is decoupled and of WKB type as $t \to 0$ with $x = \sigma + O(1)$, we expand as

$$\psi(x, t) = \phi(x, t) + \overline{A} \exp\left[-\lambda(x, t)t^{-1}\right], \tag{10.34a}$$
$$\phi(x, t) = \exp\left[-\rho(x, t)t^{-1}\right], \tag{10.34b}$$

as $t \to 0$ with $x = \sigma + O(1)$, where $0 < \lambda(x, t), \rho(x, t) = O(1)$ for all $x > \sigma$. Substitution of (10.34) into (10.33) leads to expansions for $\lambda(x, t)$ and $\rho(x, t)$ of the form

$$\lambda(x, t) = \lambda_0(x) + t\left[\lambda_1(x) + \lambda_2(x) \ln t\right] + O\left(t^2\right), \tag{10.35a}$$
$$\rho(x, t) = \rho_0(x) + t\left[\rho_1(x) + \rho_2(x) \ln t + \rho_3(x) \ln(-\ln t)\right] + O\left(t^2\right), \tag{10.35b}$$

as $t \to 0$ with $x = \sigma + O(1)$. On solving at each order in turn we find, via (10.31)-(10.35), that in region **III**

$$\alpha(x, t) = \alpha_h(t) - \phi(x, t) - \overline{A} \exp\left[-\frac{(x-\sigma)^2}{4t} - \left(\tfrac{1}{2} + \overline{B}\right) \ln(x - \sigma)\right.$$
$$\left. + \overline{B} \ln t + O(t \ln^2 t)\right], \tag{10.36a}$$

$$\beta(x, t) = \beta_h(t) + \exp\left[-\frac{(x-\sigma)^2}{4t} - \left(\tfrac{1}{2} - \frac{m}{1-m} + \overline{C}\right) \ln(x - \sigma)\right.$$
$$\left. + \overline{C} \ln t + \overline{D} - \overline{E} \ln(-\ln t) + O(t \ln^2 t)\right], \tag{10.36b}$$

as $t \to 0$ with $x = \sigma + O(1)$, where $\overline{A}, \overline{B}, \overline{C}, \overline{D}$ and $\overline{E}$ are arbitrary constants to be determined via matching (see, for example, Van Dyke [70]). Clearly, expansions (10.31) satisfy the initial conditions (10.3) in region **III**, together with the boundary conditions (10.4) as $x \to \infty$. It remains to match the expansions (10.31) in region **III** as $x \to \sigma^+$ with expansions (10.22) in region **II** as $\eta \to \infty$. Clearly asymptotic matching between regions **I** and **II** cannot be achieved (primarily because of the algebraic terms in expansions (10.31) in region **III**), and we conclude that a further region is required between **II** and **III** to accomodate the change in structure required by expansions (10.36) in region **III** from expansions (10.22) in region **II**. The existence of this further region is confirmed on noting (after consideration of higher order terms in expansion (10.22b)) that a nonuniformity develops in expansion (10.22b) as $\eta \to \infty$, before entering region **III**. In developing expansion (10.22b) in region **II**, a typical retained term from equation (10.2b) at leading order is $t^{\frac{r}{2}-1}\bar{\beta}_0$, whereas a typical neglected term is $t^{\frac{rm}{2}}\bar{\beta}_0^m$. The ratio of neglected to retained terms is then

$$R(\eta, t) = t^{1-\frac{r}{2}(1-m)}\bar{\beta}_0^{m-1}(\eta), \tag{10.37}$$

which is of $O\left(t^{1-\frac{r}{2}(1-m)}\right)$ $(R = o(1)$ since $r < \frac{2}{1-m})$ for $\eta = O(1)$ as $t \to 0$. However, when $\eta \gg 1$, $\bar{\beta}_0(\eta) \sim O\left(\eta^{-(r+1)}e^{-\eta^2/4}\right)$, via (10.28b), and so

$$R(\eta, t) \sim t^{1-\frac{r}{2}(1-m)}\eta^{(r+1)(1-m)}e^{\eta^2(1-m)/4} + \ldots \tag{10.38}$$

as $t \to 0$ and $R(\eta, t)$ becomes of $O(1)$ for $\eta$ sufficiently large, confirming the onset of a nonuniformity in expansion (10.22b) as $\eta \to \infty$. Further examination of (10.38) reveals that the nonuniformity occurs when

$$\eta^2 = \lambda_0(-\ln t) - 2(r+1)\ln[\lambda_0(-\ln t)] + \ldots \equiv C^2(t) \tag{10.39}$$

as $t \to 0$, where

$$\lambda_0 = \frac{4}{1-m}\left[1 - \frac{r}{2}(1-m)\right], \tag{10.40}$$

with, via (10.30), $\alpha = 1 - O\left(\frac{t^{\frac{1}{1-m}}}{(-\ln t)}\right)$ and $\beta = O\left(t^{\frac{1}{1-m}}\right)$. To continue the asymptotic structure we relabel region **II** as region **II(a)** and introduce region **II(b)** in which, via (10.39), $\eta \sim C(t)$ as $t \to 0$. In region **II(b)** we define the scaled coordinate $\bar{\eta} = [\eta - C(t)]\chi^{-1}(t)$, with $\bar{\eta} = O(1)$ as $t \to 0$ and the gauge functions $\chi(t), C(t) = o(1)$ as $t \to 0$, where $C(t)$ is given by (10.39), and look for expansions of the form

$$\alpha(\bar{\eta}, t) = 1 - t^{\frac{1}{1-m}}\left[\hat{\alpha}_0(\bar{\eta}) - \frac{1}{\ln t}\hat{\alpha}_1(\bar{\eta})\right] + O\left(t^{\frac{(n-m+1)}{1-m}}\right), \tag{10.41a}$$

$$\beta(\bar{\eta}, t) = t^{\frac{1}{1-m}}\hat{\beta}_0(\bar{\eta}) + O\left(t^{\frac{(n-m+1)}{1-m}}\right), \tag{10.41b}$$

as $t \to 0$. We note that the inclusion of the $O\left(t^{\frac{1}{1-m}}\right)$ term in expansion (10.41a) allows matching at leading algebraic order with region **III**. On substituting expansions (10.41) into equations (10.2) (when written in terms of $\bar\eta$ and $t$) we find that there is only one possible balance at leading order. This requires $\chi(t) = O(C^{-1}(t))$ and without loss of generality, we put $\chi(t) \equiv C^{-1}(t)$. At leading order we then obtain

$$\hat\alpha_{0\bar\eta\bar\eta} + \frac{1}{2}\hat\alpha_{0\bar\eta} = 0, \tag{10.42a}$$

$$\hat\beta_{0\bar\eta\bar\eta} + \frac{1}{2}\hat\beta_{0\bar\eta} = 0, \tag{10.42b}$$

with $-\infty < \bar\eta < \infty$ in region **IIb**. Equations (10.42) are to be solved subject to matching with region **IIa** as $\bar\eta \to -\infty$ and to region **III** as $\bar\eta \to \infty$. Equations (10.42) have the general solutions

$$\hat\alpha_0(\bar\eta) = \hat A_0 + \hat B_0 e^{-\frac{\bar\eta}{2}}, \tag{10.43a}$$

$$\hat\beta_0(\bar\eta) = \hat C_0 + \hat D_0 e^{-\frac{\bar\eta}{2}}, \tag{10.43b}$$

for some, as yet undetermined constants $\hat A_0$, $\hat B_0$, $\hat C_0$ and $\hat D_0$. Clearly, to match with region **II(a)** we require $\hat B_0 \equiv 0$ (via (10.22a) and (10.30a)) and matching to leading algebraic order $\left(O\left(t^{\frac{1}{1-m}}\right)\right)$ in region **III** requires $\hat A_0 = \hat C_0 = (1-m)^{\frac{1}{1-m}}$. We note that the remaining algebraic terms in region **III** will match to the corresponding higher order algebraic terms in expansions (10.41) in region **II(b)**.

To complete the asymptotic structure of $\beta$ as $t \to 0$, it remains to match expansion (10.41b) in region **II(b)** to expansion (10.22b) in region **II(a)** (as $\bar\eta \to -\infty$) and to expansion (10.31b) in region **III** (as $\bar\eta \to \infty$). Matching up to exponential order then gives

$$\hat D_0 = C_\infty$$

and

$$\overline{C} = \frac{2r+1}{2} - \frac{m}{1-m}, \quad \overline{D} = \ln\left[\frac{C_\infty}{\left[4\left(\frac{1}{1-m} - \frac{r}{2}\right)\right]^{\frac{m}{1-m}}}\right], \quad \overline{E} = \frac{m}{1-m}.$$

To complete the asymptotic structure for $\alpha$ as $t \to 0$, we must determine the next term, $\hat\alpha_1(\bar\eta)$, in expansion (10.41a) in region **II(b)** in order to match up to exponential order with regions **II(a)** and **III**. In particular, via (10.43), we have

$$\hat\alpha_{1\bar\eta\bar\eta} + \frac{1}{2}\hat\alpha_{1\bar\eta} = \frac{\left(-\left[(1-m)^{\frac{1}{1-m}} + C_\infty e^{-\frac{\bar\eta}{2}}\right]^m + (1-m)^{\frac{m}{1-m}}\right)}{(4/(1-m)) - 2r}, \tag{10.44}$$

for $-\infty < \bar{\eta} < \infty$. Equation (10.44) is to be solved subject to matching with region **II(a)** as $\bar{\eta} \to -\infty$ and region **III** as $\bar{\eta} \to \infty$. The matching condition with region **IIa** ($\bar{\eta} \to -\infty$) is given by

$$\hat{\alpha}_1(\bar{\eta}) \sim \frac{C_\infty^m}{m\left(1 - \frac{r}{2}(1-m)\right)} e^{-m\bar{\eta}/2} \quad \text{as} \quad \bar{\eta} \to -\infty. \tag{10.45}$$

The general solution to equation (10.44) may be written as

$$\hat{\alpha}_1(\bar{\eta}) = \hat{A}_1 + \hat{B}_1 e^{-\frac{\bar{\eta}}{2}} + P(\bar{\eta}), \quad -\infty < \bar{\eta} < \infty, \tag{10.46}$$

where $\hat{A}_1$ and $\hat{B}_1$ are arbitrary constants and $P(\bar{\eta})$ is a particular integral. It is readily established that

$$P(\bar{\eta}) \sim \begin{cases} \frac{mC_\infty}{2\left(1 - \frac{r}{2}(1-m)\right)} \bar{\eta}\, e^{-\bar{\eta}/2} & \text{as } \bar{\eta} \to \infty, \\[2ex] \frac{C_\infty^m}{m\left(1 - \frac{r}{2}(1-m)\right)} e^{-m\bar{\eta}/2} & \text{as } \bar{\eta} \to -\infty. \end{cases}$$

Condition (10.45) then requires $\hat{B}_1 = 0$. Finally, matching with region **III** up to exponential order now follows directly giving

$$\hat{A}_1 = 0, \qquad \overline{A} = \frac{C_\infty(3m-1)}{1-m}, \qquad \overline{B} = \frac{2r+1}{2}.$$

At this stage the main asymptotic structure is complete. However, we note that whilst initial conditions (10.3) are satisfied in regions **I** – **III** and boundary conditions (10.4) are satisfied in region **III**, with $g(x)$ given by (10.6b) as $x \to 0^+$, the expansions (10.20) in region **I** do not, in general, satisfy boundary conditions (10.5) at $x = 0$, and a further passive region is required in the neighbourhood of $x = 0$ as $t \to 0$, which we denote as region $\mathbf{I}_0$. It may be readily deduced that $x = O\left(t^{\frac{1}{2}}\right)$ as $t \to 0$ in region $\mathbf{I}_0$, and the appropriate expansions in this region (which satisfy initial conditions (10.3), boundary conditions (10.5) and match to region **I** for $x \gg t^{\frac{1}{2}}$) are obtained as

$$\alpha(\eta^*, t) = 1 - t[\beta_0 g_0]^m + O\left(t^{\frac{p}{2}+1}\right), \quad p \geq 1, \tag{10.47a}$$

$$\beta(\eta^*, t) = \begin{cases} \beta_0 g_0 + t^{\frac{1}{2}} \beta_0\, g_1\, \eta^*\left[1 + \frac{1}{C^*}\int_{\eta^*}^\infty \frac{e^{-s^2/4}}{s^2}\, ds\right] + O\left(t^{\frac{1}{2}}\right), \ p = 1, \\[2ex] \beta_0 g_0 + t\left[(\beta_0 g_0)^m - k(\beta_0 g_0)^n + 2\delta_{2,p}\, \beta_0\, g_2\left(1 + \frac{\eta^{*2}}{2}\right)\right] \\[1ex] + O\left(t^2\right), \quad p \geq 2 \end{cases}$$

as $t \to 0$, with $\eta^* = xt^{-\frac{1}{2}} = O(1)$ and

$$C^* = \int_0^\infty \frac{1 - e^{-s^2/4}}{s^2}\, ds,$$

and where $\delta_{2,p}$ is the Kronecker delta.

A schematic representation of the location and thickness of the asymptotic regions as $t \to 0$ in this case is given in Figure 10.4.

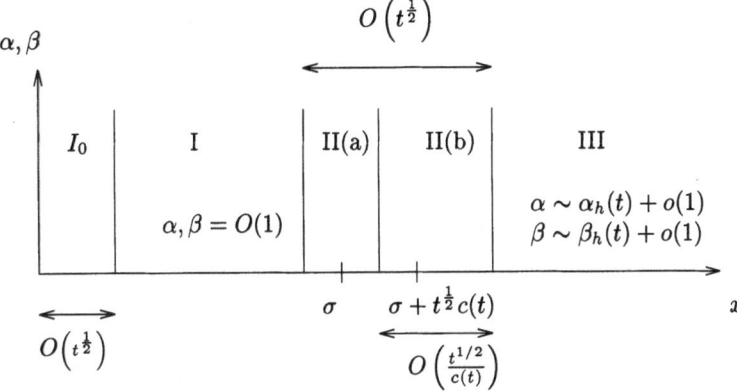

**Fig. 10.4.** Schematic representation of the location and thickness of the asymptotic regions as $t \to 0$ in the case when $m < n$ and $1 \le r < \frac{2}{(1-m)}$. Note that in region II(a): $\alpha = 1 - O\left(t^{1+rm/2}\right)$, $\beta = O\left(t^{r/2}\right)$ as $t \to 0$ while in region II(b): $\alpha = 1 - O\left(t^{1/(1-m)}\right)$, $\beta = O\left(t^{1/(1-m)}\right)$ as $t \to 0$.

**(b) $r = \frac{2}{1-m}$**

The primary asymptotic structure in this case is again given by (10.19), and in regions **I** and **I$_0$** expansions (10.20) and (10.47) hold respectively, with expansions (10.31) again being appropriate in region **III**. In this case, expansion (10.20b) again becomes nonuniform when $x = \sigma - O\left(t^{\frac{1}{2}}\right)$ and expansions (10.21) indicate that $\alpha = 1 - O\left(t^{\frac{1}{1-m}}\right)$ and $\beta = O\left(t^{\frac{1}{1-m}}\right)$ in region **II**. To examine region **II**, we introduce the scaled coordinate $\eta = (x - \sigma)t^{-\frac{1}{2}}$ and look for asymptotic expansions of the form

$$\alpha(\eta, t) = 1 - t^{\frac{1}{1-m}} \bar{\alpha}_0(\eta) + o\left(t^{\frac{1}{1-m}}\right), \tag{10.48a}$$

$$\beta(\eta, t) = t^{\frac{1}{1-m}} \bar{\beta}_0(\eta) + o\left(t^{\frac{1}{1-m}}\right) \tag{10.48b}$$

as $t \to 0$ with $\eta = O(1)$. On substitution of expansions (10.48) into equations (10.2) (when written in terms of $\eta$ and $t$) we obtain at leading order

$$\bar{\alpha}_{0\eta\eta} + \frac{\eta}{2}\bar{\alpha}_{0\eta} - \frac{\bar{\alpha}_0}{1-m} = -\bar{\beta}_0^m, \tag{10.49a}$$

$$\bar{\beta}_{0\eta\eta} + \frac{\eta}{2}\bar{\beta}_{0\eta} - \frac{\bar{\beta}_0}{1-m} + \bar{\beta}_0^m = 0, \tag{10.49b}$$

with $-\infty < \eta < \infty$ in region **II**. Equations (10.49) are to be solved subject to matching expansions (10.48) in region **II** with expansions (10.20) in region **I** as $\eta \to -\infty$, and to expansions (10.31) in region **III** as $\eta \to \infty$, that is,

$$\bar{\alpha}_0(\eta) \sim \beta_0^m g_\sigma^m (-\eta)^{\frac{2m}{1-m}} \quad \text{as} \quad \eta \to -\infty, \tag{10.50a}$$

$$\bar{\beta}_0(\eta) \sim \beta_0 g_\sigma (-\eta)^{\frac{2}{1-m}} \quad \text{as} \quad \eta \to -\infty \tag{10.50b}$$

and

$$\bar{\alpha}_0(\eta) \to (1-m)^{\frac{1}{1-m}} \quad \text{as} \quad \eta \to \infty, \tag{10.50c}$$

$$\bar{\beta}_0(\eta) \to (1-m)^{\frac{1}{1-m}} \quad \text{as} \quad \eta \to \infty, \tag{10.50d}$$

with

$$\bar{\alpha}_0, \bar{\beta}_0 \geq 0 \quad \text{for all} \quad -\infty < \eta < \infty. \tag{10.50e}$$

The nonlinear, singular boundary value problem (10.49b), (10.50b), (10.50d) and (10.50e) is examined in detail in Needham and King [57] where it is established that a unique solution exists for any $\beta_0 g_\sigma > 0$ and $0 < m < 1$ (with $m$ in the current notation replacing $p$ in the notation of Needham and King), and that the solution is monotone decreasing in $\eta$. The boundary value problem (10.49a), (10.50a), (10.50c) and (10.50e) has then a unique solution which is monotone decreasing in $\eta$ (the details are similar to those of (10.27) and are not repeated here). This completes the details of the asymptotic structure in this case.

We note in particular that the leading order terms of expansions (10.48) in region **II** are able to match directly with expansions (10.31) in region **III** as $\eta \to \infty$ and a region of type **II(b)** is not required in this case. This is due to the presence of the reaction term in the leading order equation (10.49b), which was absent in the previous case. Furthermore, subsequent terms in expansions (10.48) in region **II** match directly with expansions (10.31) in region **III** at each algebraic order. Finally we note that the constants $\overline{A}, \overline{B}, \overline{C}, \overline{D}$ and $\overline{E}$ in region **III** have not been explicitly determined in this case. The explicit determination of these constants would follow from exponentially small corrections to $\bar{\alpha}_0$ and $\bar{\beta}_0$ as $\eta \to \infty$ in the boundary value problems (10.49a), (10.50a), (10.50c) and (10.50e) and (10.49b), (10.50b), (10.50d) and (10.50e).

## (c) $r > \frac{2}{1-m}$

In this case, the primary asymptotic structure is again given by (10.19). Regions **I**, **I₀** and **III** are as before, with the respective expansions given by (10.20), (10.47) and (10.31). An examination of expansions (10.20) as $x \to \sigma^-$ reveals that expansion (10.20b) first becomes nonuniform when $x = \sigma - O\left(t^{\frac{1}{r(1-m)}}\right)$, with region **II** being thicker in this case than in (a) and (b) (with its thickness depending upon $r$ and $m$). We further observe from expansions (10.20) that $\alpha = 1 - O\left(t^{\frac{1}{1-m}}\right)$ and $\beta = O\left(t^{\frac{1}{1-m}}\right)$ in region **II**.

To examine region **II**, we introduce the scaled coordinate $\hat{\eta} = (x - \sigma)t^{-\frac{1}{r(1-m)}}$ and look for asymptotic expansions of the form

$$\alpha(\hat{\eta}, t) = 1 - t^{\frac{1}{1-m}} \hat{\alpha}_0(\hat{\eta}) + o\left(t^{\frac{1}{1-m}}\right), \tag{10.51a}$$

$$\beta(\hat{\eta}, t) = t^{\frac{1}{1-m}} \hat{\beta}_0(\hat{\eta}) + o\left(t^{\frac{1}{1-m}}\right) \tag{10.51b}$$

as $t \to 0$ with $\hat{\eta} = O(1)$. On substitution of expansions (10.51) into equations (10.2) (when written in terms of $\hat{\eta}$ and $t$) we obtain at leading order

$$\hat{\eta}\,\hat{\alpha}_{0\hat{\eta}} - r\hat{\alpha}_0 = -r(1 - m)\,\hat{\beta}_0^{\ m}, \tag{10.52a}$$

$$\hat{\eta}\,\hat{\beta}_{0\hat{\eta}} - r\hat{\beta}_0 + r(1 - m)\,\hat{\beta}_0^m = 0, \tag{10.52b}$$

with $\hat{\eta} > -\infty$ in region **II**. Equations (10.52) are to be solved subject to matching with region **I** as $\hat{\eta} \to -\infty$. The matching conditions are given by

$$\hat{\alpha}_0(\hat{\eta}) \sim \beta_0^m g_\sigma^m (-\hat{\eta})^{rm} \quad \text{as} \quad \hat{\eta} \to -\infty, \tag{10.53a}$$

$$\hat{\beta}_0(\hat{\eta}) \sim \beta_0 g_\sigma (-\hat{\eta})^r \quad \text{as} \quad \hat{\eta} \to -\infty. \tag{10.53b}$$

The solution to (10.52b), (10.53b) may be obtained directly by separation of variables as

$$\hat{\beta}_0(\hat{\eta}) = \left[(\beta_0 g_\sigma)^{1-m} (-\hat{\eta})^{r(1-m)} + (1 - m)\right]^{\frac{1}{(1-m)}}, \quad \hat{\eta} > -\infty. \tag{10.54}$$

An examination of (10.54) reveals that a weak singularity develops in $\hat{\beta}_0$ as $\hat{\eta} \to 0^-$. Now, via (10.54), the solution to (10.52a), (10.53a) is given by

$$\hat{\alpha}_0(\hat{\eta}) = r(1 - m)(-\hat{\eta})^r \int_{-\infty}^{\hat{\eta}} (-s)^{-(r+1)} \hat{\beta}_0^m(s)\mathrm{d}s, \quad -\infty < \hat{\eta} < 0, \tag{10.55}$$

and we note from (10.54) and (10.55) that $\hat{\alpha}_0(\hat{\eta})$ and $\hat{\beta}_0(\hat{\eta})$ are monotone decreasing in $\hat{\eta}$ with

$$\hat{\alpha}_0(\hat{\eta}) = (1 - m)^{\frac{1}{1-m}} + O\left[(-\hat{\eta})^{r(1-m)}\right], \tag{10.56a}$$

$$\hat{\beta}_0(\hat{\eta}) = (1 - m)^{\frac{1}{1-m}} + O\left[(-\hat{\eta})^{r(1-m)}\right], \tag{10.56b}$$

as $\hat{\eta} \to 0^-$. Thus $\hat{\alpha}_0(\hat{\eta})$ and $\hat{\beta}_0(\hat{\eta})$ develop singularities in sufficiently large derivatives as $\hat{\eta} \to 0^-$ and cannot be continued beyond $\hat{\eta} = 0$. Higher order terms in expansions (10.51) reveal a weak nonuniformity in expansions (10.51) as $\hat{\eta} \to 0^-$, occurring when $\hat{\eta} = O\left(t^{\frac{r(1-m)-2}{2r(1-m)}}\right)$ $\left(\text{that is } x = \sigma - O\left(t^{\frac{1}{2}}\right)\right)$. We therefore require a further asymptotic region between regions **II** and **III** which we label region **II**$_0$. To investigate region **II**$_0$ we introduce the scaled variable $\eta = (x - \sigma)t^{-\frac{1}{2}}$. An examination of (10.51) and (10.56) suggests expanding $\alpha$ and $\beta$ in region **II**$_0$ as

$$\alpha(\eta, t) \sim 1 - t^{\frac{1}{1-m}}\left[(1 - m)^{\frac{1}{1-m}} - \alpha_0(\eta)t^{\frac{r}{2}(1-m)-1} + \dots\right], \tag{10.57a}$$

$$\beta(\eta, t) \sim t^{\frac{1}{1-m}}\left[(1 - m)^{\frac{1}{1-m}} + \beta_0(\eta)t^{\frac{r}{2}(1-m)-1} + \dots\right], \tag{10.57b}$$

as $t \to 0$ with $\eta = O(1)$. We note that the term of $O\left(t^{\frac{r}{2}(1-m)+\frac{m}{1-m}}\right)$ in expansions (10.57) allows matching at exponential order with region **III**. Further, via (10.32), we note that in order to match expansions (10.57) in region **II₀** as $\eta \to \infty$ with the algebraic terms in expansions (10.31) of region **III** as $x \to \sigma^+$ we may, depending on the size of $r$, require further terms to be inserted prior to the $O\left(t^{\frac{r}{2}(1-m)+\frac{m}{1-m}}\right)$ term in expansions (10.57). On substitution of (10.57) into equations (10.2) (when written in terms of $\eta$ and $t$) the problems for $\alpha_0$ and $\beta_0$ (at $O\left(t^{\frac{r}{2}(1-m)+\frac{m}{1-m}}\right)$) in region **II₀** are readily obtained as

$$\alpha_{0\eta\eta} + \frac{\eta}{2}\alpha_{0\eta} - \hat{R}\alpha_0 = -\frac{m}{1-m}\beta_0, \tag{10.58a}$$

$$\beta_{0\eta\eta} + \frac{\eta}{2}\beta_{0\eta} - \frac{r(1-m)}{2}\beta_0 = 0, \tag{10.58b}$$

where $-\infty < \eta < \infty$ and $\hat{R} = \frac{r}{2}(1-m) + \frac{m}{1-m}$, with

$$\alpha_0(\eta) \sim O\left[(-\eta)^{r(1-m)}\right] \quad \text{as} \quad \eta \to -\infty, \tag{10.59a}$$

$$\beta_0(\eta) \sim O\left[(-\eta)^{r(1-m)}\right] \quad \text{as} \quad \eta \to -\infty, \tag{10.59b}$$

and

$$\alpha_0(\eta) \sim O\left(e^{-\frac{\eta^2}{4}}\right) \quad \text{as} \quad \eta \to \infty, \tag{10.60a}$$

$$\beta_0(\eta) \sim O\left(e^{-\frac{\eta^2}{4}}\right) \quad \text{as} \quad \eta \to \infty, \tag{10.60b}$$

after matching with regions **I** and **III** respectively. The boundary value problem (10.58b), (10.59b) and (10.60b) has a unique solution which may readily be written down in terms of hypergeometric (Kummer) functions, with the solution to (10.58a), (10.59a) and (10.60a) then being given, after minor modifications, by (10.27) (in part (a)). However, for brevity, we do not pursue these details any further. Clearly, expansions (10.57) in region **II₀** as $\eta \to \infty$ match directly with expansions (10.31) in **III** as $x \to \sigma^+$ at leading algebraic and exponential order. The inclusion of further terms in expansions (10.57) in region **II₀** will allow matching with region **III** to successively higher algebraic order. We do not pursue these details any further.

A schematic representation of the location and thickness of the asymptotic regions as $t \to 0$ in this case is given in Figure 10.5.

This completes the asymptotic structure of the solution to IBVP as $t \to 0$ for $m < n$. We observe in this case that $\alpha_\infty(t)$, $\beta_\infty(t)$ have been determined in the course of the analysis (by requirements in region **III**) as

$$\alpha_\infty(t) \sim \alpha_h(t), \qquad \beta_\infty(t) \sim \beta_h(t) \quad \text{as } t \to 0,$$

with $\alpha_h$, $\beta_h$ as defined in (10.31), (10.32). This confirms our conjecture in Table 10.1 (noting that $\alpha_h \equiv A(t)$ and $\beta_h \equiv B(t)$).

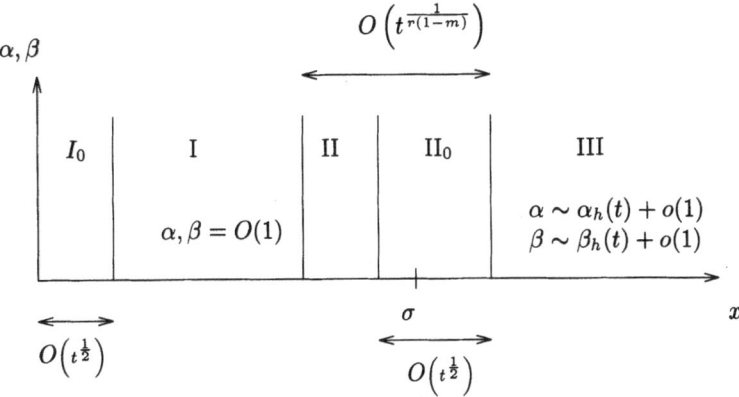

**Fig. 10.5.** Schematic representation of the location and thickness of the asymptotic regions as $t \to 0$ in the case when $m < n$ and $r > \frac{2}{(1-m)}$. Note that in regions II and II$_0$: $\alpha = 1 - O\left(t^{\frac{1}{(1-m)}}\right)$ and $\beta = O\left(t^{\frac{1}{(1-m)}}\right)$ as $t \to 0$.

### 10.2.2 $m = n$

Here there are three distinct cases to be considered, these being $k < 1$, $k = 1$ and $k > 1$. We examine these in turn. Again, we expect the asymptotic structure to have the three primary regions as in (10.19).

### (a) $k < 1$

Here the asymptotic structure as $t \to 0$ follows, after minor modifications, that given in Section 10.2.1, with $n$ now replacing $m$ throughout, and with the expansions for $\alpha$ and $\beta$ being valid up to the orders given. We note in particular that due to change in the expansions $\alpha_h(t)$, $\beta_h(t)$ (the solution to the spatially homogeneous system (10.7)) given by, in this case,

$$\alpha_h(t) \sim 1 - (1-m)^{\frac{1}{1-m}}(1-k)^{\frac{m}{1-m}}t^{\frac{1}{1-m}} + O\left(t^{\frac{2}{1-m}}\right), \quad (10.61a)$$

$$\beta_h(t) \sim (1-m)^{\frac{1}{1-m}}(1-k)^{\frac{m}{1-m}}t^{\frac{1}{1-m}} + O\left(t^{\frac{2}{1-m}}\right), \quad (10.61b)$$

as $t \to 0$, that there will be minor modifications to some of the matching constants (these modifications are straightforward and are omitted here for brevity). Again, the required asymptotic structure in this case confirms the conjectures of Table 10.1.

### (b) $k = 1$

We begin with region **I** where $0 \le x < \sigma - o(1)$ and $\alpha, \beta = O(1)$ as $t \to 0$ and, since $\alpha(x,0)$, $\beta(x,0) > 0$ are analytic in region **I**, we expand $\alpha(x,t)$ and

$\beta(x,t)$ as regular power series in $t$. After substitution into equation (10.2), equating powers of $t$ to zero, and applying initial condition (10.3), we obtain

$$\alpha(x,t) = 1 - t\left\{(\beta_0 g(x))^m\right\}$$
$$- t^2\left\{\frac{\beta_0^m}{2}[g(x)^m]'' - \frac{1}{2}(\beta_0 g(x))^{2m} + \frac{m\beta_0^m}{2}g^{m-1}(x)g''(x)\right\} + O\left(t^3\right),$$

$$\tag{10.62a}$$

$$\beta(x,t) = \beta_0 g(x) + t\left\{\beta_0 g''(x)\right\} + t^2\left\{\frac{\beta_0}{2}g^{IV}(x) - \frac{\beta_0^{2m}}{2}g^{2m}(x)\right\} + O\left(t^3\right),$$

$$\tag{10.62b}$$

as $t \to 0$ with $0 \le x < \sigma - o(1)$. Now when $(\sigma - x) \ll 1$, expansions (10.62) become

$$\alpha(x,t) = 1 - t\left\{(\beta_0 g_\sigma)^m (\sigma - x)^{rm} + \ldots\right\}$$
$$- t^2\left\{\frac{1}{2}(\beta_0 g_\sigma)^m rm(rm-1)(\sigma - x)^{rm-2}\right.$$
$$- \frac{1}{2}(\beta_0 g_\sigma)^{2m}(\sigma - x)^{2rm} + \frac{1}{2}(\beta_0 g_\sigma)^m r(r-1)(\sigma - x)^{rm-2} + \ldots\right\}$$
$$+ O\left(t^3\right), \tag{10.63a}$$

$$\beta(x,t) = \beta_0 g_\sigma(\sigma - x)^r + t\left\{\beta_0 g_\sigma r(r-1)(\sigma - x)^{r-2} + \ldots\right\}$$
$$+ t^2\left\{\frac{\beta_0}{2}r(r-1)(r-2)(r-3)(\sigma - x)^{r-4}\right.$$
$$- (\beta_0 g_\sigma)^{2m}(\sigma - x)^{2rm} + \ldots\right\} + O\left(t^3\right), \tag{10.63b}$$

as $t \to 0$, and a nonuniformity develops first in expansion (10.63b). There are four cases to consider, these being ($\frac{1}{2} \le m < 1$), ($0 < m < \frac{1}{2}; r < \frac{4}{1-2m}$), ($0 < m < \frac{1}{2}; r = \frac{4}{1-2m}$) and ($0 < m < \frac{1}{2}; r > \frac{4}{1-2m}$). We consider each of these in turn.

## (i) $\frac{1}{2} \le m < 1$

The primary structure is, again, in this case given by (10.19) with expansions (10.62) becoming nonuniform when $x = \sigma - O\left(t^{\frac{1}{2}}\right)$, when we observe, via (10.63) that $\alpha = 1 - O\left(t^{1+\frac{rm}{2}}\right)$ and $\beta = O\left(t^{\frac{1}{2}}\right)$. Hence, in region **II** we introduce the scaled coordinate $\eta = (x - \sigma)t^{-\frac{1}{2}}$ and look for asymptotic expansions of the form

$$\alpha(\eta,t) = 1 - t^{1+\frac{rm}{2}}\bar{\alpha}_0(\eta) + o\left(t^{1+\frac{rm}{2}}\right), \tag{10.64a}$$

$$\beta(\eta,t) = t^{\frac{r}{2}}\bar{\beta}_0(\eta) + o\left(t^{\frac{r}{2}}\right), \tag{10.64b}$$

as $t \to 0$ with $\eta = O(1)$. On substitution of expansions (10.64) into equations (10.2) (when written in terms of $\eta$ and $t$) we obtain at leading order

$$\bar{\alpha}_{0\eta\eta} + \frac{\eta}{2}\bar{\alpha}_{0\eta} - R\bar{\alpha}_0 = -\bar{\beta}_0^m, \tag{10.65a}$$

$$\bar{\beta}_{0\eta\eta} + \frac{\eta}{2}\bar{\beta}_{0\eta} - \frac{r}{2}\bar{\beta}_0 = 0, \tag{10.65b}$$

where $-\infty < \eta < \infty$ and $R = 1 + \frac{rm}{2}$. Equations (10.65) are to be solved subject to matching with region **I** as $\eta \to -\infty$, and region **III** as $\eta \to \infty$, that is,

$$\bar{\alpha}_0(\eta) \sim \beta_0^m g_\sigma^m (-\eta)^{rm} \quad \text{as} \quad \eta \to -\infty, \tag{10.66a}$$

$$\bar{\beta}_0(\eta) \sim \beta_0 g_\sigma (-\eta)^r \quad \text{as} \quad \eta \to -\infty, \tag{10.66b}$$

and

$$\bar{\alpha}_0(\eta), \ \bar{\beta}_0(\eta) \quad \text{remain bounded as} \quad \eta \to \infty. \tag{10.67}$$

The boundary value problems (10.65a), (10.66a) and (10.67) and (10.65b), (10.66b) and (10.67) have unique, monotone decreasing solutions, which have been studied in section 10.2.1 and are given by (10.26) and (10.27). In particular, for $\eta \gg 1$ we observe that, via (10.30), as we move into region **III**,

$$\alpha(\eta, t) \sim 1 - t^{1+\frac{rm}{2}} \frac{4 C_\infty^m}{m(1-m)} \eta^{-(2+m(r+1))} e^{-m\eta^2/4} + \ldots, \tag{10.68a}$$

$$\beta(\eta, t) \sim t^{\frac{r}{2}} C_\infty \eta^{-(r+1)} e^{-\eta^2/4} + \ldots, \tag{10.68b}$$

as $t \to 0$, with $C_\infty$ as defined in (10.29). We must now examine whether we can construct an expansion in region **III** which will match directly to (10.68) (otherwise we will have to introduce intermediate regions, as in the previous cases). It is straightforward to demonstrate (by constructing higher order terms) that, in fact, expansions (10.68) only have a very weak nonuniformity for $\eta \gg 1$ (as can also be established by noting that the ratio of neglected to retained terms in obtaining (10.65) is $R(\eta) = O\left(t^{2 - \frac{r}{2}(1-2m)} e^{\frac{\eta^2(1-2m)}{4}}\right)$ for $\eta \gg 1$ as $t \to 0$, which remains small ($\frac{1}{2} \leq m < 1$) as we move into region **III**, when $\eta = O\left(t^{-\frac{1}{2}}\right)$ ). In region **III**, where $x = \sigma + O(1)$, $\alpha = 1 - o(1)$ and $\beta = o(1)$, expressions (10.68) suggest that we look for expansions of the form

$$\alpha(x, t) = 1 - \psi(x, t), \tag{10.69a}$$

$$\beta(x, t) = \phi(x, t), \tag{10.69b}$$

as $t \to 0$ where $0 < \psi(x, t), \phi(x, t) = O\left(e^{-\frac{1}{t}}\right)$ as $t \to 0$ with $x = \sigma + O(1)$. On substitution of expansions (10.69) into equations (10.2) we may now obtain a balance at leading order, given by

$$\psi_t - \psi_{xx} = \phi^m, \tag{10.70a}$$

$$\phi_t - \phi_{xx} = 0, \tag{10.70b}$$

as $t \to 0$ with $x = \sigma + O(1)$. We first consider the uncoupled equation (10.70b) which is to be solved subject to the matching condition

$$\phi(x,t) \sim C_\infty \frac{t^{r+\frac{1}{2}}}{(x-\sigma)^{r+1}} e^{-\frac{(x-\sigma)^2}{4t}} \quad \text{as} \quad x \to \sigma^+. \tag{10.71}$$

The condition (10.71) suggests looking for a solution of (10.70b) of the form

$$\phi(x,t) = e^{-\frac{p(x,t)}{t}} \tag{10.72}$$

as $t \to 0$ with $x = \sigma + O(1)$ and $p(x,t) = O(1)$. Substitution of (10.72) into (10.70b) leads to an expansion for $p(x,t)$ of the form

$$p(x,t) = p_0(x) + t\,[p_1(x) + p_2(x)\ln t] \; + \; O\left(t^2\right) \tag{10.73}$$

as $t \to 0$ with $x = \sigma + O(1)$. On solving at each order in turn we obtain

$$p(x,t) = \frac{1}{4}(x-\sigma)^2 + t\left\{\left(D_\beta + \frac{1}{2}\right)\ln(x-\sigma) - D_\beta \ln t + A_\beta\right\} + O\left(t^2 \ln^2 t\right). \tag{10.74}$$

Now, on imposing the matching condition (10.71) we obtain

$$D_\beta = r + \frac{1}{2}, \qquad A_\beta = -\ln C_\infty,$$

giving in region **III**

$$\beta(x,t) = \exp\left[-\frac{(x-\sigma)^2}{4t} - (r+1)\ln(x-\sigma)\right.$$
$$\left. + \left(r + \tfrac{1}{2}\right)\ln t + \ln C_\infty + O\left(t\ln^2 t\right)\right] \tag{10.75}$$

as $t \to 0$ with $x = \sigma + O(1)$. Equation (10.70a) has to be solved subject to the matching condition

$$\psi(x,t) \sim \frac{4C_\infty^m}{m(1-m)} \frac{t^{2+mr+\frac{m}{2}}}{(x-\sigma)^{2+m(r+1)}} \exp\left\{-\frac{m(x-\sigma)^2}{4t}\right\} \quad \text{as} \quad x \to \sigma^+. \tag{10.76}$$

Now, (10.72), (10.73) and (10.76) suggest looking for a solution of (10.70a) of the form

$$\psi(x,t) \sim \frac{\check{A}t^{\check{B}}}{(x-\sigma)^{\check{C}}} \exp\left\{-\frac{m(x-\sigma)^2}{4t}\right\} \tag{10.77}$$

as $t \to 0$ with $x = \sigma + O(1)$. Upon substitution of (10.77) into (10.70a) we obtain

$$\check{A} = \frac{4C_\infty^m}{m(1-m)}, \qquad \check{B} = 2 + m\left(r + \frac{1}{2}\right), \qquad \check{C} = 2 + m(r+1), \tag{10.78}$$

giving in region **III**

$$\alpha(x,t) \sim 1 - \frac{4C_\infty^m}{m(1-m)} \frac{t^{2+m\left(r+\frac{1}{2}\right)}}{(x-\sigma)^{2+m(r+1)}} \exp\left\{-\frac{m(x-\sigma)^2}{4t}\right\} \qquad (10.79)$$

as $t \to 0$ with $x = \sigma + O(1)$. Clearly the expansions (10.75) and (10.79) satisfy the initial conditions in region **III** (namely $\alpha(x,t)$, $\beta(x,t) \to 0$ as $t \to 0$), and remain uniform for $x \gg 1$ ($\alpha(x,t) \to 1$, $\beta(x,t) \to 0$ as $x \to \infty$). Further, the expansions (10.75) and (10.79) in region **III** as $x \to \sigma^+$ match with expansions (10.64b), (10.64a) respectively in region **II** as $\eta \to \infty$. This completes the main asymptotic structure in this case. A schematic representation of the location and thickness of the asymptotic regions as $t \to 0$ in this case is given in Figure 10.6.

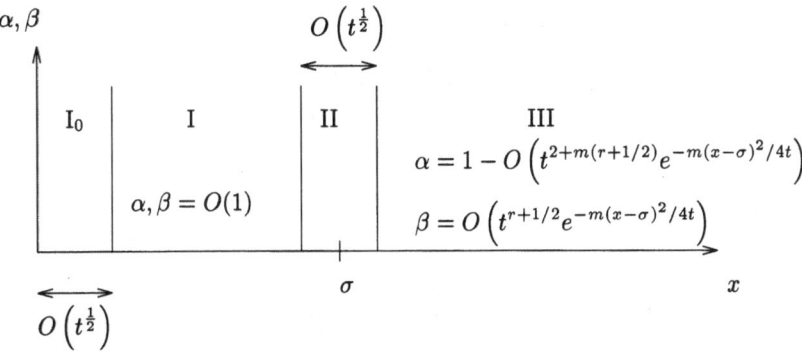

**Fig. 10.6.** Schematic representation of the location and thickness of the asymptotic regions as $t \to 0$ in the case when $m = n$, $k = 1$ and $\frac{1}{2} \le m < 1$. Note that in region II: $\alpha = 1 - O\left(t^{1+rm/2}\right)$, $\beta = O\left(t^{r/2}\right)$ as $t \to 0$.

Finally, we observe that in the course of this analysis, the asymptotic structure has required $\alpha_\infty(t) \equiv 1$ and $\beta_\infty(t) \equiv 0$, and confirms the conjecture given in Table 10.1.

**(ii) $0 < m < \frac{1}{2}$; $1 \le r < \frac{4}{1-2m}$**

In this case, regions $\mathbf{I_0}$, **I** and **II** are as given in Sections 10.2.1 (after some minor modifications), 10.2.2(b) and 10.2.2(b) part (i) respectively, with the (significant) exception that an examination of higher order terms in expansions (10.64) of region **II** (when $0 < m < \frac{1}{2}$, $1 \le r < \frac{4}{1-2m}$) indicates that a strong nonuniformity will develop in expansion (10.64b) when $\eta \gg 1$, in particular, before $\eta = O\left(t^{-\frac{1}{2}}\right)$ and we move into region **III**. This is in contrast to the previous case $\frac{1}{2} \le m < 1$ of Section 10.2.2(b) part (i) where the expansions were only weakly non–uniform for $\eta \gg 1$ on a scale of $\eta = O\left(t^{-\frac{1}{2}}\right)$. We therefore anticipate that the structure will contain a further intermediate asymptotic region between regions **II** and **III**. The existence of this additional region may be confirmed in the following way: a typical term retained from equation (10.2b) at leading order in region **II** is $t^{\frac{5}{2}-1}\eta\bar{\beta}_{0\eta}$, whereas a typical neglected term is $t^{rm+1}\bar{\alpha}_0\bar{\beta}_0^m$. The ratio of neglected to retained terms is then

$$R(\eta,t) = t^{2-\frac{r}{2}(1-2m)}\frac{\bar{\alpha}_0\bar{\beta}_0^m}{\eta\bar{\beta}_{0\eta}}, \tag{10.80}$$

which is of $o(1)$ (since $2 - \frac{r}{2}(1-2m) > 0$) for $\eta = O(1)$ as $t \to 0$. However, when $\eta \gg 1$, $\bar{\alpha}_0(\eta) \sim O\left(\eta^{-(2+m(r+1))}e^{\frac{-m\eta^2}{4}}\right)$, via (10.68a) and $\bar{\beta}_0(\eta) \sim O\left(\eta^{-(r+1)}e^{\frac{-\eta^2}{4}}\right)$, via (10.68b), and so

$$R(\eta,t) \sim t^{2-\frac{r}{2}(1-2m)}\eta^{-4+(r+1)(1-2m)}e^{\frac{\eta^2(1-2m)}{4}}, \tag{10.81}$$

and $R(\eta,t)$ becomes of $O(1)$ for $\eta$ logorithmically large, confirming the onset of a strong nonuniformity in expansion (10.64b) before moving into region **III**. Further examination of (10.81) reveals that the non-uniformity occurs when

$$\eta^2 = \lambda_0(-\ln t) + \lambda_1 \ln[\lambda_2(-\ln t)] + \ldots \equiv C^2(t) \quad \text{as} \quad t \to 0, \tag{10.82}$$

where

$$\lambda_0 = \frac{4}{1-2m}\left[2 - \frac{r}{2}(1-2m)\right], \tag{10.83}$$

and $\lambda_1$, $\lambda_2$ are constants to be fixed via the asymptotic matching procedure. To continue the asymptotic structure, we now introduce region **II(b)** (with region **II** becoming region **II(a)**) where, via (10.82), $\eta \sim C(t)$ as $t \to 0$. To examine region **II(b)** we introduce the scaled coordinate $\bar{\eta} = [\eta - C(t)]\chi^{-1}(t)$, where $\bar{\eta} = O(1)$, $\chi(t) = o(1)$, as $t \to 0$ in region **II(b)**, and look for expansions of the form

$$\alpha(\bar{\eta},t) = 1 - \phi(t)F(\bar{\eta}) + o(\phi(t)), \tag{10.84a}$$
$$\beta(\bar{\eta},t) = \psi(t)G(\bar{\eta}) + o(\psi(t)), \tag{10.84b}$$

as $t \to 0$, where the gauge functions $\phi(t), \psi(t) = o(1)$ as $t \to 0$ are to be determined. On rewriting equations (10.2) in terms of the scaled variables $F$, $G$ and $\bar{\eta}$, we find that we require in this region, to obtain the most structured leading order balance, that $\chi(t) = O\left(C(t)^{-1}\right)$, $\phi(t) = O\left((tC(t)^{-2})^{\frac{1}{(1-2m)}}\right)$ and $\psi(t) = O\left((t\phi(t)C(t)^{-2})^{\frac{1}{(1-2m)}}\right)$. Thus, without loss of generality, we put

$$\chi(t) \equiv C(t)^{-1}, \quad \phi(t) \equiv \left(\frac{t}{C(t)^2}\right)^{\frac{1}{1-2m}} \quad \text{and} \quad \psi(t) \equiv \left(\frac{t}{C(t)^2}\right)^{\frac{2}{1-2m}} \tag{10.85}$$

At leading order we then obtain the equations

$$F_{\bar{\eta}\bar{\eta}} + \frac{1}{2}F_{\bar{\eta}} = -G^m, \tag{10.86a}$$

$$G_{\bar{\eta}\bar{\eta}} + \frac{1}{2}G_{\bar{\eta}} - FG^m = 0, \tag{10.86b}$$

with $\bar{\eta} > -\infty$. Matching the expansions (10.84) in region **II(b)** to the expansions (10.64) in region **II(a)** as $\bar{\eta} \to -\infty$ follows directly, giving

$$\lambda_1 = \frac{8}{1-2m} - 2(r+1), \qquad \lambda_2 = \lambda_0, \tag{10.87}$$

and the matching conditions

$$F(\bar{\eta}) \sim \frac{4C_\infty^m}{m(1-m)} e^{-m\bar{\eta}/2} \quad \text{as} \quad \bar{\eta} \to -\infty, \tag{10.88a}$$

$$G(\bar{\eta}) \sim C_\infty e^{-\bar{\eta}/2} \quad \text{as} \quad \bar{\eta} \to -\infty. \tag{10.88b}$$

In addition to (10.88), we require, to enable matching to region **III** when $x = \sigma + O(1)$, that

$$F(\bar{\eta}), \; G(\bar{\eta}) \quad \text{are bounded as} \quad \bar{\eta} \to \infty. \tag{10.89}$$

We must now address the boundary value problem (10.86), (10.88) and (10.89). We note that the nonlinearities in equations (10.86) are not Lipschitz continuous at $G = 0$ $(0 < m < 1)$. Following the arguments for a similar scalar boundary value problem (see Appendix B) we may establish that the solution to (10.86), (10.88) and (10.89) is classical but has finite support in $G(\bar{\eta})$, that is (due to the lack of Lipschitz continuity at $G = 0$),

$$G(\bar{\eta}) \begin{cases} > 0, & \bar{\eta} < \bar{\eta}_0, \\ \equiv 0, & \bar{\eta} \geq \bar{\eta}_0, \end{cases} \tag{10.90}$$

for some $\bar{\eta}_0$ which depends upon $C_\infty$ and $m$. Here $G(\bar{\eta})$ is monotone decreasing in $-\infty < \bar{\eta} \leq \bar{\eta}_0$ with

$$G(\bar{\eta}) \sim \left[ \frac{F_0(1-m)^2}{2(1+m)} \right]^{\frac{1}{1-m}} (\bar{\eta}_0 - \bar{\eta})^{\frac{2}{1-m}} + \dots \quad \text{as} \quad \bar{\eta} \to \bar{\eta}_0^-,$$

where $F_0 = F(\bar{\eta}_0)$. The solution for $F(\bar{\eta})$ may then be written in terms of $G(\bar{\eta})$ as

$$F(\bar{\eta}) = 2e^{-\bar{\eta}/2} \int_{-\infty}^{\bar{\eta}} G^m(s)e^{s/2}\,ds + 2\int_{\bar{\eta}}^{\bar{\eta}_0} G^m(s)\,ds, \quad -\infty < \bar{\eta} < \infty. \tag{10.91}$$

Now, via (10.90) and (10.91) we have

$$G(\bar{\eta}) \equiv 0 \quad \text{as} \quad \bar{\eta} \to \infty, \tag{10.92a}$$

$$F(\bar{\eta}) \sim \bar{A}e^{-\bar{\eta}/2} \quad \text{as} \quad \bar{\eta} \to \infty, \tag{10.92b}$$

with

$$\overline{A} = \int_{-\infty}^{\overline{\eta}_0} G^m(s) e^{s/2} \, ds \quad (> 0).$$

Calculation of higher order terms confirms that the support of $\beta(\overline{\eta}, t)$ remains finite in this region. We move out of region **II(b)** when $\overline{\eta} = O\left(C^2(t)\right)$ which then gives, on putting $\overline{\eta} = C^2(t)\hat{\eta}$,

$$\eta = C(t) \left[1 + \hat{\eta}\right]$$

which, in terms of $x$, may be written

$$x = \sigma + t^{\frac{1}{2}} C(t) + t^{\frac{1}{2}} C(t)\hat{\eta}.$$

We therefore introduce a further region, labelled region **II(c)** in which $\hat{\eta} = O(1)$ as $t \to 0$. From (10.92) and (10.84) we write

$$\alpha(\hat{\eta}, t) = 1 - \left(\frac{t}{C^2(t)}\right)^{\frac{1}{1-2m}} \exp\left\{-C^2(t) f(\hat{\eta}, t)\right\}, \qquad (10.93a)$$

$$\beta(\hat{\eta}, t) \equiv 0, \qquad (10.93b)$$

as $t \to 0$ with $\hat{\eta} = O(1)$, and expand

$$f(\hat{\eta}, t) = f_0(\hat{\eta}) + o(1)$$

as $t \to 0$. On substitution into equations (10.2), when written in terms of $\hat{\eta}$ and $t$, we obtain the leading order problem as

$$f_{0\hat{\eta}}^2 - f_{0\hat{\eta}} \frac{(\hat{\eta} + 1)}{2} = 0, \quad \hat{\eta} > 0, \qquad (10.94a)$$

$$f_0(\hat{\eta}) \sim \frac{\hat{\eta}}{2} \quad \text{as} \quad \hat{\eta} \to 0. \qquad (10.94b)$$

The solution to this boundary value problem is readily obtained as

$$f_0(\hat{\eta}) = \frac{\hat{\eta}}{2} + \frac{\hat{\eta}^2}{4}, \quad \hat{\eta} > 0. \qquad (10.95)$$

Finally, we move into region **III** when $\hat{\eta} = O\left(t^{-\frac{1}{2}} C^{-1}(t)\right)$. Thus, in region **III** we write

$$x = \sigma + t^{\frac{1}{2}} C(t) + y,$$

where $y = t^{\frac{1}{2}} C(t)\hat{\eta} = O(1)$ as $t \to 0$. On examining (10.93a) and (10.95) for $\hat{\eta} \gg 1$, we expand as

$$\alpha(y, t) = 1 - \left(\frac{t}{C^2(t)}\right)^{\frac{1}{1-2m}} \exp\left\{-\frac{\tilde{\phi}(y, t)}{t}\right\}, \qquad (10.96a)$$

$$\beta(y, t) \equiv 0, \qquad (10.96b)$$

as $t \to 0$ with $y = O(1)$, and

$$\tilde{\phi}(y,t) = \tilde{\phi}_0(y) + t^{\frac{1}{2}}C(t)\tilde{\phi}_1(y) + O\left(t^{\frac{1}{2}}C^{-1}(t)\right). \qquad (10.97)$$

On substitution of (10.96) and (10.97) into equations (10.2) and solving at each order in turn, we obtain, after matching with region **II(c)**,

$$\tilde{\phi}_0(y) = \frac{y^2}{4}, \quad \tilde{\phi}_1(y) = \frac{y}{2}.$$

Thus, finally, in region **III** we have

$$\alpha(y,t) = 1 - \left(\frac{t}{C^2(t)}\right)^{\frac{1}{1-2m}} \exp\left\{-\frac{y^2}{4t} - \frac{yC(t)}{2t^{\frac{1}{2}}} + \ldots\right\}, \qquad (10.98a)$$

$$\beta(y,t) \equiv 0, \qquad (10.98b)$$

as $t \to 0$ with $y = O(1)$. Expansions (10.98) now remain uniform as $y \to \infty$, and no further regions are required. This completes the asymptotic structure. A schematic representation of the location and thickness of the asymptotic regions as $t \to 0$ in this case is given in Figure 10.7.

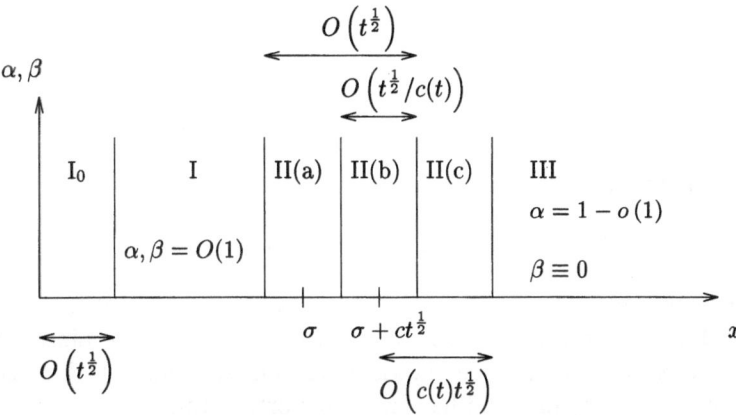

**Fig. 10.7.** Schematic representation of the location and thickness of the asymptotic regions as $t \to 0$ in the case when $m = n$, $k = 1$, $0 < m < \frac{1}{2}$ and $1 \le r < \frac{4}{1-2m}$.

We observe, in this case, that $\alpha_\infty(t) \equiv 1$ and $\beta_\infty(t) \equiv 0$, whilst the support of $\beta(x,t)$ remains finite in $t > 0$. The edge of the support of $\beta(x,t)$ is given, via (10.82), (10.83) and (10.87) as $x = s(t)$ where

$$s(t) \sim \sigma + t^{\frac{1}{2}}\left[\lambda_0^{\frac{1}{2}}(-\ln t)^{\frac{1}{2}} + \left\{\frac{\lambda_1}{2}\ln[\lambda_0(-\ln t)] + \bar{\eta}_0\right\}\lambda_0^{-\frac{1}{2}}(-\ln t)^{\frac{-1}{2}} + \ldots\right]$$

$$(10.99)$$

as $t \to 0$. We note that, in this case, the support of $\beta(x,t)$ expands initially.

**(iii)** $0 < m < \frac{1}{2}$; $r = \frac{4}{1-2m}$

Again, in this case expansion (10.62b) in region **I** becomes nonuniform first, when $x = \sigma - O\left(t^{\frac{1}{2}}\right)$ with, via (10.63), $\alpha = 1 - O\left(t^{\frac{1}{1-2m}}\right)$ and $\beta = O\left(t^{\frac{2}{1-2m}}\right)$ in region **II**. To examine region **II**, we introduce the scaled coordinate $\eta = (x - \sigma)t^{-\frac{1}{2}}$, and look for asymptotic expansions of the form

$$\alpha(\eta, t) = 1 - t^{\frac{1}{1-2m}} \bar{\alpha}_0(\eta) + o\left(t^{\frac{1}{1-2m}}\right), \tag{10.100a}$$

$$\beta(\eta, t) = t^{\frac{2}{1-2m}} \bar{\beta}_0(\eta) + o\left(t^{\frac{2}{1-2m}}\right), \tag{10.100b}$$

as $t \to 0$ with $\eta = O(1)$. On substitution of expansions (10.100) into equations (10.2) (when written in terms of $\eta$ and $t$) we obtain at leading order

$$\bar{\alpha}_{0\eta\eta} + \frac{\eta}{2}\bar{\alpha}_{0\eta} - \frac{\bar{\alpha}_0}{1-2m} = -\bar{\beta}_0^m, \tag{10.101a}$$

$$\bar{\beta}_{0\eta\eta} + \frac{\eta}{2}\bar{\beta}_{0\eta} - \frac{2\bar{\beta}_0}{1-2m} - \bar{\alpha}_0\bar{\beta}_0^m = 0, \tag{10.101b}$$

with $\eta > -\infty$. Equations (10.101) are to be solved subject to matching expansions (10.100) in region **II** with expansions (10.62) in region **I** as $\eta \to -\infty$, that is,

$$\bar{\alpha}_0(\eta) \sim (\beta_0 g_\sigma)^m (-\eta)^{\frac{4m}{1-2m}} \quad \text{as} \quad \eta \to -\infty, \tag{10.102a}$$
$$\bar{\beta}_0(\eta) \sim \beta_0 g_\sigma(-\eta)^{\frac{4}{1-2m}} \quad \text{as} \quad \eta \to -\infty, \tag{10.102b}$$

whilst matching to region **III** requires that

$$\bar{\alpha}_0(\eta), \ \bar{\beta}_0(\eta) \quad \text{remain bounded as} \quad \eta \to \infty. \tag{10.103}$$

We must now address the nonlinear boundary value problem (10.101), (10.102) and (10.103). Again the nonlinear terms in (10.101) fail to be Lipschitz continuous at $\bar{\beta}_0 = 0$. It may be shown, following Appendix C, that $\bar{\beta}_0(\eta)$ is monotone decreasing in $\eta$ and has finite support, so that

$$\bar{\beta}_0(\eta) \begin{cases} > 0, \ -\infty < \eta < \eta_0, \\ \equiv 0, \ \eta \geq \eta_0, \end{cases} \tag{10.104}$$

with $\eta_0$ depending upon $g_\sigma$ and $m$. In particular,

$$\bar{\beta}_0(\eta) \sim \left[\frac{a_0(1-m)^2}{2(1+m)}\right]^{\frac{1}{1-m}} (\eta_0 - \eta)^{\frac{2}{1-m}} \tag{10.105}$$

as $\eta \to \eta_0^-$. Here $a_0 = \bar{\alpha}_0(\eta_0) \ (> 0)$. In addition, it can be shown that $\bar{\alpha}_0(\eta)$ is monotone decreasing in $\eta$ and has

$$\bar{a}_0(\eta) \sim O\left(e^{-\frac{\eta^2}{4}}\right) \quad \text{as} \quad \eta \to \infty. \tag{10.106}$$

Consideration of higher order terms in region **II** confirms that the support of $\beta(\eta, t)$ remains finite. Thus we may now move into region **III** in which $x = \sigma + O(1)$ and

$$\alpha(x, t) = 1 - \exp\left\{-\frac{\phi(x, t)}{t}\right\}, \tag{10.107a}$$

$$\beta(x, t) \equiv 0, \tag{10.107b}$$

as $t \to 0$. We expand

$$\phi(x, t) = \phi_0(x) + t\left[\phi_1(x) + \phi_2(x)\ln t\right] + \dots \tag{10.108}$$

as $t \to 0$ with $x = \sigma + O(1)$. After substitution of (10.107) and (10.108) into equations (10.2) and solving at each order in turn we obtain

$$\phi_0(x) = \frac{(x + A)^2}{4}, \qquad \phi_1(x) = (\frac{1}{2} - B)\ln(x + A) + C, \qquad \phi_2(x) = B,$$

and so

$$\alpha(x, t) = 1 - \exp\left\{-\frac{(x + A)^2}{4t} - (\frac{1}{2} - B)\ln(x + A) - C - B\ln t + \dots\right\},$$
$$\tag{10.109a}$$

$$\beta(x, t) \equiv 0, \tag{10.109b}$$

as $t \to 0$ with $x = \sigma + O(1)$. Here $A$, $B$ and $C$ are arbitrary constants to be fixed on matching; however, for brevity, we do not pursue these details. Expansions (10.109) now remain uniform for $x \gg 1$, and the asymptotic structure is complete. A schematic representation of the location and thickness of the asymptotic regions as $t \to 0$ in this case is given in Figure 10.8. Again, in this case, we have determined (via (10.109)) that $\alpha_\infty(t) \equiv 1$ and $\beta_\infty(t) \equiv 0$ in the course of the analysis. Moreover, the support of $\beta(x, t)$ remains finite in $t > 0$, with the edge of the support being given by $x = s(t)$, where now

$$s(t) \sim \sigma + t^{\frac{1}{2}}\eta_0 + \dots \tag{10.110}$$

as $t \to 0$. The initial expansion or contraction of the support of $\beta(x, t)$ depends upon the sign of $\eta_0$, which we have been unable to determine explicitly.

**(iv) $0 < m < \frac{1}{2}$; $r > \frac{4}{1-2m}$**

Expansion (10.62b) in region **I** now becomes nonuniform when $x = \sigma - O\left(t^{\frac{2}{r(1-2m)}}\right)$ with, via (10.63), $\alpha = 1 - O\left(t^{\frac{1}{1-2m}}\right)$ and $\beta = O\left(t^{\frac{2}{1-2m}}\right)$ in region **II**. To examine region **II**, we introduce the scaled coordinate $\hat{\eta} = (x - \sigma)t^{-\frac{2}{r(1-2m)}}$, and look for asymptotic expansions of the form

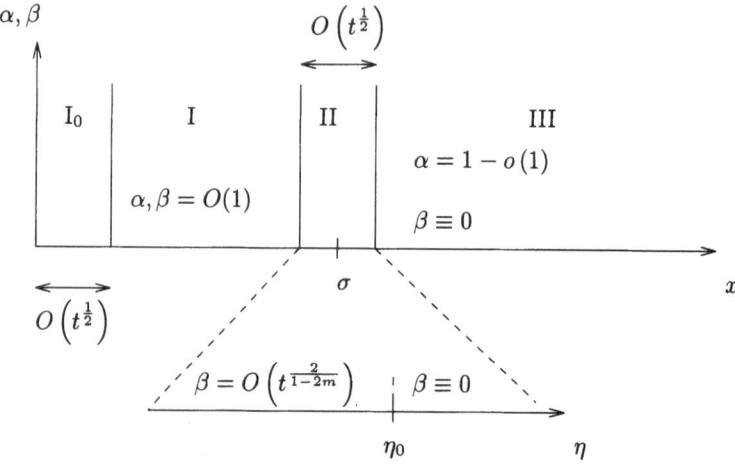

**Fig. 10.8.** Schematic representation of the location and thickness of the asymptotic regions as $t \to 0$ in the case when $m = n, k = 1, 0 < m < \frac{1}{2}$ and $r = \frac{4}{1-2m}$. Note that $\alpha = 1 - O\left(t^{1/(1-2m)}\right)$ in region II.

$$\alpha(\hat{\eta}, t) = 1 - t^{\frac{1}{1-2m}} \hat{\alpha}_0(\hat{\eta}) + o\left(t^{\frac{1}{1-2m}}\right), \tag{10.111a}$$

$$\beta(\hat{\eta}, t) = t^{\frac{2}{1-2m}} \hat{\beta}_0(\hat{\eta}) + o\left(t^{\frac{2}{1-2m}}\right), \tag{10.111b}$$

as $t \to 0$ with $\hat{\eta} = O(1)$. On substitution of expansions (10.111) into equations (10.2) (when written in terms of $\hat{\eta}$ and $t$) we obtain the coupled leading order problems for $\hat{\alpha}_0(\hat{\eta})$ and $\hat{\beta}_0(\hat{\eta})$ as

$$\hat{\eta} \hat{\alpha}_{0\hat{\eta}} - \frac{r}{2}\hat{\alpha}_0 = -\frac{r}{2}(1 - 2m) \hat{\beta}_0^m, \quad \hat{\eta} > -\infty, \tag{10.112a}$$

$$\hat{\alpha}_0(\hat{\eta}) \sim (\beta_0 g_\sigma)^m (-\hat{\eta})^{rm} \quad \text{as} \quad \hat{\eta} \to -\infty \tag{10.112b}$$

and

$$\hat{\eta} \hat{\beta}_{0\hat{\eta}} - r\hat{\beta}_0 - \frac{r}{2}(1 - 2m) \hat{\alpha}_0 \hat{\beta}_0^m = 0, \quad \hat{\eta} > -\infty, \tag{10.113a}$$

$$\hat{\beta}_0(\hat{\eta}) \sim \beta_0 g_\sigma (-\hat{\eta})^r \quad \text{as} \quad \hat{\eta} \to -\infty. \tag{10.113b}$$

Conditions (10.112b) and (10.113b) arise from matching with region **I** as $\hat{\eta} \to -\infty$. On defining new variables

$$F(s) = \hat{\alpha}_0(\hat{\eta}), \quad G(s) = \hat{\beta}_0(\hat{\eta}), \quad \hat{\eta} = -e^{-s}, \tag{10.114}$$

equations (10.112a) and (10.113a) are reduced to the second-order autonomous system

$$F_s = \frac{r}{2}(1 - 2m) G^m - \frac{r}{2}F, \tag{10.115a}$$

$$G_s = -\frac{r}{2}(1 - 2m) FG^m - rG, \tag{10.115b}$$

where $-\infty < s < \infty$. We require a solution to equations (10.115) (with $F, G \geq 0$ for $-\infty < s < \infty$) subject to the boundary conditions (10.112b) and (10.113b), which become

$$F(s) \sim (\beta_0 g_\sigma)^m e^{-rms}, \quad G(s) \sim \beta_0 g_\sigma e^{-rs} \quad \text{as} \quad s \to -\infty. \quad (10.116)$$

The system of equations (10.115) has just one finite equilibrium point at $(0,0)$ in the positive quadrant of the $(F,G)$ phase plane (this quadrant being a positive invariant region for (10.115)). We note that this is not hyperbolic (which is a consequence of the non-Lipschitz nature of the problem at the origin) and can not be classified via linearization. However, we observe from equations (10.115) that

$$\frac{F^2(s)}{2} + G(s) = \hat{k} e^{-rs}, \quad (10.117)$$

where $\hat{k}$ is a positive constant, and further that $V(F,G) = \frac{F^2}{2} + G$ provides a strong Liapunov function for the equilibrium point $(0,0)$ on $(F,G) \in (-\infty, \infty) \times [0, \infty)$ (since $V(0,0) = 0$, $V(F,G) > 0$ and $V_s(F,G) = -rke^{-rs} < 0$ for all $(F,G) \in (-\infty, \infty) \times [0, \infty) \setminus (0,0)$) from which it follows that the origin is asymptotically stable in this region. An integral path, $\ell_0$, which satisfies conditions (10.116) must originate in the region bounded by the vertical and horizontal isoclines in the positive quadrant of the $(F,G)$ phase plane given by $G = \left(\frac{F}{1-2m}\right)^{\frac{1}{m}}$ and $G = 0$ respectively, and approach the origin as $s \to \infty$. On this phase path (10.117) holds with $\hat{k} = \beta_0 g_\sigma$. The integral path $\ell_0$ reaches the $F$ axis (tangentially) in finite s, say as $s \to s_0^-$. Thereafter, $\ell_0$ proceeds along the $F$ axis, aproaching the origin as $s \to \infty$. Thus the solution $(F(s), G(s))$ has finite support in $G(s)$. We readily conclude that $G(s)$ is monotone decreasing in $s$, with

$$G(s) \begin{cases} > 0, & -\infty < s < s_0, \\ \equiv 0, & s \geq s_0, \end{cases}$$

with $s_0$ depending upon $\beta_0, g_\sigma, r$ and $m$. In particular,

$$G(s) \begin{cases} \sim \left[\frac{\sqrt{2\beta_0 g_\sigma}}{2} r(1-m)(1-2m)\right]^{\frac{1}{1-m}} e^{-\frac{r}{2(1-m)}s}(s_0 - s)^{\frac{1}{1-m}}, & \text{as } s \to s_0^-, \\ \equiv 0, & \text{for } s \geq s_0. \end{cases}$$
$$(10.118)$$

Correspondingly, via (10.117) we have

$$F(s) = \sqrt{2} \left[\beta_0 g_\sigma e^{-rs} - G(s)\right]^{\frac{1}{2}}, \quad -\infty < s < \infty, \quad (10.119)$$

so that $F(s)$ is monotone decreasing in $s$, with

$$F(s) = \sqrt{2\beta_0 g_\sigma} \, e^{-\frac{rs}{2}}, \quad \text{for} \quad s \geq s_0. \quad (10.120)$$

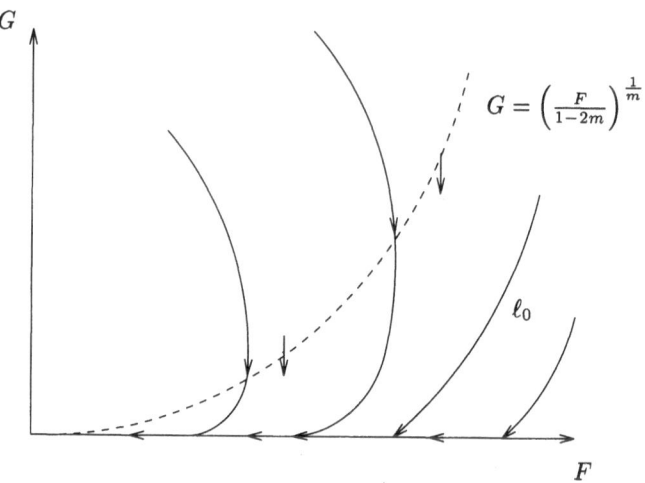

$$G = \left(\frac{F}{1-2m}\right)^{\frac{1}{m}}$$

$\ell_0$

$F$

**Fig. 10.9.** A sketch of the phase portrait for the system of equations (10.115).

The phase portrait of the system of equations (10.115) is sketched in Figure 10.9. On returning to the original variables, we have that

$$\hat{\alpha}_0(\hat{\eta}) \sim \sqrt{2\hat{k}}\,(-\hat{\eta})^{\frac{r}{2}}, \tag{10.121a}$$

$$\hat{\beta}_0(\hat{\eta}) \sim \left[\frac{\sqrt{2\hat{k}}}{2}r(1-m)(1-2m)(-\hat{\eta}_0)^{\frac{r}{2}}\right]^{\frac{1}{1-m}}\left[\ln\left(\frac{\hat{\eta}}{\hat{\eta}_0}\right)\right]^{\frac{1}{1-m}} \tag{10.121b}$$

as $\hat{\eta} \to \hat{\eta}_0^-$, where $\hat{\eta}_0 = -e^{-s_0} < 0$. The support of $\beta(\hat{\eta}, t)$ ends at $\hat{\eta} = \hat{\eta}_0$ in this region. However, since $\ln\left(\frac{\hat{\eta}}{\hat{\eta}_0}\right) = \frac{\hat{\eta}_0 - \hat{\eta}}{(-\hat{\eta}_0)} + O\left((\hat{\eta}_0 - \hat{\eta})^2\right)$ as $\hat{\eta} \to \hat{\eta}_0^-$, we note that the degree of $(\hat{\eta}_0 - \hat{\eta})$ in (10.121b) as $\hat{\eta} \to \hat{\eta}_0^-$ (which is $\frac{1}{1-m}$), is too weak, and consideration of further terms in expansion (10.111b) reveals a weak nonuniformity as $\hat{\eta} \to \hat{\eta}_0^-$. Therefore a further region is required, in which $\hat{\eta} = \hat{\eta}_0 + o(1)$ as $t \to 0$, and diffusion effects are retained at leading order to enable the appropriate behaviour to be achieved at the edge of the support of $\beta$. We label this region, region **II(b)**, with the previous region now being referred to as **II(a)**. We introduce the scaled coordinate $\tilde{\eta}$ by

$$\hat{\eta} = \hat{\eta}_0 + t^{\gamma}\tilde{\eta}, \tag{10.122}$$

with $\gamma > 0$ to be determined, and $\tilde{\eta} = O(1)$ as $t \to 0$ in region **II(b)**. An examination of (10.112a), (10.113a) and (10.121) then determines that $\alpha = 1 - O\left(t^{\frac{1}{1-2m}}\right)$ and $\beta = O\left(t^{\frac{2}{1-2m}+\frac{7}{1-m}}\right)$ in region **II(b)**. Thus we expand $\alpha$ and $\beta$ as

$$\alpha(\tilde{\eta}, t) = 1 - t^{\frac{1}{1-2m}}\bar{\alpha}_0(\tilde{\eta}) + o\left(t^{\frac{1}{1-2m}}\right), \tag{10.123a}$$

$$\beta(\tilde{\eta}, t) = t^{\frac{2}{1-2m}+\frac{7}{1-m}}\bar{\beta}_0(\tilde{\eta}) + o\left(t^{\frac{2}{1-2m}+\frac{7}{1-m}}\right), \tag{10.123b}$$

as $t \to 0$, with $\tilde{\eta} = O(1)$. On substituting expansions (10.123) into equations (10.2) (when written in terms of $\tilde{\eta}$ and $t$), to retain diffusion terms at leading order requires that

$$\gamma = 1 - \frac{4}{r(1-2m)} > 0, \tag{10.124}$$

after which we note that the leading order problem for $\bar{\alpha}_0(\tilde{\eta})$ in region **II(b)** (which is decoupled from $\bar{\beta}_0(\tilde{\eta})$) is given by

$$\bar{\alpha}_{0\tilde{\eta}\tilde{\eta}} + \frac{2\hat{\eta}_0}{r(1-2m)}\bar{\alpha}_{0\tilde{\eta}} = 0, \quad \tilde{\eta} > -\infty, \tag{10.125}$$

which is to be solved subject to matching with region **II(a)** as $\tilde{\eta} \to -\infty$, that is,

$$\bar{\alpha}_0(\tilde{\eta}) \sim \sqrt{2\hat{k}}\,(-\hat{\eta}_0)^{\frac{r}{2}} \quad \text{as} \quad \tilde{\eta} \to -\infty. \tag{10.126}$$

The solution of (10.125), (10.126) is

$$\bar{\alpha}_0(\tilde{\eta}) = \sqrt{2\hat{k}}\,(-\hat{\eta}_0)^{\frac{r}{2}}, \quad -\infty < \tilde{\eta} < \infty. \tag{10.127}$$

The leading order problem for $\bar{\beta}_0(\tilde{\eta})$ in region **II(b)** is then given by

$$\bar{\beta}_{0\tilde{\eta}\tilde{\eta}} + \frac{2\hat{\eta}_0}{r(1-2m)}\bar{\beta}_{0\tilde{\eta}} - \sqrt{2\hat{k}}(-\hat{\eta}_0)^{\frac{r}{2}}\bar{\beta}_0^m = 0, \quad -\infty < \tilde{\eta} < \infty, \tag{10.128a}$$

$$\bar{\beta}_0(\tilde{\eta}) \sim \left[\frac{\sqrt{2\hat{k}}}{2}r(1-m)(1-2m)(-\hat{\eta}_0)^{\frac{r}{2}-1}\right]^{\frac{1}{1-m}}(-\tilde{\eta})^{\frac{1}{1-m}} \quad \text{as} \quad \tilde{\eta} \to -\infty, \tag{10.128b}$$

$$\bar{\beta}_0(\tilde{\eta}) \geq 0 \quad \text{for all} \quad -\infty < \tilde{\eta} < \infty, \tag{10.128c}$$

$$\bar{\beta}_0(\tilde{\eta}) \quad \text{bounded as} \quad \tilde{\eta} \to \infty. \tag{10.128d}$$

The solution to the boundary value problem (10.128) has finite support in $\tilde{\eta}$ (see Appendix D). With $\tilde{\eta}_0$ being the edge of the support of $\bar{\beta}_0(\tilde{\eta})$, we have (via Appendix D) that the solution to (10.128) is monotone decreasing in $-\infty < \tilde{\eta} < \tilde{\eta}_0$, and further, from (10.128a) that

$$\bar{\beta}_0(\tilde{\eta}) \begin{cases} \sim \left[\frac{\sqrt{2\hat{k}}(-\hat{\eta}_0)^{\frac{r}{2}}(1-m)^2}{2(1+m)}\right]^{\frac{1}{1-m}}(\tilde{\eta}_0 - \tilde{\eta})^{\frac{2}{1-m}} \quad \text{as} \quad \tilde{\eta} \to \tilde{\eta}_0^-, \\ \equiv 0, \qquad\qquad\qquad\qquad\qquad\qquad\qquad \text{for} \quad \tilde{\eta} \geq \tilde{\eta}_0, \end{cases} \tag{10.129}$$

which has the required decay rate in $(\tilde{\eta}_0 - \tilde{\eta})$ as the edge of the support is approached. Consideration of further terms in this region now shows that expansion (10.123b) remains uniform as $\tilde{\eta} \to \tilde{\eta}_0^-$. In particular, we observe that the edge of the support of $\beta(x, t)$, $x = s(t)$, is given by

$$s(t) \sim \sigma + t^{\frac{2}{r(1-2m)}} \hat{\eta}_0 + t^{\frac{r(1-2m)-2}{r(1-2m)}} \tilde{\eta}_0 + \dots \tag{10.130}$$

as $t \to 0$. We note from (10.130) that, since $\hat{\eta}_0 < 0$, the edge of the support is contracting initially.

As noted above, the asymptotic structure of $\alpha(x,t)$ as $t \to 0$ does not end in this region and we are left to introduce the final region, region **III**, where $x = s(t) + O(1)$, $\beta(x,t) \equiv 0$ and $\alpha(x,t) = 1 - o(1)$. To examine region **III** we introduce the scaled coordinate $y = x - s(t)$ (where $s(t)$ is given by (10.130)) and look for an expansion of the form

$$\alpha(y,t) = 1 - \Phi(y,t) \tag{10.131}$$

as $t \to 0$ where $\Phi(y,t) = o(1)$ and $y = O(1)$. On substitution of (10.131) into equation (10.2a) (when written in terms of $y$ and $t$ with $\beta \equiv 0$) we obtain

$$\Phi_t = \Phi_{yy} + \dot{s}(t)\Phi_y, \tag{10.132}$$

with $t \ll 1$ and $0 < y < \infty$. Moreover, the structure of expansion (10.123a) as $\tilde{\eta} \to \infty$ (using (10.127)) suggests that we write in region **III**

$$\Phi(y,t) = t^{\frac{1}{1-2m}} \tilde{F}(y,t) \tag{10.133}$$

as $t \to 0$, $y = O(1)$. On substitution from (10.133) into (10.132) we obtain a WKB type equation for $\tilde{F}(y,t)$, and so we write

$$\tilde{F}(y,t) = \sqrt{2\hat{k}} \, (-\hat{\eta}_0)^{\frac{r}{2}} e^{-\frac{\tilde{\phi}(y,t)}{t}}, \tag{10.134}$$

as $t \to 0$ where $\tilde{\phi} = O(1)$, $\tilde{\phi}(y,t) > 0$ for all $y > 0$. Substitution of (10.134), via (10.133), into (10.132) leads to an expansion of $\tilde{\phi}(y,t)$ of the form

$$\tilde{\phi}(y,t) = \tilde{\phi}_0(y) + t^{\frac{2}{r(1-2m)}} \tilde{\phi}_1(y) + O\left(\tilde{C}(t)\right), \tag{10.135}$$

where

$$\tilde{C}(t) = \begin{cases} t^{\frac{4}{r(1-2m)}}, & r \geq \frac{6}{1-2m}, \\ t^{1-\frac{2}{r(1-2m)}}, & \frac{4}{(1-2m)} < r < \frac{6}{1-2m}, \end{cases}$$

as $t \to 0$, with $y = O(1)$. On substitution of (10.134) and (10.135) into equation (10.132) and solving at each order in turn we obtain, via (10.131), that

$$\alpha(y,t) = 1 - t^{\frac{1}{1-2m}} \sqrt{2\hat{k}} \, (-\hat{\eta}_0)^{\frac{r}{2}} \exp\left\{ -\frac{y^2}{4t} - \frac{y\hat{\eta}_0}{2} t^{\frac{2}{r(1-2m)}-1} + o\left(t^{\frac{2}{r(1-2m)}-1}\right) \right\} \tag{10.136}$$

as $t \to 0$ with $y = O(1)$. Clearly (10.136) remains uniform for $y \gg 1$ and no further far field regions are required. However, we note that expansion (10.136) develops a weak nonuniformity as $y \to 0^+$ (when $y = O\left(t^{\frac{2}{r(1-2m)}}\right)$)

and we require an inner region close to $y = 0$. We label this inner region **III(a)** and relabel region **III** as **III(b)**. The structure of expansion (10.136) when $y = O\left(t^{\frac{2}{r(1-2m)}}\right)$ suggests that in region **III(a)** we introduce the scaled coordinate $z = yt^{-\frac{2}{r(1-2m)}}$, and expand as

$$\alpha(z,t) \sim 1 - t^{\frac{1}{1-2m}} \sqrt{2\hat{k}} \, (-\hat{\eta}_0)^{\frac{1}{2}} \exp\left\{ -f(z) \, t^{\frac{4}{r(1-2m)}-1} \right\} \tag{10.137}$$

as $t \to 0$ with $0 < z < \infty$. Substitution of (10.137) into (10.2a) (with $\beta \equiv 0$ and when written in terms of $z$ and $t$), gives at leading order

$$f_z^2 - \frac{2(z+\hat{\eta}_0)}{r(1-2m)} f_z + \left( \frac{4}{r(1-2m)} - 1 \right) f = 0, \tag{10.138}$$

with $0 < z < \infty$ in region **III(a)**. The monotone increasing solution of (10.138) which allows matching to region **III(b)** as $z \to \infty$, is given by

$$f(z) \begin{cases} = \frac{(z+\hat{\eta}_0)^2}{4}, & -\hat{\eta}_0 < z < \infty, \\ \equiv 0, & 0 \le z \le -\hat{\eta}_0. \end{cases} \tag{10.139}$$

Therefore, via (10.137) and (10.139), the expansion for $\alpha$ in region **III(a)** is given by

$$\alpha(z,t) \sim \begin{cases} 1 - t^{\frac{1}{1-2m}} \sqrt{2\hat{k}} \, (-\hat{\eta}_0)^{\frac{1}{2}} \exp\left\{ -\frac{(z+\hat{\eta}_0)^2}{4} t^{\frac{4}{r(1-2m)}-1} \right\}, & -\hat{\eta}_0 < z < \infty, \\ 1 - t^{\frac{1}{1-2m}} \sqrt{2\hat{k}} \, (-\hat{\eta}_0)^{\frac{1}{2}}, & 0 \le z \le -\hat{\eta}_0, \end{cases} \tag{10.140}$$

as $t \to 0$ with $0 \le z < \infty$. Finally, we note that expansion (10.140) of region **III(a)** (as $z \to 0$) matches classicaly to expansion (10.123a) of region **II(b)** (as $\tilde{\eta} \to \infty$). This completes the small-$t$ asymptotic structure in this case. Note that expansions (10.62) in region **I** do not, in general, satisfy the boundary conditions (10.5) at $x = 0$ and a further passive region is required in the neighbourhood of $x = 0$ as $t \to 0$. The details of this region follow, after minor modifications, those given in region $\mathbf{I}_0$ of Section 10.2.1. A schematic representation of the location and thickness of the asymptotic regions as $t \to 0$ in this case is given in Figure 10.10, with a blow out of region **II** being given in Figure 10.11.

We note that again in this case the asymptotic structure has determined $\alpha_\infty(t) \equiv 1$ and $\beta_\infty(t) \equiv 0$, with the support of of $\beta(x,t)$ remaining finite in $t > 0$.

## (c) $k > 1$

We begin in region **I** where $0 \le x < \sigma - o(1)$ and $\alpha, \beta = O(1)$ as $t \to 0$ and, since $\alpha(x,0), \beta(x,0) > 0$ are analytic in region **I**, we expand $\alpha(x,t)$ and

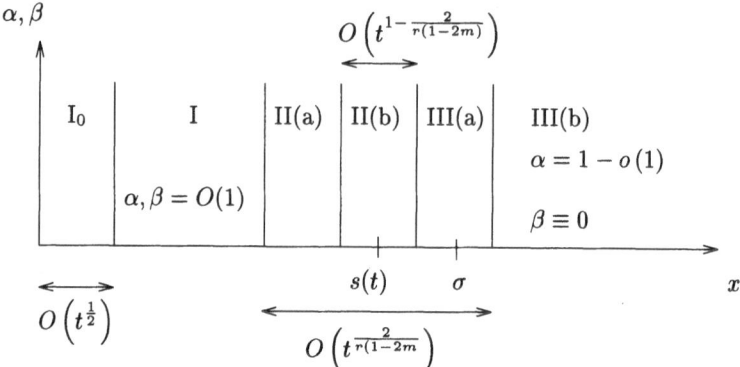

**Fig. 10.10.** Schematic representation of the location and thickness of the asymptotic regions as $t \to 0$ in the case when $m = n$, $k = 1$, $0 < m < \frac{1}{2}$ and $r > \frac{4}{1-2m}$.

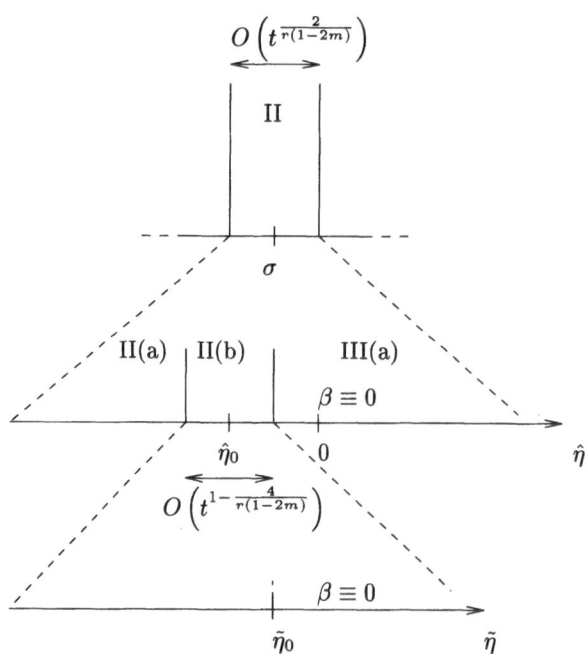

**Fig. 10.11.** A blow out of region II.

$\beta(x,t)$ as regular power series in $t$. After substitution into equation (10.2), equating powers of $t$ to zero, and applying initial condition (10.3), we obtain

$$\alpha(x,t) = 1 - t\left\{(\beta_0 g(x))^m\right\} + O\left(t^2\right), \tag{10.141a}$$

$$\beta(x,t) = \beta_0 g(x) + t\left\{\beta_0 g''(x) - (k-1)(\beta_0 g(x))^m\right\} + O\left(t^2\right), \tag{10.141b}$$

as $t \to 0$ with $0 \le x < \sigma - o(1)$. Now when $(\sigma - x) \ll 1$, expansions (10.141) become

$$\alpha(x,t) = 1 - t\left\{(\beta_0 g_\sigma)^m (\sigma - x)^{rm} + \ldots\right\} + O\left(t^2\right), \tag{10.142a}$$

$$\beta(x,t) = \left\{\beta_0 g_\sigma (\sigma - x)^r + \ldots\right\} + t\left\{\beta_0 g_\sigma r(r-1)(\sigma - x)^{r-2}\right.$$
$$\left. - (k-1)(\beta_0 g_\sigma)^m (\sigma - x)^{rm} + \ldots\right\} + O\left(t^2\right), \tag{10.142b}$$

as $t \to 0$, and a nonuniformity develops first in expansion (10.142b). There are three cases to consider.

## (i) $1 \le r < \frac{2}{1-m}$

In this case, expansion (10.141b) in region **I** becomes nonuniform when $x = \sigma - O\left(t^{\frac{1}{2}}\right)$, when we observe, via (10.142), that $\alpha = 1 - O\left(t^{1+\frac{rm}{2}}\right)$ and $\beta = O\left(t^{\frac{r}{2}}\right)$. Hence, in region **II** we introduce the scaled coordinate $\eta = (x-\sigma)t^{-\frac{1}{2}}$ and look for asymptotic expansions of the form

$$\alpha(\eta,t) = 1 - t^{1+\frac{rm}{2}}\,\bar\alpha_0(\eta) + o\left(t^{1+\frac{rm}{2}}\right), \tag{10.143a}$$

$$\beta(\eta,t) = t^{\frac{r}{2}}\,\bar\beta_0(\eta) + o\left(t^{\frac{r}{2}}\right) \tag{10.143b}$$

as $t \to 0$ with $\eta = O(1)$. On substitution of expansions (10.143) into equations (10.2) (when written in terms of $\eta$ and $t$) we obtain at leading order

$$\bar\alpha_{0\eta\eta} + \frac{\eta}{2}\bar\alpha_{0\eta} - R\bar\alpha_0 = -\bar\beta_0^m, \tag{10.144a}$$

$$\bar\beta_{0\eta\eta} + \frac{\eta}{2}\bar\beta_{0\eta} - \frac{r}{2}\bar\beta_0 = 0, \tag{10.144b}$$

where $-\infty < \eta < \infty$ and $R = 1 + \frac{rm}{2}$. Equations (10.144) are to be solved subject to matching with region **I** as $\eta \to -\infty$, and region **III** as $\eta \to \infty$, that is,

$$\bar\alpha_0(\eta) \sim \beta_0^m g_\sigma^m (-\eta)^{rm} \quad \text{as} \quad \eta \to -\infty, \tag{10.145a}$$

$$\bar\beta_0(\eta) \sim \beta_0 g_\sigma (-\eta)^r \quad \text{as} \quad \eta \to -\infty \tag{10.145b}$$

and

$$\bar\alpha_0(\eta),\ \bar\beta_0(\eta) \quad \text{remain bounded as} \quad \eta \to \infty. \tag{10.146}$$

The boundary value problems (10.144a), (10.145a) and (10.146) and (10.144b), (10.145b) and (10.146) have unique, monotone decreasing solutions, which are studied in Section 10.2.1 and are given by (10.26) and (10.27). In particular, for $\eta \gg 1$ we observe that, via (10.30), as we move into region **III**,

$$\alpha(\eta, t) \sim 1 \, - \, t^{1 + \frac{rm}{2}} \frac{4 \, C_\infty^m}{m(1-m)} \, \eta^{-(2+m(r+1))} \, e^{-m\eta^2/4} + \dots \text{(10.147a)}$$

$$\beta(\eta, t) \sim t^{\frac{r}{2}} \, C_\infty \, \eta^{-(r+1)} \, e^{-\eta^2/4} + \dots, \tag{10.147b}$$

as $t \to 0$, with $C_\infty$ as defined in (10.29). However, an examination of higher order terms in expansions (10.143) of region **II** indicates that a non–uniformity will develop in expansion (10.143b) for $\eta \gg 1$ and before we reach region **III**. We therefore anticipate that the structure will contain a further asymptotic region. The existence of this additional region may be confirmed in the following way: a typical term retained from equation (10.2b) at leading order in region **II** is $t^{\frac{r}{2}-1}\bar{\beta}_0$, whereas a typical neglected term is $t^{\frac{rm}{2}}\bar{\beta}_0^m$. The ratio of neglected to retained terms is then

$$R(\eta, t) \;=\; t^{1 - \frac{r}{2}(1-m)} \bar{\beta}_0^{m-1}(\eta), \tag{10.148}$$

which is of $O\left(t^{1-\frac{r}{2}(1-m)}\right)$ $(R = o(1)$ since $r < \frac{2}{1-m})$ for $\eta = O(1)$ as $t \to 0$. However, when $\eta \gg 1$, $\bar{\beta}_0(\eta) \sim O\left(\eta^{-(r+1)}e^{-\eta^2/4}\right)$, via (10.147b), and so

$$R(\eta, t) \;\sim\; t^{1 - \frac{r}{2}(1-m)} \eta^{(r+1)(1-m)} e^{\eta^2(1-m)/4} + \dots \tag{10.149}$$

as $t \to 0$ and $R(\eta, t)$ becomes of $O(1)$ for $\eta$ sufficiently large, confirming the onset of a nonuniformity in expansion (10.143b) as $\eta \to \infty$. Further examination of (10.149) reveals that the nonuniformity occurs when

$$\eta^2 \;=\; \lambda_0(-\ln t) \, + \, \lambda_1 \ln\left[\lambda_2(-\ln t)\right] + \dots \; \equiv \; C^2(t) \tag{10.150}$$

as $t \to 0$, where

$$\lambda_0 = \frac{4}{1-m}\left[1 - \frac{r}{2}(1-m)\right], \tag{10.151}$$

and $\lambda_1$, $\lambda_2$ are constants to be fixed via the asymptotic matching procedure. To continue the asymptotic structure, we now introduce region **II(b)** (with region **II** becoming region **II(a)**), where, via (10.150), $\eta \sim C(t)$ as $t \to 0$. To examine region **II(b)** we introduce the scaled coordinate $\bar{\eta} = [\eta - C(t)] \chi^{-1}(t)$, where $\bar{\eta} = O(1)$, $\chi(t) = o(1)$, as $t \to 0$ in region **II(b)**, and look for expansions of the form

$$\alpha(\bar{\eta}, t) = 1 - \phi(t)F(\bar{\eta}) + o\left(\phi(t)\right), \tag{10.152a}$$

$$\beta(\bar{\eta}, t) = \psi(t)G(\bar{\eta}) + o\left(\psi(t)\right), \tag{10.152b}$$

as $t \to 0$, where the gauge functions $\phi(t), \psi(t) = o(1)$ as $t \to 0$, are to be determined. On rewriting equations (10.2) in terms of the scaled variables $F$, $G$ and $\bar{\eta}$, we find that in this region the most structured balance

at leading order requires $\chi(t) = O\left(C^{-1}(t)\right)$, $\phi(t) = O\left((tC^{-2}(t))^{\frac{1}{1-m}}\right)$ and
$\psi(t) = O\left((tC^{-2}(t))^{\frac{1}{1-m}}\right)$. Thus, without loss of generality, we put

$$
\chi(t) \equiv C^{-1}(t), \quad \phi(t) \equiv \left(\frac{t}{C^2(t)}\right)^{\frac{1}{1-m}} \quad \text{and} \quad \psi(t) \equiv \left(\frac{t}{C^2(t)}\right)^{\frac{1}{1-m}}.
$$
$$(10.153)$$

At leading order we then obtain the equations

$$
F_{\bar{\eta}\bar{\eta}} + \frac{1}{2}F_{\bar{\eta}} = -G^m, \tag{10.154a}
$$

$$
G_{\bar{\eta}\bar{\eta}} + \frac{1}{2}G_{\bar{\eta}} - (k-1)G^m = 0, \tag{10.154b}
$$

with $\bar{\eta} > -\infty$. Matching the expansions (10.152) in region **II(b)** to the expansions (10.143) in region **II(a)** as $\bar{\eta} \to -\infty$ follows directly giving

$$
\lambda_1 = \frac{4}{1-m} - 2(r+1), \qquad \lambda_2 = \lambda_0, \tag{10.155}
$$

and the matching conditions

$$
F(\bar{\eta}) \sim \frac{4C_\infty^m}{m(1-m)} e^{-\frac{m\bar{\eta}}{2}} \quad \text{as} \quad \bar{\eta} \to -\infty,
$$

$$
G(\bar{\eta}) \sim C_\infty e^{-\frac{\bar{\eta}}{2}} \quad \text{as} \quad \bar{\eta} \to -\infty.
$$

Moreover, to match with region **III** we require

$$
F(\bar{\eta}), \; G(\bar{\eta}) \quad \text{remain bounded as} \quad \bar{\eta} \to \infty. \tag{10.156}
$$

Therefore at leading order in region **III** we obtain the uncoupled boundary value problems

$$
G_{\bar{\eta}\bar{\eta}} + \frac{1}{2}G_{\bar{\eta}} - (k-1)G^m = 0, \quad -\infty < \bar{\eta} < \infty, \tag{10.157a}
$$
$$
G(\bar{\eta}) > 0, \quad -\infty < \bar{\eta} < \infty, \tag{10.157b}
$$
$$
G(\bar{\eta}) \sim C_\infty e^{-\frac{\bar{\eta}}{2}}, \quad \text{as} \quad \bar{\eta} \to -\infty, \tag{10.157c}
$$
$$
G(\bar{\eta}) \quad \text{bounded as} \quad \bar{\eta} \to \infty, \tag{10.157d}
$$

and

$$
F_{\bar{\eta}\bar{\eta}} + \frac{1}{2}F_{\bar{\eta}} = -G^m, \quad -\infty < \bar{\eta} < \infty, \tag{10.158a}
$$
$$
F(\bar{\eta}) > 0, \quad -\infty < \bar{\eta} < \infty, \tag{10.158b}
$$
$$
F(\bar{\eta}) \sim \frac{4C_\infty^m}{m(1-m)} e^{-\frac{m\bar{\eta}}{2}} \quad \text{as} \quad \bar{\eta} \to -\infty, \tag{10.158c}
$$
$$
F(\bar{\eta}) \quad \text{bounded as} \quad \bar{\eta} \to \infty. \tag{10.158d}
$$

The uncoupled boundary value problem (10.157) is studied in Appendix B. The solution to this has $G(\bar{\eta})$ with finite support, so that

$$G(\bar{\eta}) \begin{cases} > 0, & -\infty < \bar{\eta} < \bar{\eta}_0, \\ \equiv 0, & \bar{\eta} \geq \bar{\eta}_0, \end{cases}$$

with $\bar{\eta}_0$ being the edge of the support (and depending upon $C_\infty$ and $k$) and $G(\bar{\eta})$ being monotone decreasing in $\bar{\eta}$. The corresponding solution to the boundary value problem (10.158) is then obtained as

$$F(\bar{\eta}) = 2e^{-\bar{\eta}/2} \int_{-\infty}^{\bar{\eta}} G^m(s)e^{s/2}\,\mathrm{d}s + 2 \int_{\bar{\eta}}^{\bar{\eta}_0} G^m(s)\,\mathrm{d}s, \quad -\infty < \bar{\eta} < \infty.$$
$$(10.159)$$

In addition we have

$$G(\bar{\eta}) \sim \left[ \frac{(1-m)^2(k-1)}{2(1+m)} \right]^{\frac{1}{1-m}} (\bar{\eta}_0 - \bar{\eta})^{\frac{2}{1-m}} \quad \text{as } \bar{\eta} \to \bar{\eta}_0^- \quad (10.160)$$

and

$$F(\bar{\eta}) = \overline{A}e^{-\frac{\bar{\eta}}{2}}, \quad \bar{\eta} \geq \bar{\eta}_0, \quad (10.161)$$

with

$$\overline{A} = \int_{-\infty}^{\bar{\eta}_0} G^m(s)e^{s/2}\,\mathrm{d}s.$$

A consideration of higher order terms shows that the support of $\beta(x,t)$ remains finite and so the asymptotic structure of $\beta$ as $t \to 0$ ends in this region with the edge of the support of $\beta(x,t)$, say $x = s(t)$ given by

$$s(t) \sim \sigma + t^{\frac{1}{2}}C(t) + \frac{t^{\frac{1}{2}}}{C(t)}\bar{\eta}_0 + \dots$$

as $t \to 0$, which becomes, on using (10.150)

$$s(t) \sim \sigma + t^{\frac{1}{2}}\left[ \lambda_0^{\frac{1}{2}}(-\ln t)^{\frac{1}{2}} + \left\{ \frac{\lambda_1}{2}\ln[\lambda_0(-\ln t)] + \bar{\eta}_0 \right\} \lambda_0^{-\frac{1}{2}}(-\ln t)^{-\frac{1}{2}} + \dots \right]$$
$$(10.162)$$

as $t \to 0$, with $\lambda_0$ and $\lambda_1$ as given in (10.151) and (10.155). We observe that, in this case, the edge of the support expands initially. The asymptotic structure of $\alpha(x,t)$ as $t \to 0$ does not end in this region and we are left to introduce the final region, region **III**, where $x = s(t) + O(1)$ and $\alpha(x,t) = 1 - o(1)$ with $\beta(x,t) \equiv 0$. The details of this final region follow, after minor modifications, those of Section 10.2.2 (b)(ii). Finally, we note that expansions (10.141) in region **I** do not, in general, satisfy the boundary conditions (10.5) at $x = 0$ and a further passive region is required in the neighbourhood of $x = 0$ as $t \to 0$. The details of the region follow, after minor modifications, those given in region $\mathbf{I}_0$ of Section 10.2.1.

In this case, the asymptotic structure has again determined that $\alpha_\infty(t) \equiv 1$ and $\beta_\infty(t) \equiv 0$, with the support of $\beta(x, t)$ remaining finite in $t > 0$.

## (ii) $r = \frac{2}{1-m}$

Again, in this case expansion (10.141b) in region **I** becomes nonuniform first, when $x = \sigma - O\left(t^{\frac{1}{2}}\right)$ with, via (10.142), $\alpha = 1 - O\left(t^{\frac{1}{1-m}}\right)$ and $\beta = O\left(t^{\frac{1}{1-m}}\right)$ in region **II**. To examine region **II**, we introduce the scaled coordinate $\eta = (x - \sigma)t^{-\frac{1}{2}}$, and look for asymptotic expansions of the form

$$\alpha(\eta, t) = 1 - t^{\frac{1}{1-m}} \bar{\alpha}_0(\eta) + o\left(t^{\frac{1}{1-m}}\right), \qquad (10.163a)$$

$$\beta(\eta, t) = t^{\frac{1}{1-m}} \bar{\beta}_0(\eta) + o\left(t^{\frac{1}{1-m}}\right) \qquad (10.163b)$$

as $t \to 0$ with $\eta = O(1)$. On substitution of expansions (10.163) into equations (10.2) (when written in terms of $\eta$ and $t$) we obtain at leading order

$$\bar{\alpha}_{0\eta\eta} + \frac{\eta}{2}\bar{\alpha}_{0\eta} - \frac{\bar{\alpha}_0}{1-m} = -\bar{\beta}_0^m, \qquad (10.164a)$$

$$\bar{\beta}_{0\eta\eta} + \frac{\eta}{2}\bar{\beta}_{0\eta} - \frac{\bar{\beta}_0}{1-m} - (k-1)\bar{\beta}_0^m = 0, \qquad (10.164b)$$

with $\eta > -\infty$. Equations (10.164) are to be solved subject to matching expansions (10.163) in region **II** with expansions (10.141) in region **I** as $\eta \to -\infty$, that is,

$$\bar{\alpha}_0(\eta) \sim (\beta_0 g_\sigma)^m (-\eta)^{\frac{2m}{1-m}} \quad \text{as} \quad \eta \to -\infty, \qquad (10.165a)$$

$$\bar{\beta}_0(\eta) \sim \beta_0 g_\sigma (-\eta)^{\frac{2}{1-m}} \quad \text{as} \quad \eta \to -\infty, \qquad (10.165b)$$

whilst region **III** requires

$$\bar{\alpha}_0(\eta), \bar{\beta}_0(\eta) \quad \text{remain bounded as} \quad \eta \to \infty. \qquad (10.166)$$

We also require the conditions

$$\bar{\alpha}_0, \bar{\beta}_0 \geq 0 \quad \text{for all} \quad -\infty < \eta < \infty. \qquad (10.167)$$

The uncoupled boundary value problem (10.164b), (10.165b), (10.166) and (10.167) is considered in detail in Appendix C, where it is shown that a monotone decreasing solution to this problem exists, which has finite support, that is,

$$\bar{\beta}_0(\eta) \begin{cases} > 0, & -\infty < \eta < \eta_0, \\ \equiv 0, & \eta \geq \eta_0. \end{cases}$$

Further, it is established that the sign of $\eta_0$ depends on the single parameter

$$\lambda = \beta_0 g_\sigma \left[\frac{2(1+m)}{(k-1)(1-m)^2}\right]^{\frac{1}{1-m}},$$

in the sense that, $\eta_0 > 0$ when $\lambda > 1$, $\eta_0 = 0$ when $\lambda = 1$, $\eta_0 < 0$ when $0 < \lambda < 1$. We note via (10.164b) that

$$\bar{\beta}_0(\eta) \sim \left[\frac{(k-1)(1-m)^2}{2(1+m)}\right]^{\frac{1}{1-m}} (\eta_0 - \eta)^{\frac{2}{1-m}} \quad \text{as } \eta \to \eta_0^-, \quad (10.168)$$

which has the required decay rate in $(\eta_0 - \eta)$ as the edge of the support of $\beta$ is approached. We note in this case that the edge of the support of $\beta(x,t)$, say $x = s(t)$, is given by

$$s(t) \sim \sigma + t^{\frac{1}{2}}\eta_0 + \dots \quad (10.169)$$

as $t \to 0$. We note that the initial behaviour of the edge of the support of $\beta$ depends upon the sign of $\eta_0$. The solution to (10.164a), $\bar{\alpha}_0(\eta)$, is monotone decreasing in $-\infty < \eta < \infty$, with

$$\bar{\alpha}_0(\eta) \sim O\left(e^{-\frac{\eta^2}{4}}\right) \quad \text{as } \eta \to \infty. \quad (10.170)$$

The asymptotic structure of $\alpha(x,t)$ as $t \to 0$ does not end in this region and we are left to introduce the final region, region **III**, where $x = s(t) + O(1)$ and $\alpha(x,t) = 1 - o(1)$ with $\beta(x,t) \equiv 0$. The details of this final region depend on the sign of $\eta_0$. If $\eta_0 > 0$ ($< 0$) the edge of the support of $\beta$ expands (contracts) initially with the asymptotic structure being given, after minor modifications, in Section 10.2.2(b)(ii) (Section 10.2.2 (b)(iv)) respectively. If $\eta_0 = 0$, further corrections to expansion (10.169) are required to determine the initial behaviour of the edge of the support of $\beta$. We do not pursue these details here. Finally, we note that expansions (10.141) in region **I** do not, in general, satisfy the boundary conditions (10.5) at $x = 0$ and a further passive region is required in the neighbourhood of $x = 0$ as $t \to 0$. The details of the region follow, after minor modifications, those given in region $I_0$ of Section 10.2.1.

In this case again, the asymptotic structure has determined that $\alpha_\infty(t) \equiv 1$ and $\beta_\infty(t) \equiv 0$, whilst $\beta(x,t)$ has finite support in $t > 0$.

**(iii)** $r > \frac{2}{1-m}$

Expansion (10.141b) in region **I** now becomes nonuniform when $x = \sigma - O\left(t^{\frac{1}{r(1-m)}}\right)$ with, via (10.142), $\alpha = 1 - O\left(t^{\frac{1}{1-m}}\right)$ and $\beta = O\left(t^{\frac{1}{1-m}}\right)$ in region **II**. Hence in region **II**, we introduce the scaled coordinate $\hat{\eta} = (x-\sigma)t^{-\frac{1}{r(1-m)}}$, and look for asymptotic expansions of the form

$$\alpha(\hat{\eta},t) = 1 - t^{\frac{1}{1-m}} \hat{\alpha}_0(\hat{\eta}) + o\left(t^{\frac{1}{1-m}}\right), \quad (10.171a)$$

$$\beta(\hat{\eta},t) = t^{\frac{1}{1-m}} \hat{\beta}_0(\hat{\eta}) + o\left(t^{\frac{1}{1-m}}\right), \quad (10.171b)$$

as $t \to 0$ with $\hat{\eta} = O(1)$. On substitution of expansions (10.171) into equations (10.2) (when written in terms of $\hat{\eta}$ and $t$) we obtain the coupled leading order problems for $\hat{\alpha}_0(\hat{\eta})$ and $\hat{\beta}_0(\hat{\eta})$ as

$$\hat{\eta}\,\hat{\alpha}_{0\hat{\eta}} - r\hat{\alpha}_0 \; = \; -r(1-m)\,\hat{\beta}_0^m, \quad \hat{\eta} > -\infty, \qquad (10.172a)$$
$$\hat{\alpha}_0(\hat{\eta}) \sim (\beta_0 g_\sigma)^m \, (-\hat{\eta})^{rm} \quad \text{as} \quad \hat{\eta} \to -\infty \qquad (10.172b)$$

and

$$\hat{\eta}\,\hat{\beta}_{0\hat{\eta}} - r\hat{\beta}_0 \; - \; r(1-m)(k-1)\,\hat{\beta}_0^m \; = \; 0, \quad \hat{\eta} > -\infty, \quad (10.173a)$$
$$\hat{\beta}_0(\hat{\eta}) \sim \beta_0 g_\sigma(-\hat{\eta})^r \quad \text{as} \quad \hat{\eta} \to -\infty. \qquad (10.173b)$$

Conditions (10.172b) and (10.173b) arise from matching with region **I** as $\hat{\eta} \to -\infty$. The solution to (10.173) may be obtained directly by separation of variables as

$$\hat{\beta}_0(\hat{\eta}) = \left[(\beta_0 g_\sigma)^{1-m} \, (-\hat{\eta})^{r(1-m)} - (k-1)(1-m)\right]^{\frac{1}{(1-m)}}, \quad \hat{\eta} > -\infty. \tag{10.174}$$

An examination of (10.174) reveals that a weak singularity develops in $\hat{\beta}_0$ as $\hat{\eta} \to \hat{\eta}_0^-$ where

$$\hat{\eta}_0 = - \left[\frac{(k-1)(1-m)}{(\beta_0 g_\sigma)^{1-m}}\right]^{\frac{1}{r(1-m)}}. \tag{10.175}$$

Now, via (10.174), the solution to (10.172), is given in terms of $\hat{\beta}_0$ by

$$\hat{\alpha}_0(\hat{\eta}) \; = \; r(1-m)(-\hat{\eta})^r \int_{-\infty}^{\hat{\eta}} (-s)^{-(r+1)}\hat{\beta}_0^m(s)\mathrm{d}s, \quad -\infty < \hat{\eta} < \hat{\eta}_0, \tag{10.176}$$

and we note from (10.174) and (10.176) that $\hat{\alpha}_0$ and $\hat{\beta}_0$ are monotone decreasing in $\hat{\eta}$, with

$$\hat{\alpha}_0(\hat{\eta}) = \hat{\kappa}\,(-\hat{\eta})^r \; + \; O\left((\hat{\eta}_0 - \hat{\eta})^{\frac{1}{1-m}}\right), \tag{10.177a}$$

$$\hat{\beta}_0(\hat{\eta}) = \left[\frac{r(1-k)(1-m)^2}{(-\hat{\eta}_0)}\right]^{\frac{1}{1-m}} (\hat{\eta}_0 - \hat{\eta})^{\frac{1}{1-m}}$$
$$+ \; O\left((\hat{\eta}_0 - \hat{\eta})^{\frac{2-m}{1-m}}\right) \tag{10.177b}$$

as $\hat{\eta} \to \hat{\eta}_0^-$ with

$$\hat{\kappa} = r(1-m) \int_{-\infty}^{\hat{\eta}_0} (-s)^{-(r+1)}\hat{\beta}_0^m(s)\mathrm{d}s. \tag{10.178}$$

The support of $\beta(\hat{\eta}, t)$ ends at $\hat{\eta} = \hat{\eta}_0$ in this region. However, in (10.177b) the degree of $(\hat{\eta}_0 - \hat{\eta})$ as $\hat{\eta} \to \hat{\eta}_0^-$ (which is $\frac{1}{1-m}$), is too weak, and consideration of further terms in expansion (10.171b) reveals a weak nonuniformity as $\hat{\eta} \to \hat{\eta}_0^-$. Therefore a further region is required to complete the structure, in which $\hat{\eta} = \hat{\eta}_0 + o(1)$ as $t \to 0$, and diffusion effects are retained at leading order to enable the appropriate behaviour to be achieved at the edge of the support of $\beta$. We label this region, region **II(b)** (with region **II** becoming region **II(a)**), and introduce the scaled coordinate $\tilde{\eta}$ by

$$\hat{\eta} = \hat{\eta}_0 + t^\gamma \tilde{\eta}, \tag{10.179}$$

with $\gamma > 0$ to be determined, and $\tilde{\eta} = O(1)$ as $t \to 0$ in region **II(b)**. An examination of (10.177) and expansions (10.171) then determines that $\alpha = 1 - O\left(t^{\frac{1}{1-m}}\right)$ and $\beta = O\left(t^{\frac{1+\gamma}{1-m}}\right)$ in region **II(b)**. Thus we expand $\alpha$ and $\beta$ as

$$\alpha(\tilde{\eta}, t) = 1 - t^{\frac{1}{1-m}} \bar{\alpha}_0(\tilde{\eta}) + o\left(t^{\frac{1}{1-m}}\right), \tag{10.180a}$$

$$\beta(\tilde{\eta}, t) = t^{\frac{1+\gamma}{1-m}} \bar{\beta}_0(\tilde{\eta}) + o\left(t^{\frac{1+\gamma}{1-m}}\right), \tag{10.180b}$$

as $t \to 0$, with $\tilde{\eta} = O(1)$. On substituting expansions (10.180) into equations (10.2) (when written in terms of $\tilde{\eta}$ and $t$), to retain diffusion terms at leading order requires that

$$\gamma = 1 - \frac{2}{r(1-m)} > 0, \tag{10.181}$$

after which the leading order uncoupled boundary value problems for $\bar{\alpha}_0$ and $\bar{\beta}_0$ are given by

$$\bar{\alpha}_{0\tilde{\eta}\tilde{\eta}} + \frac{\hat{\eta}_0}{r(1-m)} \bar{\alpha}_{0\tilde{\eta}} = 0, \quad -\infty < \tilde{\eta} < \infty, \tag{10.182a}$$

$$\bar{\alpha}_0(\tilde{\eta}) > 0 \quad \text{for all} \quad -\infty < \tilde{\eta} < \infty, \tag{10.182b}$$

$$\bar{\alpha}_0(\tilde{\eta}) \sim \hat{\kappa} (-\hat{\eta}_0)^r \quad \text{as} \quad \tilde{\eta} \to -\infty, \tag{10.182c}$$

$$\bar{\alpha}_0(\tilde{\eta}) \quad \text{bounded as} \quad \tilde{\eta} \to \infty \tag{10.182d}$$

and

$$\bar{\beta}_{0\tilde{\eta}\tilde{\eta}} + \frac{\hat{\eta}_0}{r(1-m)} \bar{\beta}_{0\tilde{\eta}} - (k-1)\bar{\beta}_0^m = 0, \quad -\infty < \tilde{\eta} < \infty, \tag{10.183a}$$

$$\bar{\beta}_0(\tilde{\eta}) > 0 \quad \text{for all} \quad -\infty < \tilde{\eta} < \infty, \tag{10.183b}$$

$$\bar{\beta}_0(\tilde{\eta}) \sim \left[\frac{r(k-1)(1-m)^2}{(-\hat{\eta}_0)}\right]^{\frac{1}{1-m}} (-\tilde{\eta})^{\frac{1}{1-m}} \quad \text{as} \quad \tilde{\eta} \to -\infty, \tag{10.183c}$$

$$\bar{\beta}_0(\tilde{\eta}) \quad \text{bounded as} \quad \tilde{\eta} \to \infty. \tag{10.183d}$$

Conditions (10.182c) and (10.183c) arise from matching with region **II(a)** as $\tilde{\eta} \to -\infty$. It can be shown (after minor modifications to Appendix D) that the problem (10.183) has a unique solution on $-\infty < \tilde{\eta} < \infty$, which has finite support, so that

$$\bar{\beta}_0(\tilde{\eta}) \begin{cases} > 0, & -\infty < \tilde{\eta} < \tilde{\eta}_0, \\ \equiv 0, & \tilde{\eta} \ge \tilde{\eta}_0. \end{cases}$$

We observe (via Appendix D) that the solution to (10.183) is monotone decreasing in $-\infty < \tilde{\eta} < \tilde{\eta}_0$, and further, from (10.183a) that

$$\bar{\beta}_0(\tilde{\eta}) \sim \left[ \frac{(k-1)(1-m)^2}{2(1+m)} \right]^{\frac{1}{1-m}} (\tilde{\eta}_0 - \tilde{\eta})^{\frac{2}{1-m}} \quad \text{as} \quad \tilde{\eta} \to \tilde{\eta}_0^-, \quad (10.184)$$

which has the required decay rate in $(\tilde{\eta}_0 - \tilde{\eta})$ as the edge of the support is approached. Consideration of further terms in this region shows that expansion (10.180b) continues to have finite support. In particular, we observe that the edge of the support of $\beta(x,t)$, $x = s(t)$, is given by

$$s(t) \sim \sigma + t^{\frac{1}{r(1-m)}} \hat{\eta}_0 + t^{\frac{r(1-m)-1}{r(1-m)}} \tilde{\eta}_0 + \ldots \quad (10.185)$$

as $t \to 0$. We note from (10.185) that, since $\hat{\eta}_0 < 0$, the edge of the support is contracting initially. The solution of (10.182a) on $-\infty < \tilde{\eta} < \infty$ subject to (10.182c) is given by

$$\bar{\alpha}_0(\tilde{\eta}) \equiv \hat{\kappa}(-\hat{\eta}_0)^r. \quad (10.186)$$

We note that the asymptotic structure of $\alpha(x,t)$ as $t \to 0$ does not end in this region and we are left to introduce the final region, region **III**, where $x = s(t) + O(1)$ and $\alpha(x,t) = 1 - o(1)$ with $\beta(x,t) \equiv 0$. The details of this final region follow, after minor modifications, those of Section 10.2.2(b)(iv) (since the support of $\beta$ contracts initially). Finally, we note that expansions (10.141) in region **I** do not, in general, satisfy the boundary conditions (10.5) at $x = 0$ and a further passive region is required in the neighbourhood of $x = 0$ as $t \to 0$. The details of the region follow, after minor modifications, those given in region $\mathbf{I}_0$ of Section 10.2.1.

Here again, the asymptotic structure has determined that $\alpha_\infty(t) \equiv 1$ and $\beta_\infty(t) \equiv 0$, with the support of $\beta(x,t)$ remaining finite in $t > 0$.

### 10.2.3 $m > n$

In this case the asymptotic structure as $t \to 0$ follows, after minor modifications, that given in Section 10.2.2 (c). Again we have $\alpha_\infty(t) \equiv 1$ and $\beta_\infty(t) \equiv 0$, with the support of $\beta(x,t)$ remaining finite in $t > 0$ and given by

$$s(t) \sim \begin{cases} \sigma + t^{\frac{1}{2}} C(t) + \frac{t^{\frac{1}{2}}}{C(t)} \tilde{\eta}_0 + \ldots, & 0 < r < \frac{2}{1-n}, \\ \sigma + t^{\frac{1}{2}} \eta_0 + \ldots, & r = \frac{2}{1-n}, \\ \sigma + t^{\frac{1}{r(1-n)}} \hat{\eta}_0 + t^{\frac{r(1-n)-1}{r(1-n)}} \tilde{\eta}_0 + \ldots, & 0 < r < \frac{2}{1-n} \end{cases}$$

as $t \to 0$, with $C(t)$ as given in (10.150) and where $\bar{\eta}_0$, $\eta_0$, $\hat{\eta}_0$ and $\tilde{\eta}_0$ are real constants with $\hat{\eta}_0 < 0$.

## 10.3 Conclusions

The results of this chapter are summarized as follows:

(i) $(m < n), (m = n; k < 1)$

"Lifting at infinity". The support of $\beta$ becomes unbounded in infinitesimal $t$, with $\beta(x, t)$ being bounded above zero as $x \to \infty$, $0 < t \ll 1$. The far field functions are given by $\alpha_\infty(t) \sim 1 - (1 - m)^{\frac{1}{1-m}} t^{\frac{1}{1-m}} + \ldots$, $\beta_\infty(t) \sim (1 - m)^{\frac{1}{1-m}} t^{\frac{1}{1-m}} + \ldots$ $(\alpha_\infty(t) \sim 1 - (1 - m)^{\frac{1}{1-m}} (1 - k)^{\frac{m}{1-m}} t^{\frac{1}{1-m}} + \ldots$, $\beta_\infty(t) \sim 1 - (1 - m)^{\frac{1}{1-m}} (1 - k)^{\frac{1}{1-m}} t^{\frac{1}{1-m}} + \ldots)$ for $0 < t \ll 1$, when $m < n$ $(m = n; k < 1)$ respectively.

(ii) $(\frac{1}{2} \leq m = n < 1; k = 1)$

The support of $\beta$ becomes unbounded in infinitesimal $t$ with $\beta(x, t) \to 0$ as $x \to \infty$, $0 < t \ll 1$. The far field functions are $\alpha_\infty(t) \equiv 1$, $\beta_\infty(t) \equiv 0$ for $0 < t \ll 1$.

(iii) $(0 < m = n < \frac{1}{2}; k = 1)$ $(m = n; k > 1)$ and $(m > n)$

The support of $\beta$ remains finite for $0 < t \ll 1$, with the edge of the support initially expanding (contracting) for $r < r_c$ ($> r_c$) respectively, where $r_c = \frac{4}{1-2m}$ in the case $(0 < m = n < \frac{1}{2}; k = 1)$ and $r_c = \frac{2}{1-m}$ in the cases $(m = n; k > 1)$ and $(m > n)$. The behaviour of the edge of the support of $\beta$ when $r = r_c$ is a more delicate situation and is described in the above analysis. In all cases expressions describing the initial motion of the edge of the support have been determined. The far field functions in these cases are $\alpha_\infty(t) \equiv 1$, $\beta_\infty(t) \equiv 0$ for $0 < t \ll 1$.

Numerical solutions of IBVP (see McCabe *et al* [41]) corroborate the above results and indicate that travelling waves will occur only when $m > n$ and $k$ is sufficiently small, which is in line with the predictions of Chapters 7 and 8. It has been established that the fractional order autocatalytic step ((6.1) with $0 < m < 1$) alone cannot support permanent form travelling waves (see King and Needham [28]). However, the results of Chapters 7 and 8, the numerical simulations [41], the travelling wave theory [27] and the analysis presented in this chapter strongly suggest that the inclusion of the fractional order termination step ((6.2) with $0 < n < 1$) restores, in the long time, the capability of the fractional order autocatalytic step (6.1) to support finite speed travelling waves.

We propose the following conjectures on the global behaviour of the coupled system:

(i) In the case $m < n$, the concentrations $\alpha(x, t)$ and $\beta(x, t)$ will be bounded above zero and below unity, respectively, for all $0 < t < \infty$ ("Lifting at infinity"). The reactant concentration, $\alpha(x, t) \to 0$ monotonically as $t \to \infty$ uniformly in $x$ whilst the concentration of autocatalyst, $\beta(x, t)$, will increase to a maximum, $\beta_{\max}(x)$ at some $t = t_{\beta\max}$ before $\beta(x, t) \to 0$ as $t \to \infty$ uniformly in $x$.

(ii) In the case ($m = n; k < 1$), the concentrations $\alpha(x,t)$ and $\beta(x,t)$ will be bounded above zero and below unity, respectively, for all $0 < t < \infty$ ("Lifting at infinity"). The reactant concentration $\alpha(x,t) \to \alpha_s$ monotonically as $t \to t_c$ uniformly in $x$ with $\alpha_s$ a positive, constant, residual concentration. The concentration of autocatalyst, $\beta(x,t)$, will increase to a maximum, $\beta_{max}(x)$, at some $t = t_{\beta\,max}$ before $\beta(x,t) \to 0$ as $t \to t_c$ uniformly in $x$ for some finite $t = t_c$.

(iii) In the case ($\frac{1}{2} \le m = n < 1; k = 1$), the support of $\beta$ will be unbounded for all $t > 0$. The concentration of autocatalyst $\beta(x,t) \to 0$ monotonically, uniformly in $x$ as $t \to 0$. The reactant concentration, $\alpha(x,t)$, will first decrease (with a minimum at $x = 0$ and $\alpha(x,t) \to 1$ as $x \to \infty$) then approach unity as $t \to \infty$, uniformly in $x$, through diffusion.

(iv) In the cases ($0 < m = n < \frac{1}{2}; k = 1$) and ($m = n; k > 1$), the support of $\beta(x,t)$ will remain finite for all $t > 0$. The concentration of autocatalyst, $\beta(x,t)$, and its support, will collapse to zero in finite $t$. The reactant concentration, $\alpha(x,t)$ will first decrease (with a minimum at $x = 0$ and $\alpha(x,t) \to 1$ as $x \to \infty$) then approach unity as $t \to \infty$, uniformly in $x$, through diffusion.

(v) In the case $m > n$ there are two possibilities.

(a) If $k$ is sufficiently small and $\beta_0$ sufficiently large ($\beta_0 > k^{\frac{1}{m-n}}$ is necessary), permanent form travelling waves will develop in $\alpha(x,t)$ and $\beta(x,t)$ as $t \to \infty$. The wave in $\alpha$ will be monotone increasing in $x$ with $\alpha \to 1$ ahead of the wave and $\alpha \equiv \alpha_s$ behind the wave with $\alpha_s$ a positive, constant, residual concentration. The wave in $\beta$ will have a single crest with $\beta(x,t) \equiv 0$ both behind and ahead of the wave. That is, the wave in $\beta$ will be compactly supported for large $t$, and the formation of a dead core occurs.

(b) If $k$ is large or $\beta_0$ small, then the system will undergo the finite time extinction behaviour as described in (iv).

The authors are at present working to provide rigorous results to support the above conjectures.

# A

## Construction of a Global Nonnegative Solution to the Scalar Equation $w_t = w_{xx} + \mu^* w^n$

In this appendix, we construct a global non-negative solution to the scalar equation

$$w_t = w_{xx} + \mu^* w^m, \quad 0 \le x < x_s(t), \ t > 0, \tag{A.1}$$

where $\mu^* < 0$, which has finite support, with

$$w(x_s(t), t) = 0, \qquad w_x(x_s(t), t) = 0, \tag{A.2}$$

and

$$w(x, t) > 0, \quad 0 \le x < x_s(t), \ t \ge 0. \tag{A.3}$$

Here $x_s(t)$ is the edge of the support, and decays to zero in finite time, $t_0$. To this end we introduce the similarity variables

$$w(z, t) = (t_0 - t)^\alpha H(z), \qquad z = x(t_0 - t)^{-\beta}, \tag{A.4}$$

with $\alpha, \beta > 0$. Substitution of (A.4) into equation (A.1) (when written in terms of $z$) requires for a non-trivial balance that,

$$\alpha = \frac{1}{1 - m}, \qquad \beta = \frac{1}{2}, \tag{A.5}$$

after which (A.1) becomes

$$H_{zz} - \frac{z}{2} H_z + \frac{1}{1 - m} H + \mu^* H^m = 0, \quad 0 \le z < z_0, \tag{A.6}$$

which is to be solved subject to the conditions

$$H(z_0) = 0, \qquad H_z(z_0) = 0, \tag{A.7}$$

and

$$H(z) > 0 \quad \text{for} \quad 0 \le z < z_0. \tag{A.8}$$

Here $z_0$ is the edge of the support of $H(z)$. We consider the solution for which $z_0 \ll 1$, and introduce the scaled variable

$$y = 1 - \frac{z}{z_0}, \tag{A.9}$$

so that $0 \le y \le 1$, and look for a solution of (A.6)–(A.8) in the form

$$H(y) = z_0^{\frac{2}{1-m}} \hat{H}(y) + O\left(z_0^{\frac{2(2-m)}{1-m}}\right) \quad \text{as} \quad z_0 \to 0, \tag{A.10}$$

where $\hat{H}(y) = O(1)$ as $z_0 \to 0$. On substituting (A.10) into (A.6) (when written in terms of $y$) we obtain at leading order as $z_0 \to 0$,

$$\hat{H}_{yy} + \mu^* \hat{H}^m = 0, \quad 0 < y \le 1, \tag{A.11}$$

which is to be solved subject to the conditions

$$\hat{H}(0) = 0, \qquad \hat{H}_y(0) = 0, \tag{A.12}$$

and

$$\hat{H}(y) > 0 \quad \text{for} \quad 0 < y \le 1. \tag{A.13}$$

The solution of (A.11)–(A.13) may be obtained directly by integration, to give

$$\hat{H}(y) = \left[\frac{-\mu^*(1-m)^2}{2(1+m)}\right]^{\frac{1}{1-m}} y^{\frac{2}{1-m}}, \quad 0 \le y \le 1, \tag{A.14}$$

which via (A.9) and (A.10) gives

$$H(z) \sim \left[\frac{-\mu^*(1-m)^2}{2(1+m)}\right]^{\frac{1}{1-m}} (z_0 - z)^{\frac{2}{1-m}}, \quad 0 \le z \le z_0 \tag{A.15}$$

as $z_0 \to 0$. Finally, via (A.4) we obtain, in terms of the original variables,

$$w(x,t) \begin{cases} \sim \left[\frac{-\mu^*(1-m)^2}{2(1+m)}\right]^{\frac{1}{1-m}} \left(z_0^*(t_0-t)^{\frac{1}{2}} - x\right)^{\frac{2}{1-m}}, & 0 \le x < z_0^*(t_0-t)^{\frac{1}{2}}, \\ \equiv 0, & x \ge z_0^*(t_0-t)^{\frac{1}{2}}, \end{cases} \tag{A.16}$$

where we have fixed $z_0 = z_0^*$ sufficiently small. We observe in (A.16) that $t_0$, the extinction time, is a free parameter.

# B

## Asymptotic Solutions to the Eigenvalue Problem (8.68)-(8.71) as $m \to 0^+$ and $m \to 1^-$

In this appendix we examine the asymptotic solutions to the eigenvalue problem (8.68)-(8.71) as $m \to 0^+$ and $m \to 1^-$. We first introduce the scaled variables

$$y(\hat{x}) = \alpha^{\frac{1}{m-1}} G(\bar{\eta}), \quad \hat{x} = \bar{\eta} - \bar{\eta}_0. \tag{B.1}$$

Under this transformation the eigenvalue problem (8.68)-(8.71) becomes

$$y'' + \frac{1}{2}y' - y^m = 0, \quad -\infty < \hat{x} < 0, \tag{B.2}$$

$$y(0) = 0, \quad y'(0) = 0, \tag{B.3}$$

$$y(\hat{x}) \sim F_+ e^{-\hat{x}/2} \quad \text{as} \quad \hat{x} \to -\infty, \tag{B.4}$$

where a prime denotes $\frac{d}{d\hat{x}}$ and $F_+ = C_\infty \alpha^{\frac{1}{m-1}} e^{\frac{-\bar{\eta}_0}{2}}$. Note that the far field constant $F_+$ can be determined numerically by solving the initial value problem (B.2)-(B.3) and once $F_+$ has been determined in this way, we obtain an expression for the position of the edge of the support, $\bar{\eta}_0$, given by

$$\bar{\eta}_0 = -2 \ln \left( \frac{F_+ \alpha^{\frac{1}{1-m}}}{C_\infty} \right). \tag{B.5}$$

We begin by considering the asymptotic solution of (B.2)-(B.4) as $m \to 0^+$. Equation (B.2), at leading order, is given by

$$y'' + \frac{1}{2}y' - 1 = 0, \tag{B.6}$$

which when solved subject to the initial conditions (B.3) gives

$$y(\hat{x}) \sim 4\left[ e^{-\hat{x}/2} - 1 \right] + 2\hat{x} \quad \text{as} \quad m \to 0^+. \tag{B.7}$$

Now as $\hat{x} \to -\infty$, (B.7) becomes

$$y(\hat{x}) \sim 4e^{-\hat{x}/2}, \tag{B.8}$$

which gives a value for the far field constant of $F_+ = 4$ as $m \to 0^+$, and so from (B.5),

$$\bar{\eta}_0 = -2\ln\left(\frac{4\alpha}{C_\infty}\right) + o(1) \quad \text{as} \quad m \to 0^+. \tag{B.9}$$

Finally, we consider the asymptotic solution of (B.2)-(B.4) as $m \to 1^-$. We first write $m = 1 - \epsilon$ in equation (B.2) and consider the limit $\epsilon \to 0^+$. It is straightforward to show that the local solution to (B.2)-(B.4) as $\hat{x} \to 0^-$, is given by

$$y(\hat{x}) \sim e^{\frac{2}{\epsilon}\ln(-\epsilon\hat{x}/2)}. \tag{B.10}$$

This suggests that, to obtain a solution for $\epsilon \ll 1$, we should rescale by $\bar{x} = \epsilon\hat{x}$ and look for a solution of the form

$$y(\bar{x}) = e^{\frac{2}{\epsilon}(X(\bar{x})+o(1))} \quad \text{as} \quad \epsilon \to 0, \tag{B.11}$$

with $\bar{x} = O(1)$. Under this rescaling equation (B.2) becomes, at leading order,

$$4(X')^2 + X' - e^{-2X} = 0, \quad \bar{x} < 0, \tag{B.12}$$

and is solved subject to the boundary conditions

$$X(\bar{x}) \sim \ln(-\bar{x}/2) \quad \text{as} \quad \bar{x} \to 0^-, \tag{B.13}$$

$$X(\bar{x}) \sim -\frac{1}{4}\bar{x} \quad \text{as} \quad \bar{x} \to -\infty, \tag{B.14}$$

where a prime denotes $\frac{d}{d\bar{x}}$. On solving the quadratic in (B.12) we obtain

$$X'(\bar{x}) = \frac{-1 - [1 + 16e^{-2X}]^{\frac{1}{2}}}{8} \equiv f(X), \quad \bar{x} < 0, \tag{B.15}$$

where

$$f(X) = \begin{cases} -\frac{1}{2}e^{-X} & \text{as } X \to -\infty, \\ -\frac{1}{4} + O\left(e^{-2X}\right) & \text{as } X \to \infty. \end{cases} \tag{B.16}$$

It is readily shown that (B.15) has the solution, which satisfies (B.13),

$$\bar{x} = \int_{-\infty}^{X} [f(s)]^{-1} \, ds. \tag{B.17}$$

Considering (B.17) as $X \to \infty$, we find that

$$X(\bar{x}) \sim -\frac{1}{4}\bar{x} + c_1 \quad \text{as} \quad \bar{x} \to -\infty, \tag{B.18}$$

where

$$c_1 = \frac{1}{4}\left\{\int_0^\infty \left(\frac{1}{f(s)} + 4\right)\, ds + \int_{-\infty}^0 \frac{1}{f(s)}\, ds\right\} = \ln 2 - \frac{1}{2}, \qquad \text{(B.19)}$$

which on substituting back into (B.11) gives

$$y(\hat{x}) = e^{2c_1/\epsilon}\, e^{-\hat{x}/2} \qquad \text{as} \quad \hat{x} \to -\infty, \qquad \text{(B.20)}$$

which when compared to the condition (B.4) gives the far field constant $F_+ \sim e^{2c_1/\epsilon}$ as $\epsilon \to 0^+$. Therefore, we have, from (B.5),

$$\bar{\eta}_0 \sim \frac{-2}{1-m}\ln\left(\frac{4e^{-1}\alpha}{C_\infty^{1-m}}\right) \qquad \text{as} \quad m \to 1^-. \qquad \text{(B.21)}$$

# C

# Analysis of Boundary Value Problem (8.76)-(8.78)

In this appendix we analyze the boundary value problem (8.76)-(8.78). We can write this as

$$\bar{F}_{\eta\eta} + \frac{\eta}{2}\bar{F}_\eta + G(\bar{F}) = 0, \quad -\infty < \eta < \infty, \tag{C.1}$$

$$\bar{F}(\eta) \sim \lambda(-\eta)^{\frac{2}{1-m}} \quad \text{as} \quad \eta \to -\infty, \tag{C.2}$$

$$\bar{F}(\eta_0) = 0, \quad \bar{F}_\eta(\eta_0) = 0, \tag{C.3}$$

$$\bar{F} \quad \begin{cases} > 0, -\infty < \eta < \eta_0, \\ = 0, \eta > \eta_0. \end{cases} \tag{C.4}$$

Here,

$$\lambda = u_0 g_\sigma > 0, \tag{C.5}$$

and

$$G(X) = \begin{cases} -\left[\frac{1}{1-m}X + \alpha X^m\right], & X \geq 0, \\ -X, & X < 0, \end{cases} \tag{C.6}$$

whilst $\eta_0$ is an eigenvalue to be fixed by solution of (C.1)-(C.4). First we may establish the following proposition.

**Proposition C.1.** *Let $\bar{F}(\eta)$ be a solution of (C.1)-(C.4), then $\bar{F}(\eta)$ is monotone decreasing in $-\infty < \eta \leq \eta_0$.*

*Proof.* Suppose $\bar{F}(\eta)$ is not monotone decreasing in $-\infty < \eta \leq \eta_0$, then, via conditions (C.2) and (C.3), there exists a value $\eta = \eta^*$ at which $\bar{F}(\eta)$ has a positive maximum, that is,

$$\bar{F}(\eta^*) > 0, \quad \bar{F}_\eta(\eta^*) = 0, \tag{C.7}$$

whilst

$$\bar{F}_{\eta\eta}(\eta^*) \leq 0. \tag{C.8}$$

However, from equation (C.1) we have,

$$\bar{F}_{\eta\eta}(\eta^*) = -G(\bar{F}(\eta^*)) > 0, \tag{C.9}$$

via (C.7) and (C.6), which contradicts (C.8). The result follows.    □

We next have the following proposition.

**Proposition C.2.** *Let $\bar{F}_1(\eta)$ and $\bar{F}_2(\eta)$ be solutions of (C.1)-(C.4) corresponding to $\lambda = \lambda_1$ and $\lambda = \lambda_2$ respectively, with $\lambda_2 > \lambda_1 > 0$. Then $\bar{F}_1(\eta) \le \bar{F}_2(\eta)$ for all $-\infty < \eta < \infty$.*

*Proof.* Let $\bar{F}_1(\eta)$ and $\bar{F}_2(\eta)$ be as above, and define

$$w(\eta) = \bar{F}_1(\eta) - \bar{F}_2(\eta), \quad -\infty < \eta < \infty. \tag{C.10}$$

It follows, from equation (C.1), that

$$w_{\eta\eta} + \frac{\eta}{2}w_\eta + h(\eta)w = 0, \quad -\infty < \eta < \infty, \tag{C.11}$$

where

$$h(\eta) = \begin{cases} \frac{G(\bar{F}_1(\eta)) - G(\bar{F}_2(\eta))}{[\bar{F}_1(\eta) - \bar{F}_2(\eta)]}, & \bar{F}_1(\eta) \ne \bar{F}_2(\eta), \\ 0, & \bar{F}_1(\eta) = \bar{F}_2(\eta). \end{cases} \tag{C.12}$$

Moreover, from (C.2) and (C.4) we have

$$w(\eta) < 0 \quad \text{for all } \eta < -\tau, \tag{C.13}$$
$$w(\eta) \equiv 0 \quad \text{for all } \eta \ge \max[\eta_{01}, \eta_{02}], \tag{C.14}$$

where $\tau > 0$ is sufficiently large and $\eta_{01}, \eta_{02}$ are the values of $\eta_0$ corresponding to $\lambda = \lambda_1$ and $\lambda = \lambda_2$ respectively. Now suppose $w(\eta) \not\le 0$ for $-\infty < \eta < \infty$. Then there exists a value $-\infty < \eta^* < \max[\eta_{01}, \eta_{02}]$ such that $w(\eta)$ has a positive maximum at $\eta = \eta^*$, that is,

$$w(\eta^*) > 0, \quad w_\eta(\eta^*) = 0, \tag{C.15}$$

whilst

$$w_{\eta\eta}(\eta^*) \le 0. \tag{C.16}$$

However, equation (C.11) requires

$$w_{\eta\eta}(\eta^*) = -h(\eta^*)w(\eta^*). \tag{C.17}$$

Since $\bar{F}_1(\eta^*) > \bar{F}_2(\eta^*)$, then via (C.12) we have $h(\eta^*) < 0$, and so (C.17) gives

$$w_{\eta\eta}(\eta^*) > 0,$$

contradicting (C.16). The result follows.    □

*Remark C.3.* Let $\bar{F}_1(\eta)$ and $\bar{F}_2(\eta)$ be as given in Proposition C.2. Then $\bar{F}_1(\eta) < \bar{F}_2(\eta)$ when $-\infty < \eta < \min[\eta_{01}, \eta_{02}]$ *(this follows since* $\bar{F}_1(\eta)$, $\bar{F}_2(\eta) > 0$ *on* $-\infty < \eta < \min[\eta_{01}, \eta_{02}]$ *and so if there is a point* $\eta^*$ *such that* $\bar{F}_1(\eta^*) = \bar{F}_2(\eta^*)$, *then via Proposition C.2,* $\bar{F}_{1\eta}(\eta^*) = \bar{F}_{2\eta}(\eta^*)$, *whilst* $\eta = \eta^*$ *is a regular point of equation (C.1). Uniqueness in* $-\infty < \eta < \eta^*$ *then contradicts (C.2)).* □

We now observe that when

$$\lambda = \lambda_c = \left[\frac{\alpha(1-m)^2}{2(1+m)}\right]^{\frac{1}{1-m}},$$

then the solution to (C.1)-(C.4) is

$$\bar{F}_c(\eta) = \begin{cases} \lambda_c(-\eta)^{\frac{2}{1-m}}, & -\infty < \eta \le 0, \\ 0, & \eta > 0, \end{cases}$$

and

$$\eta_{0c} = 0.$$

We therefore conclude, via Proposition C.2, that

$$\eta_0 \begin{cases} \ge 0 \text{ for } \lambda > \lambda_c, \\ \le 0 \text{ for } \lambda < \lambda_c. \end{cases} \tag{C.18}$$

Moreover, numerical solutions of (C.1)-(C.4) suggest that *strict* inequalities hold in (C.18).

# D

## Analysis of Boundary Value Problem (8.90)-(8.92)

In this appendix we establish that the boundary value problem, given by (8.90)-(8.92), has a unique solution for each fixed $\tilde{\eta}_0 \in \mathbb{R}$. Equation (8.90) can be reduced to the second order autonomous dynamical system

$$H_{\tilde{\eta}} = Q, \quad Q_{\tilde{\eta}} = \alpha H^m - \frac{\eta_c}{r(1-m)} Q, \tag{D.1}$$

where $\eta_c < 0$. The dynamical system (D.1) has only one equilibrium point at $(0,0)$ in the $(H, Q)$ phase plane. The equilibrium point $(0,0)$ is not hyperbolic, and cannot be classified via linearization or centre manifold theory and a more detailed analysis is required. This has been given in detail in Chapter 7, where it is established that the equilibrium point $(0,0)$ has the structure of a saddle point in $H \geq 0$, with a unique stable manifold and a unique unstable manifold. The global $(H, Q)$ phase portrait in $H \geq 0$ can now be readily sketched, and is illustrated in Figure D.1, where the stable manifold at $(0,0)$ is labelled $S_1$. The integral path in the phase plane corresponding to the stable manifold $S_1$, has $Q = Q_s(H)$ in $H \geq 0$, where

$$Q_s(H) < 0 \quad \text{for all } H > 0, \tag{D.2}$$

$$Q_s(H) \quad \sim \quad \begin{cases} -\sqrt{\frac{2\alpha}{m+1}} H^{\frac{m+1}{2}} & \text{as } H \to 0^+, \\ -\frac{r(1-m)\alpha}{|\eta_c|} H^m & \text{as } H \to \infty. \end{cases} \tag{D.3}$$

Therefore, the solution of equations (D.1) corresponding to the stable manifold $S_1$ may be written as

$$H = F_s(\tilde{\eta} - \tilde{\eta}_0), \quad Q = Q_s[F_s(\tilde{\eta} - \tilde{\eta}_0)] \tag{D.4}$$

for any fixed $\tilde{\eta}_0 \in \mathbb{R}$, where $F_s(x)$ is the unique solution to the problem

$$F_{sX} = Q_s(F_s), \quad X < 0, \tag{D.5}$$

$$F_s(X) \to 0 \quad \text{as} \quad X \to 0^-, \tag{D.6}$$

$$F_s(X) \quad \sim \quad \left\{ \frac{\alpha r(1-m)^2}{|\eta_c|} \right\}^{\frac{1}{1-m}} (-X)^{\frac{1}{1-m}} \quad \text{as } X \to -\infty. \tag{D.7}$$

It now follows that, for any fixed $\tilde{\eta}_0 \in \mathbb{R}$, (D.4) provides a solution to the boundary value problem (8.90)–(8.92), and an examination of the phase portrait in Figure D.1 demonstrates that this is the only solution.

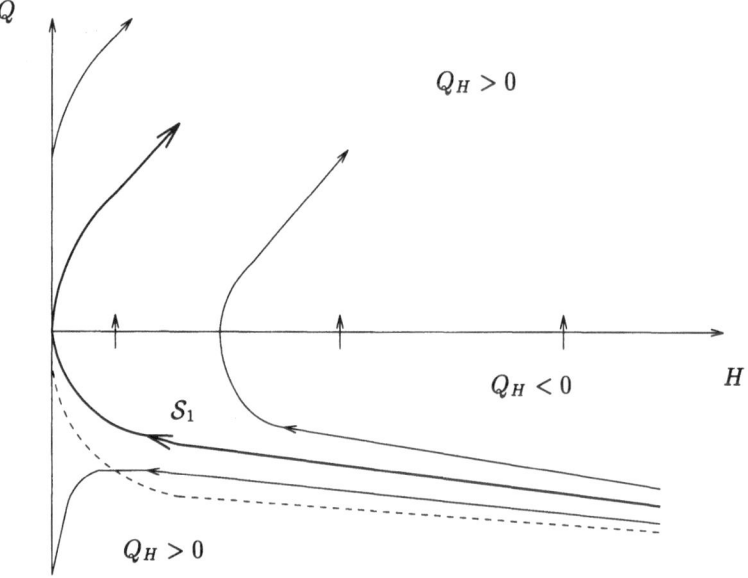

**Fig. D.1.** A sketch of the phase portrait of the system of equations (D.1). Note the dashed line represents the horizontal isocline.

# References

1. Abramowitz, M., Stegun, I.: Handbook of Mathematical Functions. Dover (1965)
2. Aguirre, A.: A Cauchy problem for $u_t - \Delta u = u^p$ with $0 < p < 1$. Asymptotic behaviour of solutions. Ann. Fac. Sci. Toulouse Math., (5), 8, 175–203 (1986)
3. Bandle, C., Stakgold, I,: The formation of the dead core in parabolic reaction-diffusion equations. Trans. Amer. Math. Soc., 28, 275–293 (1984)
4. Barnes, A.N.: Reaction-diffusion waves in an isothermal chemical system with a general order of autocatalysis. MSc Dissertation, University of Reading.
5. Billingham, J., Needham, D.J.: A note on the properties of a family of travelling wave solutions arising in cubic autocatalysis. Dynamics and stability of systems, 6, 33–49 (1991)
6. Billingham, J., Needham, D.J.: The development of travelling waves in quadratic and cubic autocatalysis with unequal diffusion rates. I. Permanent form travelling waves. Phil. Trans. Roy. Soc. Lond. Ser. A **334**:1–24 (1991)
7. Billingham, J., Needham, D.J.: The development of travelling waves in quadratic and cubic autocatalysis with unequal diffusion rates. II. An initial value problem with an immobilized or nearly immobilized autocatalyst. Phil. Trans. Roy. Soc. Lond. Ser. A **336**:497–539 (1991)
8. Billingham, J., Needham, D.J.: The development of travelling waves in quadratic and cubic autocatalysis with unequal diffusion rates III: Large time development in quadratic autocatalysis. Q. Appl. Math., 2, 343–372 (1992)
9. Bramson, M.D.: Maximal displacement of branching Brownian motion. Comm. Pure Appl. Math. 31, 531–81 (1978)
10. Britton, N.F.: Reaction-diffusion Equations and their applications to Biology. Academic, London (1986)
11. Burkhill, J.C.: The Theory of Ordinary Differential Equations. London, Dickens.
12. Coddington, E.A., Levinson, N.: Theory of Ordinary Differential Equations. McGraw-Hill, New York (1955)
13. Ebert, U., Saarloos, W. Van.: Front propagation into unstable states: universal algebraic convergence towards uniformly translating pulled fronts. Physica D, 146, 1–99 (2000)
14. Fife, P.C.: Mathematical Aspects of Reacting and Diffusing Systems. Springer, Berlin (1979)
15. Field, R.J. Noyes, R.M.: Oscillations in chemical systems. IV Limit cycle behaviour in a model of a real chemical reaction. Journal of Chemical Physics **60** :1877-1884 1974

16. Fisher, R.A.: The wave of advance of advantageous genes. Ann. Eugenics, **7**, 335–369 (1937)
17. Forstova, L., Sevcikova, H., Marek, M., Merkin, J.H.: Electric field effects on the selectivity of reactions within propagating reaction fronts. Chemical Engineering Science, 55:233-243 (2000)
18. Georgescu, A.: Asymptotic Treatment of Differential Equations. Applied Mathematics and Mathematical Computation 9. Chapman & Hall (1995)
19. Gowland, R.J., Stedman, G. A novel moving boundary reaction involving hydroxylamine and nitric acid. Journal of the Chemical Society, Chemical Communications, 1038-1039 (1983)
20. Gray, P., Merkin, D.J., Needham, D.J., Scott, S.K.: The development of travelling waves in a simple isothermal chemical system III. Cubic and mixed autocatalysis. Proc. Roy. Soc. Lond. Ser. A, **430**, 509–524 (1990)
21. Gray, P., Scott, S.K.: Chemical Oscillations and Instabilities: Nonlinear Chemical Kinetics. International series of monographs on chemistry. Oxford (1994)
22. Gray, P., Scott, S.K., Showalter, K.: The influence of the form of autocatalysis on the speed of chemical waves Proc. Roy. Soc. Lond. Ser. A, **337**, 249 (1991)
23. Grudy, R.E., Peletier, L.A.: Travelling fronts in nonlinear diffusion equations. Proc. Roy. Soc. Edinburgh Ser. A, 107, 271–288 (1987)
24. Hadeler, K.P., Rothe, F.: Travelling fronts in nonlinear diffusion equations. J. Math. Biol., 2, 251–102 (1975)
25. Hanna, A., Saul, A., Showalter, K.: Detailed studies of propagating fronts in the iodate oxidation of arsenous acid. J. Am. Chem. Phys. **104**:3838–3844 (1982)
26. Hinch, E.J.: Perturbation Methods. Cambridge (1991)
27. Kay, A.L., Needham, D.J., Leach, J.A.: Travelling waves for a coupled, singular reaction-diffusion system arising from a model of fractional order autocatalysis with decay. I. Permanent form travelling waves. Nonlinearity, Vol. 16, No. 2, 735-770 (2003)
28. King, A.C., Needham, D.J.: On a singular initial-boundary-value problem for a reaction-diffusion equation arising from a simple model of isothermal chemical autocatalysis. Proc. Roy. Soc. Lond. Ser. A, **437**, 657–671 (1992)
29. Kolmogorov, A., Petrovsky, I., Piscounov, N.: Moscow University Bulletin of Mathematics, 1, 1–25, (1937)
30. Lagerstrom, P.A., Casten, R.G.: Basic concepts underlying singular perturbation techniques. SIAM Review, Vol. 14, No. 1, 63–120 (1972)
31. Larson, D.A.: Transient bounds and time-asymptotic behaviour of solutions to nonlinear equations of Fisher type. SIAM J. Appl. Math. Vol. 34, 93–103 (1978)
32. Leach, J.A., Needham, D.J.: The evolution of travelling waves in generalized Fisher equations via matched asymptotic expansions: Algebraic corrections. Q. Jl Mech. Appl. Math. 54 (1) 157-175 (2001)
33. Leach, J.A., Needham, D.J.: The evolution of travelling waves in generalized Fisher equations via matched asymptotic expansions: Exponential corrections. To appear: Zeitschrift für angewandte Mathematik & Physik (ZAMP)
34. Leach, J.A., Needham, D.J.: A review of the evolution of reaction-diffusion waves in a class of scalar reaction-diffusion equations. International Journal of Differential Equations and Applications, Vol. 5, No. 2, 159-169 (2002)
35. Leach, J.A., Needham, D.J., Kay, A.L.: The evolution of reaction-diffusion waves in a class of scalar reaction-diffusion equations: Algebraic decay rates. Physica D, Nonlinear Phenomena, 167 (3-4) 153-182 (2002)

36. Leach, J.A., Needham, D.J., Kay, A.L.: The evolution of reaction-diffusion waves in generalized Fisher equations: Exponential decay rates. Dynamics of Continuous, Discrete and Impulsive Systems, Ser. A, Vol. 10, No. 3, 417-430 (2003)

37. Leach, J.A., Needham, D.J., Kay, A.L.: The evolution of reaction-diffusion waves in a class of scalar reaction-diffusion equations: Initial data with compact support and exponential decay rates. Q. Jl Mech. Appl. Math. 56 (2) 217-249 (2003)

38. McCabe, P.M., Leach, J.A., Needham, D.J.: The evolution of travelling waves in fractional order autocatalysis with decay. I. Permanent form travelling waves. SIAM J. Appl. Math. Vol. 59, No. 3, 870–899 (1998)

39. McCabe, P.M., Leach, J.A., Needham, D.J.: The evolution of travelling waves in fractional order autocatalysis with decay. II. The initial boundary value problem. SIAM J. Appl. Math. Vol. 60, No. 5, 1707–1748 (2000)

40. McCabe, P.M., Leach, J.A., Needham, D.J.: A note on the non-existence of permanent form travelling wave solutions in a class of singular reaction-diffusion problems. Dynamical Systems, Vol 17, No. 2, 131–135 (2002)

41. McCabe, P.M., Leach, J.A., Needham, D.J.: On an initial-boundary-value problem for a coupled, singular reaction-diffusion system arising from a model of fractional order chemical autocatalysis and decay. Q. Jl Mech. Appl. Math. 55 (4), 511-560 (2002)

42. McCabe, P.M., Leach, J.A., Needham, D.J.: A note on the small time development of the solution to a scalar, non-linear, singular reaction-diffusion equation. To appear: Zeitschrift für angewandte Mathematik & Physik (ZAMP)

43. McKean, H.P.: Application of Brownian motion to the equation of Kolmogorov-Petrovskii-Piskunov. Comm. Pure Appl. Math. 28, 323 (1975)

44. Merkin, J.H., Needham, D.J.: Propagating reaction-diffusion waves in a simple isothermal quadratic chemical system. J. Engrg. Math., 23, 343–356 (1989)

45. Merkin, J.H., Needham, D.J.: Reaction-diffusion waves in an isothermal chemical system with general orders of autocatalysis and spatial dimension. J. Appl. Math. Phys. (ZAMP) A44, 707–721 (1993)

46. Merkin, J.H., Needham, D.J., Scott, S.K.: A simple model for sustained oscillations in isothermal branched-chain or autocatalytic reactions in a well-stirred, open system. Proc. Roy. Soc. Ser. A **398**: 81-116 (1985)

47. Merkin, J.H., Needham, D.J., Scott, S.K.: The development of travelling waves in a simple isothermal chemical system. I. Quadratic autocatalysis with linear decay. Proc. Roy. Soc. Ser. A **424**: 187-209 (1989)

48. Merkin, J.H., Needham, D.J., Scott, S.K.: The development of travelling waves in a simple isothermal chemical system. II. Cubic autocatalysis with quadratic and linear decay. Proc. Roy. Soc. Ser. A **430**: 315-345 (1990)

49. Merkin, J.H., Needham, D.J., Scott, S.K.: The development of travelling waves in a simple isothermal chemical system. IV. Quadratic autocatalysis with quadratic decay. Proc. Roy. Soc. Ser. A **434**: 531-554 (1991)

50. Merkin, J.H., Sevcikova, H.: Reaction fronts in an ionic autocatalytic system with an applied electric field. Journal of Mathematical Chemistry, 25:111-132 (1999)

51. Merkin, J.H., Sevcikova, H., Snita, D.: The effects of an electric field on the local stoichemistry of front waves in an ionic chemical system. IMA Journal of Applied Mathematics, 64:157-188 (2000)

52. Murray, J.D.: Mathematical Biology, Biomathematics Texts, Springer-Verlag, New York (1989)

53. Nayfeh, A.H.: Introduction to Perturbation Techniques. A Wiley-Interscience Publication. (1993)

54. Needham, D.J.: A formal theory concerning the generation and propagation of travelling wavefronts in reaction-diffusion equations. Q. Jl Mech. Appl. Math. 45, 469–498 (1992)

55. Needham, D.J.: On the global existence of solutions to a singular semilinear parabolic equation arising from the study of autocatalytic chemical kinetics. Z. Angew. Math. Phys., 43, 471–480 (1992)

56. Needham, D.J., Barnes, A.: Reaction-diffusion and phase waves occurring in a class of scalar reaction-diffusion equations. Nonlinearity, 12, 41–58 (1999)

57. Needham, D.J., King, A.C.: On the existence and uniqueness of solutions to a singular nonlinear boundary value problem arising in isothermal autocatalytic chemical kinetics. J. Edinburgh Math. Soc. 36, 479-500 (1992)

58. Needham, D.J., Merkin, J.H.: The development of travelling waves in a simple isothermal chemical system with general orders of autocatalysis and decay. Philos. Trans. Roy. Soc. London Ser. A 337, 261–274 (1991)

59. Perko, L.: Differential Equations and Dynamical Systems. Texts in Applied Mathematics 7. Springer (2001)

60. Samarskii, A.A., Galaktionov, V.A., Kurdyumov, S.P., Mikhailov, A.P.: Blow Up in Quasilinear Parabolic Equations. De Gruyter, Berlin (1995)

61. Saul, A., Showalter, K.: Propagating reaction-diffusion fronts. *Oscillations and Travelling Waves in Chemical Systems* (eds R.J. Field and M. Burger; Wiley, New York).

62. Sel'Kov, E.E.: Self-oscillations in glycosis, 1. A simple kinetic model. European Journal of Biochemistry,4: 79-86 (1968)

63. Sherratt, J.A.: On the transition from initial data to travelling waves in the Fisher-KKP equation. Dyn. Stab. Syst. 13, 167 (1998)

64. Sherratt, J.A., Marchant, B.P.: Algebraic decay and variable speeds in wavefront solutions of a scalar reaction-diffusion equation. IMA Journal of Applied Mathematics, 56, 289–302 (1996)

65. Smith, S., Needham, D.J., Leach, J.A.: The evolution of travelling waves in a simple model for an ionic autocatalytic system. Submitted to: Proc. Roy. Soc. Ser. A

66. Smoller, J.: Shock Waves and Reaction-Diffusion Equations. Springer, Berlin (1989)

67. Snita, D., Seveikova, M., Marek, M., Merkin, J.H.: Travelling waves in an ionic autocatalytic chemical system with an imposed electric field. Proc. Roy. Soc. Lond. Ser. A 453, 2325–2351 (1997)

68. Stackgold, I.: Greens Functions and Boundary Value Problems. Wiley Interscience, New York (1990)

69. Stokes, A.N.: On Two Types of Moving Front in Quasilinear Diffusion. Math. Biosci. 31 307-315 (1976)

70. Van Dyke, M.: Perturbation Methods in Fluid Dynamics. Parabolic Press, Stanford, CA (1975)

71. Volpert, A.I., Volpert, V.A., Volpert, V.A.: Travelling Wave Solutions of Parabolic Systems. American Mathematical Society, Providence, Rhode Island (1994)

72. Wiggins, S.: Introduction to Applied Nonlinear Dynamical Systems and Chaos, Springer-Verlag, New York (1990)

73. Winfree, A.T.: The Geometry of Biological Time. Berlin, Springer (1980)
74. Xin. J.: Front propagation in heterogeneous media. SIAM Review Vol. 42, No. 2, 161–230 (2000)
75. Zailin, A.N., Zhabotinskii, A.M.: Concentration wave propagation in two dimensional liquid-phase self-organising systems. Nature, **225**:535–537 (1970)

# Index